METHODS IN MOLECULAR BIOLOGY

For further volumes:
http://www.springer.com/series/7651

Synthetic mRNA

Production, Introduction Into Cells, and Physiological Consequences

Edited by

Robert E. Rhoads

Department of Biochemistry and Molecular Biology
Louisiana State University Health Sciences Center
Shreveport, LA, USA

Humana Press

Editor
Robert E. Rhoads
Department of Biochemistry and Molecular Biology
Louisiana State University Health Sciences Center
Shreveport, LA, USA

ISSN 1064-3745 ISSN 1940-6029 (electronic)
Methods in Molecular Biology
ISBN 978-1-4939-3623-6 ISBN 978-1-4939-3625-0 (eBook)
DOI 10.1007/978-1-4939-3625-0

Library of Congress Control Number: 2016936683

Cover illustration: Bioluminescence image of a living mouse after intranodal injection with synthetic mRNA encoding firefly luciferase (as described by Kreiter *et al.* in Chapter 11)

Printed on acid-free paper

This Humana Press imprint is published by Springer Nature
The registered company is Springer Science+Business Media LLC New York

Preface

The dream of altering genetic expression to alleviate human disease states—gene therapy—has existed from the dawn of the molecular biology era. Early attempts were made with both DNA and RNA as carriers of genetic information, but the inherent instability of RNA compared with DNA attracted by far the larger share of researchers, vectors, protocols, and success stories. Yet DNA, regardless of whether it is introduced by viral or nonviral modalities, carries the potential of integration into the host genome at unintended sites, which can lead to unwelcome and permanent consequences for the patient. Also, there are some outcomes of gene therapy that are best achieved by transient rather than permanent introduction of new genetic information, e.g., lineage conversion of cell fates, genome editing, and activation of the immune system against pathogens or cancer. For these applications, RNA, particularly mRNA, can be preferable to DNA. The past decade has witnessed new discoveries on how exogenous mRNA activates the host cell's innate immune response to shut down protein synthesis and destroy the RNA, but importantly, there have also been new discoveries on how this can be managed by modifying the structure of mRNA. Progress has also been made on methods to introduce mRNA into cells, to stabilize it in the cell, and to enhance its translational efficiency. New synthetic techniques have been developed that allow for structural features to be built into mRNA which provide investigational tools, such as fluorescence emission, click chemistry, and photochemical crosslinking. Finally, there have been significant advances in the use of synthetic mRNA for protein replacement therapy, immunotherapy against cancer and infectious diseases, creation of pluripotent stem cells from somatic cells, and genome editing. The chapters in this volume present detailed laboratory protocols for (1) synthesis of mRNA with favorable properties, (2) introduction of the synthetic mRNA into a variety of cell types by a variety of techniques, and (3) use of synthetic mRNA to achieve a range of physiological outcomes.

Shreveport, LA, USA *Robert E. Rhoads*

Contents

Contributors

FRÉDÉRIC AMANT • *Laboratory of Gynecologic Oncology, Department of Oncology, KU Leuven, Leuven Cancer Institute, Leuven, Belgium; Department of Gynecology and Obstetrics, UZ Leuven, Leuven, Belgium*

DAPHNÉ BENTEYN • *Laboratory of Molecular and Cellular Therapy, Department of Immunology–Physiology and the Dendritic Cell Bank, Vrije Universiteit Brussel, Brussels, Belgium*

ZWI N. BERNEMAN • *Laboratory of Experimental Hematology, Vaccine and Infectious Disease Institute (VAXINFECTIO), Faculty of Medicine and Health Sciences, Antwerp University Hospital, University of Antwerp, Edegem, Belgium*

ANTONIO BERTOLETTI • *Emerging Infectious Diseases Program, Duke-NUS Graduate Medical School, Singapore, Singapore*

SOLENNE BIRE • *LBTM, Institute of Biotechnology, UNIL-EPFL, Lausanne, Switzerland*

AUDE BONEHILL • *Laboratory of Molecular and Cellular Therapy, Department of Immunology–Physiology and the Dendritic Cell Bank, Vrije Universiteit Brussel, Brussels, Belgium*

TROELS HOLZ BORCH • *Department of Haematology, Center for Cancer Immune Therapy, Copenhagen University Hospital Herlev, Herlev, Denmark; Department of Oncology, Copenhagen University Hospital Herlev, Herlev, Denmark*

GÁBOR BOROS • *Department of Dermatology, Faculty of Medicine, University of Debrecen, Debrecen, Hungary*

JESSICA A. BROWN • *Department of Molecular Biophysics and Biochemistry, Howard Hughes Medical Institute, Yale University School of Medicine, New Haven, CT, USA*

CHRISTIE C. CHOW • *Li Ka Shing Institute of Health Sciences, The Chinese University of Hong Kong, Princes of Wales Hospital, Shatin, Hong Kong, SAR, China*

NATHALIE COOLS • *Laboratory of Experimental Hematology, Vaccine and Infectious Disease Institute (VAXINFECTIO), Faculty of Medicine and Health Sciences, University of Antwerp, Antwerp University Hospital, Edegem, Belgium*

AN COOSEMANS • *Laboratory of Gynecologic Oncology, Department of Oncology, KU Leuven, Leuven Cancer Institute, Leuven, Belgium; Department of Gynecology and Obstetrics, UZ Leuven, Leuven, Belgium*

JURGEN CORTHALS • *Laboratory of Molecular and Cellular Therapy (LMCT), Vrije Universiteit Brussel, Brussels, Belgium*

JENS DANNULL • *Department of Surgery, Duke University School of Medicine, Durham, NC, USA*

JUDITH DERDELINCKX • *Laboratory of Experimental Hematology, Vaccine and Infectious Disease Institute (VAXINFECTIO), Faculty of Medicine and Health Sciences, University of Antwerp, Antwerp University Hospital, Edegem, Belgium*

GAYATHRI R. DEVI • *Division of Surgical Sciences, Department of Surgery, Women's Cancer Program, Duke Cancer Institute, Duke University School of Medicine, Durham, NC, USA*

MUSTAFA DIKEN • *TRON-Translational Oncology at the University Medical Center, Johannes Gutenberg University gGmbH, Mainz, Germany*

ARTEM V. DOMASHEVSKIY • *Department of Sciences, John Jay College of Criminal Justice, City University of New York, New York, NY, USA*

ESZTER EMRI • *Department of Dermatology, Faculty of Medicine, University of Debrecen, Debrecen, Hungary*
GABRIELLA EMRI • *Department of Dermatology, Faculty of Medicine, University of Debrecen, Debrecen, Hungary*
CHRISTINA L. GARDNER • *Department of Microbiology and Molecular Genetics, Center for Vaccine Research, University of Pittsburgh, Pittsburg, PA, USA*
DIXIE J. GOSS • *Department of Chemistry, Hunter College, City University of New York, New York, NY, USA*
DILANTHA GUNAWARDANA • *Department of Biochemistry and Molecular Biology, Bio21 Molecular Science and Biotechnology Institute, University of Melbourne, Parkville, VIC, Australia; Department of Botany, University of Sri Jayewardenepura, Nugegoda, Sri Lanka*
THOMAS STEEN HANSEN • *Department of Micro- and Nanotechnology, DTU Nanotech, Technical University of Denmark, Kgs. Lyngby, Denmark*
CSABA HEGEDŰS • *Department of Dermatology, Faculty of Medicine, University of Debrecen, Debrecen, Hungary*
CARLO HEIRMAN • *Laboratory of Molecular and Cellular Therapy, Department of Immunology–Physiology and the Dendritic Cell Bank, Vrije Universiteit Brussel, Brussels, Belgium*
JOSEPHIN M. HOLSTEIN • *Westfälische Wilhelms-Universität Münster, Institute of Biochemistry, Muenster, Germany*
MANJA IDORN • *Center for Cancer Immune Therapy (CCIT), Department of Hematology, Copenhagen University Hospital Herlev, Herlev, Denmark*
NICOLE ISHAC • *LNOX, GICC UMR CNRS 7292, UFR de Médecine, Tours, France*
JACEK JEMIELITY • *Center of New Technologies, University of Warsaw, Warsaw, Poland*
KATALIN KARIKÓ • *Department of Neurosurgery, University of Pennsylvania, Philadelphia, PA, USA; BioNTech RNA Pharmaceuticals GmbH, Mainz, Germany*
WILLIAM B. KLIMSTRA • *Department of Microbiology and Molecular Genetics, Center for Vaccine Research, University of Pittsburgh, Pittsburg, PA, USA*
SARENE KOH • *Singapore Institute for Clinical Sciences, Agency for Science, Technology and Research, (A*STAR), Singapore, Singapore*
JOANNA KOWALSKA • *Division of Biophysics, Institute of Experimental Physics, Faculty of Physics, University of Warsaw, Warsaw, Poland*
SEBASTIAN KREITER • *TRON-Translational Oncology at the University Medical Center, Johannes Gutenberg University gGmbH, Mainz, Germany*
NIELS B. LARSEN • *Department of Micro- and Nanotechnology, DTU Nanotech, Technical University of Denmark, Kgs. Lyngby, Denmark*
PO SING LEUNG • *School of Biomedical Sciences, The Chinese University of Hong Kong, Shatin, Hong Kong, SAR, China*
SONG LU • *Department of Chemical Pathology and Li Ka Shing Institute of Health Sciences, The Chinese University of Hong Kong, Princes of Wales Hospital, Shatin, Hong Kong, SAR, China*
KATHY O. LUI • *Department of Chemical Pathology and Li Ka Shing Institute of Health Sciences, The Chinese University of Hong Kong, Princes of Wales Hospital, Shatin, Hong Kong, SAR, China*
AZITA JOSEFINE MAHINY • *BioNTech RNA Pharmaceuticals GmbH, Mainz, Germany*
FRANCK MARTIN • *Architecture et Réactivité de l'ARN, Université de Strasbourg, CNRS, Institut de Biologie Moléculaire et Cellulaire, Strasbourg, CEDEX, France*

William F. Marzluff • *Department of Biochemistry and Biophysics, University of North Carolina, Chapel Hill, NC, USA*

Özcan Met • *Department of Haematology, Center for Cancer Immune Therapy, Copenhagen University Hospital Herlev, Herlev, Denmark; Department of Oncology, Copenhagen University Hospital Herlev, Herlev, Denmark*

Edit Miko • *Department of Dermatology, Faculty of Medicine, University of Debrecen, Debrecen, Hungary; MTA-DE Lendület Laboratory of Cellular Metabolism Research Group of the Hungarian Academy of Sciences, University of Debrecen, Debrecen, Hungary*

Kim Morias • *Department of Human Genetics, KU Leuven, Leuven, Belgium*

Hiromi Muramatsu • *Department of Neurosurgery, University of Pennsylvania, Philadelphia, PA, USA*

Smita K. Nair • *Department of Surgery, Duke University School of Medicine, Durham, NC, USA*

Sritama Nath • *Division of Surgical Sciences, Department of Surgery, Women's Cancer Program, Duke Cancer Institute, Duke University School of Medicine, Durham, NC, USA*

Jutta Petschenka • *TRON-Translational Oncology at the University Medical Center, Johannes Gutenberg University gGmbH, Mainz, Germany*

Éva Remenyik • *Department of Dermatology, Faculty of Medicine, University of Debrecen, Debrecen, Hungary*

Andrea Rentmeister • *Institute of Biochemistry, Westfälische Wilhelms-Universität Münster, Muenster, Germany; Cells-in-Motion Cluster of Excellence (EXC 1003-CiM), University of Muenster, Muenster, Germany*

Robert E. Rhoads • *Department of Biochemistry and Molecular Biology, Louisiana State University Health Sciences Center, Shreveport, LA, USA*

David J. Rodriguez • *Department of Sciences, John Jay College of Criminal Justice, City University of New York, New York, NY, USA*

Florence Rouleux-Bonnin • *LNOX, GICC UMR CNRS 7292, UFR de Médecine, Tours, France*

Kate D. Ryman • *Department of Microbiology and Molecular Genetics, Center for Vaccine Research, University of Pittsburgh, Pittsburg, PA, USA*

Ugur Sahin • *TRON-Translational Oncology at the University Medical Center, Johannes Gutenberg University gGmbH, Mainz, Germany*

David Selmeczi • *Department of Micro- and Nanotechnology, DTU Nanotech, Technical University of Denmark, Kgs. Lyngby, Denmark*

Abderraouf Selmi • *TRON-Translational Oncology at the University Medical Center, Johannes Gutenberg University gGmbH, Mainz, Germany*

Noriko Shimasaki • *Department of Pediatrics, Yong Loo Lin School of Medicine, National University of Singapore, Singapore, Singapore*

Michael K. Slevin • *Good Start Genetics, Cambridge, MA, USA*

Joan A. Steitz • *Department of Molecular Biophysics and Biochemistry, Howard Hughes Medical Institute, Yale University School of Medicine, New Haven, CT, USA*

Per thor Straten • *Department of Hematology, Center for Cancer Immune Therapy (CCIT), Copenhagen University Hospital Herlev, Herlev, Denmark*

Daniela Stummer • *Westfälische Wilhelms-Universität Münster, Institute of Biochemistry, Muenster, Germany; Cells-in-Motion Cluster of Excellence (EXC 1003-CiM), University of Muenster, Muenster, Germany*

Wei Su • *Department of Neurology, University of Washington, Seattle, WA, USA*

INGE MARIE SVANE • *Department of Haematology, Center for Cancer Immune Therapy, Copenhagen University Hospital Herlev, Herlev, Denmark; Department of Oncology, Copenhagen University Hospital Herlev, Herlev, Denmark*

KRIS THIELEMANS • *Laboratory of Molecular and Cellular Therapy, Department of Immunology–Physiology and the Dendritic Cell Bank, Vrije Universiteit Brussel, Brussels, Belgium*

DEREK W. TROBAUGH • *Department of Microbiology and Molecular Genetics, Center for Vaccine Research, University of Pittsburgh, Pittsburg, PA, USA*

STEPHEN K. TSUI • *School of Biomedical Sciences, The Chinese University of Hong Kong, Shatin, Hong Kong, SAR, China*

ÖZLEM TÜRECI • *TRON-Translational Oncology at the University Medical Center, Johannes Gutenberg University gGmbH, Mainz, Germany*

SANDRA TUYAERTS • *Department of Oncology, Laboratory of Gynecologic Oncology, KU Leuven, Leuven Cancer Institute, Leuven, Belgium*

STEFAAN W. VAN GOOL • *Laboratory of Pediatric Immunology, Department of Microbiology and Immunology, KU Leuven, Leuven, Belgium*

IGNACE VERGOTE • *Department of Oncology, Laboratory of Gynecologic Oncology, KU Leuven, Leuven Cancer Institute, Leuven, Belgium; Department of Gynecology and Obstetrics, UZ Leuven, Leuven, Belgium*

JUNWEI ZHOU • *School of Biomedical Sciences, The Chinese University of Hong Kong, Shatin, Hong Kong, SAR, China*

Part I

Introduction

Chapter 1

Synthetic mRNA: Production, Introduction into Cells, and Physiological Consequences

Robert E. Rhoads

Abstract

Recent advances have made it possible to synthesize mRNA in vitro that is relatively stable when introduced into mammalian cells, has a diminished ability to activate the innate immune response against exogenous (virus-like) RNA, and can be efficiently translated into protein. Synthetic methods have also been developed to produce mRNA with unique investigational properties such as photo-cross-linking, fluorescence emission, and attachment of ligands through click chemistry. Synthetic mRNA has been proven effective in numerous applications beneficial for human health such as immunizing patients against cancer and infections diseases, alleviating diseases by restoring deficient proteins, converting somatic cells to pluripotent stem cells to use in regenerative medicine therapies, and engineering the genome by making specific alterations in DNA. This introductory chapter provides background information relevant to the following 20 chapters of this volume that present protocols for these applications of synthetic mRNA.

Key words Cap analogs, Immunotherapy, mRNA stability, Nucleoporation, Electroporation, Cationic lipids, Innate immunity, Protein expression, Poly(A), Translational efficiency

1 Introduction

The vision of gene therapy—correcting medical defects by introducing genetic information into the patient—was present at the dawn of the molecular biology age. In fact, some of the researchers involved in deciphering of the genetic code [1] were among the first to begin investigating the possibility of gene therapy [2, 3]. An initial question was what form of genetic material should be used to implement gene therapy. Even though support has come from many quarters that an "RNA world" preceded the current "DNA world" [4], RNA is vastly less stable to hydrolysis than DNA, both in the laboratory and the cell, due the presence of a 2′-OH group in RNA that makes formation of an internal phosphodiester intermediate possible. Therefore, DNA was used for initial attempts at gene therapy [3], and today DNA-based vectors for gene therapy are actively under development, including

Robert E. Rhoads (ed.), *Synthetic mRNA: Production, Introduction Into Cells, and Physiological Consequences*, Methods in Molecular Biology, vol. 1428, DOI 10.1007/978-1-4939-3625-0_1,

nonreplicating adenoviruses, adeno-associated viruses, lentiviruses, retroviruses (the latter being RNA viruses that go through a DNA phase), and others. Yet DNA-based gene therapy protocols have sometimes resulted in serious adverse effects in patients [5], and DNA, regardless of whether it enters the cell by a viral or nonviral route, has the potential of integrating into the host genome, causing insertional mutagenesis and activating oncogenes.

The use of mRNA as a genetic vector is theoretically attractive because mRNA does not integrate into the genome, is immediately available for translation to make protein, and provides a transient signal, a feature that is desirable for some applications. The demonstration that mRNA could, in fact, function in this capacity came in 1990, not long after the initial attempts at DNA-based gene therapy, when Wolff et al. injected naked mRNA encoding chloramphenicol acetyl transferase into the skeletal muscle of mice and observed specific protein expression [6]. This was followed in 1992 by a study in which injection of a naked, synthetic mRNA encoding arginine vasopressin into the hypothalami of Brattleboro rats cured the chronic diabetes insipidus suffered by this strain [7]. Then in 1993, Martinon et al. showed that subcutaneous injection of liposome-encapsidated mRNA encoding the influenza virus nucleoprotein induced anti-influenza cytotoxic T lymphocytes [8]. But despite these early successes, there was little progress in the use of mRNA for therapeutic purposes for a decade while development of DNA-based strategies progressed rapidly. Yet today, mRNA-based approaches are being taken to solve a wide range of biomedical problems—from vaccination against cancer and infectious diseases to generation of induced pluripotent stem cells (iPSCs) to genome editing. This change in the landscape of gene therapy is due to a number of discoveries, insights, and technical advances, discussed briefly below and more extensively in the 20 following chapters of this volume.

This introductory chapter is intended to provide background information to place the protocol chapters in context. It is not a complete treatment of the subject of synthetic mRNA and mRNA-based therapeutics. For that, the reader is referred to several recent, comprehensive, and scholarly reviews [9–12]. Also, there are RNA-based methods of altering genetic expression that do not involve mRNA, e.g., Sendai virus and miRNA, that are not covered in this volume but are reviewed elsewhere [13].

2 Basic Methodology for the In Vitro Synthesis of mRNA

2.1 The RNA Chain

mRNA fulfills its role of being a transient carrier of genetic information by being unstable in the cell, and this instability extends to the laboratory as well, where special precautions are needed to prevent mRNA hydrolysis by omnipresent ribonucleases. Yet it is possible to synthesize RNA in the laboratory both chemically and enzymatically. For chemical synthesis, repetitive yields of >99 % are needed to make RNA chains of only 100 nt, but physiological mRNAs are typically 1000 - 10,000 nt [14]. For this reason, highly processive bacterial and bacteriophage RNA polymerases such as T7 [15] and SP6 [16] must be used. A plasmid is typically used as transcription template for the 5′-untranslated region (UTR), coding region, 3′-UTR, and sometimes poly(A) tract, with transcription driven by a promoter specific for the RNA polymerase.

2.2 The Cap

An essential feature for eukaryotic mRNA is the cap, a 5′-terminal 7-methylguanosine attached by a 5′–5′ triphosphate bridge to the RNA [17]. The cap is needed for both high translational efficiency and high stability in the cell, and these two factors separately increase the yield of protein synthesized. However, RNAs made by bacterial and bacteriophage polymerases are not capped unless extra steps are taken. Two methods are widely used to cap mRNAs, and both are employed in the chapters of the current volume. In the first method, capping is posttranscriptional and is accomplished by a multifunctional enzyme from vaccinia virus that first incorporates GTP into the RNA chain in a 5′-to-5′ triphosphate linkage, with concomitant production of P_i and PP_i, followed by 7-methylation of the 5′-terminal guanosine with S-adenosylmethionine [16, 18]. In the second method, a cap dinucleotide of the form m^7GpppG is added along with the other four NTPs during RNA synthesis [19–21]. The RNA polymerase incorporates m^7GpppG in the place of GTP at the 5′-end of the RNA. To increase the percentage of capping, the concentration of GTP is lowered relative to the other NTPs and the concentration of cap dinucleotide is raised. Each method has advantages and disadvantages. The capping efficiency is greater with the vaccinia method, but the cap dinucleotide method is co-transcriptional, so capping and in vitro transcription are performed in a single step. Furthermore, the cap dinucleotide method permits incorporation of chemically modified caps that confer additional properties to synthetic mRNAs (see below). In the current volume, several chapters use the vaccinia method (Boros et al., Holstein et al., Koh et al., and Domashevskiy et al.) and several use the cap dinucleotide method (Dannull and Nair, Borch et al., Idorn et al., Benteyn et al., Kreiter et al., Bire et al., Su et al., Devi and Nath, and Kowalska et al.).

2.3 The Poly(A) Tract

Another critical element for translational efficiency [22] and stability [23] of mRNA is the 3′-terminal poly(A) tract. Synthetic mRNA containing a poly(A) tract may be generated either by including a poly(dT) tract in the plasmid transcribed by the RNA polymerase or by posttranscriptional polyadenylation with a poly (A) polymerase [24]. Again, there are advantages and disadvantages of each method. Longer poly(A) tracts can be obtained and modified nucleoside residues can be incorporated with poly(A) polymerase, but transcription from a DNA template yields RNA with a defined poly(A) tract length and constitutes a one-step procedure.

3 Improving the Stability and Translational Efficiency of Synthetic mRNA

The inherent lability of mRNA is a major obstacle to using synthetic mRNA for altering cellular function because a short half-life of mRNA in the cell means a short window for protein expression. However, some natural mRNAs can have half-lives measured in days rather than minutes, e.g., ovalbumin [25], globin [26], casein [27], and histone [28], and stabilization of these mRNAs is regulated by factors such as the cell cycle and hormonal signaling. Understanding the mechanisms that govern stability has allowed various investigators over the years to introduce structural changes to synthetic mRNA to improve its stability. The chapter by Su et al. in this volume describes synthesis of mRNA that exhibits the regulated stability properties of histone mRNA. Poor translational efficiency also diminishes protein expression, leading investigators to optimize this parameter in synthetic mRNA as well. Interestingly, the two most important determinants of stability and translational efficiency, the cap and poly(A) tract, act synergistically [29–33]. Structural modifications to increase stability and translational efficiency can be divided into the various domains of mRNA: cap, UTRs, coding region, poly(A) tract, and 3′-end.

3.1 The Cap

3.1.1 ARCAs

A disadvantage of the cap dinucleotide method not mentioned above is that m^7GpppG is incorporated in either orientation [34]. This is because the α-phosphate of the first nucleotide of the growing RNA chain can be attacked by the 3′-OH of either the guanosine or 7-methylguanosine moiety of m^7GpppG. A normal linkage, m^7GpppGpG…, results from attack by the 3′-OH of the guanosine moiety, but a reversed linkage, Gpppm^{7}GpG…, results from attack by the 3′-OH of the 7-methylguanosine moiety. Thus, roughly half of the caps are incorporated backwards. The manner in which caps interact with proteins involved in mRNA processing, translation, and turnover [35–37] indicates that mRNAs with the structure Gpppm^{7}GpG… would not be recognized as being capped. This problem has been solved by synthesis of a variety of cap dinucleotides in which either the 2′ or 3′ position of the 7-methylguanosine moiety is modified to prevent incorporation in the reverse orientation [38–43]. The use of these “anti-reverse cap analogs” (ARCAs) allows synthesis of mRNAs with superior translational properties both in vitro [38, 39, 44] and after introduction into various cell types [33, 44, 45]. ARCAs also permit the unambiguous location of modifications within the cap with respect to the mRNA body and, hence, within the active sites of cap-binding proteins involved in splicing, translation, and turnover of mRNA. For example, a modification of the phosphate moiety of the triphosphate bridge that is proximal to the 7-methylguanosine moiety would always be distal to the mRNA body, provided there is an ARCA modification to prevent reverse incorporation.

3.1.2 Caps that Are Resistant to Decapping Enzymes

M-ARCAs—mRNA is degraded principally by 3′-to-5′ and 5′-to-3′ pathways [46, 47], the latter requiring prior decapping. There are two types of decapping enzymes in eukaryotic cells, DcpS and the Dcp1-Dcp2 complex. DcpS is a scavenger decapping enzyme and acts on the short capped oligonucleotides remaining after 3′-to-5′ hydrolysis by the exosome, cleaving the cap between the β- and γ-phosphates to produce m^7GMP [48]. Dcp1 and Dcp2 are the central components of an RNA-dependent decapping complex that requires an RNA body of at least 25 nt and cleaves between the α- and β-phosphate moieties to produce m^7GDP [49–53]. Substitution of the β–γ bridging O atom with CH_2 in the cap dinucleotide to produce $m_2^{7,3'\text{-}O}GpCH_2ppG$ prevents cleavage of this bond in vitro by DcpS [54]. Substitution of the α-β bridging O atom with CH_2 in the cap dinucleotide to produce $m_2^{7,3'\text{-}O}GppCH_2pG$ prevents cleavage of this bond in vitro by Dcp2 and stabilizes the mRNA to intracellular degradation in cultured cells [44].

S-ARCAs—Unfortunately, mRNAs capped with $m_2^{7,3'\text{-}O}GppCH_2pG$ are poorly translated in cultured cells because $m_2^{7,3'\text{-}O}GppCH_2pG$ has a lower affinity for the translational cap-binding protein eIF4E than the corresponding unmodified ARCA [44]. However, substituting S for the non-bridging O on the β-phosphate to make the phosphorothioate cap dinucleotide $m_2^{7,2'\text{-}O}Gpp_SpG$ allows one to synthesize mRNA that is resistant to in vitro decapping by Dcp2, is more stable in cultured cells, and is translated efficiently in cultured cells [55, 56]. One diastereomer of $m_2^{7,2'\text{-}O}Gpp_SpG$ produces mRNA that is actually translated *more* efficiently in cultured mouse mammary epithelial cells than the control (non-phosphorothioate) ARCA [55]. Injection of luciferase mRNA containing a phosphorothioate-modified cap into the lymph nodes of mice produces more luciferase than the same mRNA capped with a standard ARCA, and injection of an antigen-encoding mRNA containing a phosphorothioate-modified cap induces a more potent immune response than the control mRNA [57].

B-ARCAs—Despite the fact that mRNA capped with the most cleavage-resistant phosphorothioate analog is resistant to cleavage by Dcp2 in vitro, it is still slowly cleaved [58]. Substitution of the β non-bridging O with BH_3 to form the two-headed borano analog, $m^7Gpp_{BH3}pm^7G$ (BTH), yields mRNA that is more resistant to cleavage by Dcp2 in vitro than mRNA capped with the phosphorothioate analog, yet it is still translated as efficiently in HeLa cells [58].

3.1.3 Other Modified Cap Analogs

Fluorescent Cap Analogs for Biophysical Studies—All phases of mRNA metabolism require the specific interaction of mRNA with proteins: synthesis by RNA polymerases, splicing, modification of nucleoside residues, transport from nucleus to cytosol, targeting to specific intracellular sites of translation, translation, movement into and out of P bodies, and finally degradation. Frequently the portion of mRNA recognized by these proteins is the cap, since this is a structural feature unique to mRNA. (The caps of snRNAs are *N*-trimethylated and not recognized by most proteins involved in translation). Fluorescence spectroscopy is widely used to study protein-RNA interactions. It is therefore useful to synthesize mRNAs containing a fluorophore in or near the cap structure. In one such study, mRNA labeled at the ribose of the 7-methylguanosine moiety with either anthraniloyl or *N*-methylanthraniloyl moieties was synthesized by the one-step cap-dinucleotide method described above [59]. In a different study, anthraniloyl labeled mRNA was synthesized by the two-step vaccinia method [60]. The chapter by Domashevskiy et al. in this volume provides a detailed protocol for the latter synthesis.

Fluorophosphate-containing Caps for NMR Studies—Compounds containing ^{19}F are useful for NMR studies. Baranowski et al. [61] prepared a series of cap analogs and synthetic mRNAs in which there is an O-to-F substitution at the γ-position of the triphosphate chain and have used them to study two proteins that interact with the mRNA cap, DcpS and eIF4E.

Caps Amenable to Click Chemistry—Click chemistry describes a powerful collection of synthetic chemical reactions characterized by high yield and rapid delivery of a single product under ambient conditions, often involving azide–alkyne additions [62]. Addition of a 4-vinlybenzyl moiety to the N2 position of 7-methylguanosine in the mRNA cap allows one to create a "clickable" recipient for introducing reporter groups to mRNA to study various steps of mRNA metabolism and intracellular location [63]. The chapter by Holstein et al. in this volume provides a detailed protocol for this synthesis.

6-Thioguanosine-containing Caps for Photo-cross-linking Studies—Another way to study the interaction of mRNAs with other macromolecules is through photo-cross-linking experiments. Nowakowska et al. created cap analogs in which the guanosine moiety of m^{7}GpppG is changed to 6-thioguanosine to create a photo-cross-linking reagent [64]. The cap dinucleotide is synthesized as an ARCA to ensure correct orientation and is incorporated into mRNA co-transcriptionally. The chapter by Kowalska et al. in this volume provides a detailed protocol for this synthesis.

3.2 The 5′- and 3′-UTR

The UTRs of mRNA can have a profound effect on both stability and translational efficiency [65–68]. Various investigators have compared 5′- and 3′-UTRs to achieve maximum expression of proteins encoded by synthetic mRNA introduced into mammalian cells. Karikó et al. [69] found that replacing the 5′- and 3′-UTRs of urokinase-type plasminogen activator receptor mRNA with those of *Xenopus* β-globin mRNA produced a ~10-fold increase in expression of the encoded protein when the mRNA was introduced by cationic lipid-mediated delivery into several cultured mammalian cell lines. Holtkamp et al. [70] found that sequential α-globin 3′-UTRs present in a head-to-tail orientation between the coding region and the poly(A) tract each independently enhanced stability and translational efficiency of mRNA when introduced into immature human dendritic cells (DCs) by electroporation. Although delivered by a DNA-based vector, mRNA encoding the hepatitis B virus surface antigen caused the highest level of secretion of IFN-γ by splenocytes isolated from mice when it contained the 3′-UTR of rabbit β-globin mRNA [71].

3.3 The Coding Region

Two of the chapters in the current volume utilize codon-optimized mRNAs (Boros et al. and Bire et al.). Expression of proteins from synthetic mRNA can be diminished if the mRNA contains rare codons or rate-limiting regulatory sequences. Redesign of the mRNA by using synonymous but more frequently used codons can increase the rate of translation and hence, translational yield [72]. Also, recognition of the termination codon is influenced by adjacent nucleotide sequences [73]. An interesting observation was that a "silent" polymorphism in the MDR1 gene encoding P-glycoprotein is not really silent since it changes the activity and substrate specificity of the protein, suggesting that a rare codon can affect the timing of co-translational folding and insertion of proteins into the plasma membrane [74]. Mauro and Chappelle [75] have recently reviewed potential drawback of codon optimization such as this, including increased immunogenicity of proteins made from codon-optimized mRNA, and have cautioned that codon optimization may not provide the best strategy for increasing protein production.

3.4 The Poly(A) Tract

It was recognized early that the 3′-terminal poly(A) tract of mRNA is a stability factor and that deadenylation precedes 3′-to-5′ degradation of mRNA [23]. Poly(A) is also a strong stimulator of translational initiation due the fact that translation factor eIF4G associates with both the cap-binding protein eIF4E and the poly(A)-binding protein Pab1 [76]. As a result, the cap structure and the poly(A) tract synergize and direct ribosome entry to the 5′-end to drive efficient translation [77, 78]. This is referred to as the "closed loop" model for translation [79–81]. Even though mRNAs emerge from the nucleus with a long (200–300 nt) poly(A) tract, it must be maintained in the cytosol for continued stability and translation, and mechanisms exist for cytoplasmic polyadenylation [82, 83]. The positive correlation between poly(A) length and protein expression was shown in a particularly dramatic case involving ARCA-capped mRNA used to transfect an immortalized mouse DC line, where increasing the poly(A) length from 64 to 100 nt increased protein expression by 35-fold [33]. In another study where immature human DCs were transfected with ARCA-capped mRNAs, an optimal poly(A) length of 120 nt was determined [70]. Brown et al. [84] identified a triple-helical element in the 3′-end of some RNAs that prevents deadenylation. When added to an intronless β-globin reporter RNA it increased intracellular levels. The chapter by Brown and Steitz in the current volume presents a protocol for construction and use of this stabilizing element.

3.5 The 3′-End

Structures other than poly(A) at the 3′-end can stabilize mRNA. The best understood mRNAs of this type encode the canonical replicative histones [85]. Synthesis of these histones (H2A, H2B, H3, and H4) occurs only when DNA is being synthesized and is regulated primarily by changes in mRNA levels, which increase 35-fold as cells enter S-phase. At the end of S-phase or when DNA synthesis is inhibited, histone mRNAs are rapidly degraded. The regulated stabilization of these mRNAs is conferred by a conserved 25- to 26-nt stem-loop (SL) at the 3′-end instead of poly(A). The SL is sufficient to confer regulated stability to a reporter mRNA that is introduced into mammalian cells by electroporation [86], and such a system can be used to gain insight into the mechanisms of mRNA turnover. Degradation in the cell is initiated by the addition of U residues to the 3′-end by a poly(U) polymerase [87]. Adding a cordycepin residue (3′-deoxyadenosine) to the 3′-end of synthetic SL-containing mRNA prevents oligouridylation and stabilizes the mRNA. The chapter by Su et al. in this volume presents a detailed protocol for this system.

4 Delivery of Exogenous mRNA to the Cell

A variety of methods have been developed over the years for introducing mRNA into cells. These will be listed here briefly, but several very complete reviews provide more information [9, 10, 88, 89].

4.1 Direct Uptake of Naked mRNA

The first report of protein expression driven by uptake of exogenous synthetic mRNA was by Wolff et al. [6], who injected naked mRNA into mouse skeletal muscle. This mode of delivery is extremely inefficient, but Diken et al. [90] noted that immature DCs are different from most cells in this regard because they take up mRNA by macropinocytosis, being specialized in sampling their environment by engulfing extracellular fluid. This property made it possible for Kreiter et al. [91] to observe a strong antigen-specific T-cell immunity upon injection of naked antigen-encoding mRNA into mouse lymph nodes. This method can be optimized by co-administration of the DC-activating Fms-like tyrosine kinase 3 (FLT3) ligand as an adjuvant [92]. This protocol is presented in the chapter by Kreiter et al. in the current volume. The antigen-specific response to internodal delivery of naked mRNA is also improved when an mRNA encoding a tumor-associated antigen is injected along with a mixture of mRNAs, termed TriMix, encoding CD40 ligand, constitutively active Toll-like receptor 4, and CD70 [93]. TriMix is utilized in the chapters by Benteyn et al. and Coosemans et al. in the current volume. Rittig et al. [94] performed intradermal injection of naked synthetic mRNA encoding the tumor-associated antigens mucin 1, carcinoembryonic, human epidermal growth factor receptor 2, telomerase, survivin, and melanoma-associated antigen 1 in 30 patients. A clinical benefit was observed in a subset of the patients. Finally, Petsch et al. [95] performed intradermal injection of naked synthetic mRNA encoding the influenza virus HA protein in mice, ferrets, and pigs and observed long-lived and protective immunity to influenza A virus infections.

4.2 Cationic Liposome-Mediated RNA Transfection

Even though uptake of naked mRNA is effective for some applications and cell types, the use of complexing agents generally facilitates uptake and protects the mRNA from nucleases. Malone et al. [96] first used synthetic cationic lipids incorporated into a liposome to transfect mouse NIH 3T3 cells in culture with in vitro-synthesized luciferase mRNA and obtained a linear dose–response of luciferase activity. Conry et al. [97] used this idea to test liposome-protected mRNA as a vector for a tumor vaccine. They synthesized an mRNA encoding human carcinoembryonic antigen, injected mice with it intramuscularly, and found it produced an antigen-specific immune response. Granstein et al. [98] applied the idea of RNA-mediated immunization to the skin. They isolated RNA from S1509a spindle cell tumors, encapsidated it with DOTAP, a commercial liposomal transfection reagent, and used the preparation to pulse CAF1 epidermal cells. When they injected these cells subcutaneously into mice and then challenged the mice with living S1509a cells, they found significantly reduced tumor growth. Kormann et al. [99] used nucleoside modified mRNA (see below) encased in the cationic lipid preparation Lipofectamine 2000 to express therapeutic proteins. Intramuscular injection of erythropoietin mRNA raised the hematocrit in mice. Application of an aerosol of mRNA encoding surfactant protein B restored the missing protein and permitted survival in a mouse model of a lethal congenital lung disease.

In the current volume, cationic lipids are utilized in the protocols by Boros et al., Bire et al., Devi and Nath, and Lu et al.

4.3 Protamine-Complexed mRNA

RNA can also be protected against degradation by complexing with the polycationic protein protamine. Hoerr et al. [100] synthesized β-galactosidase mRNA and injected the liposome-encapsulated RNA-protamine complex into mice. This led to protein expression, activation of specific cytotoxic T lymphocytes, and production of IgG antibodies against β-galactosidase. Both naked and protected mRNA elicited a specific immune response, but the protected mRNA persisted longer. The protamine-stabilization approach without lipofection reagents was also used in a clinical immunotherapy trial [101]. Protamine-stabilized mRNAs coding for Melan-A, tyrosinase, gp100, Mage-A1, Mage-A3, and survivin in 21 were injected into metastatic melanoma patients. One of seven patients showed a complete response.

4.4 Electroporation and Nucleoporation

Electroporation is another technique that can be used effectively to introduce mRNA into cells ex vivo. Su et al. [102] first introduced mRNA encoding telomerase reverse transcriptase (hTERT) into immature DCs by electroporation and then administered these cells to 20 patients with metastatic prostate cancer. In 19 of them, expansion of hTERT-specific $CD8^+$ T cells was observed in the peripheral blood. Vaccination was further associated with a reduction of prostate-specific antigen velocity and clearance of circulating micrometastases. Since then, introduction of mRNAs into DCs by electroporation has been widely used [9]. It has been possible to achieve large volume electroporation of DCs via continuous flow techniques [103]. Several chapters in this volume use electroporation for various cell types. Idorn et al. use electroporation for tumor infiltrating lymphocytes, Gardner et al. use the technique for fibroblastic and myeloid cells, and Dannull and Nair use it for monocytes. Chapters that use electroporation for DCs are discussed in Subheading 6.2.1.

Nucleoporation is similar to electroporation but uses proprietary nucleofection reagents and introduces nucleic acids into both the cell nucleus and cytosol [104]. It is a gentler procedure than electroporation; cells recover more quickly than for electroporation, and mRNA translation begins as early as measurements can be taken ($<$10 min) [55, 58, 86]. The chapters in this volume by Su et al. and Koh et al. utilize this technique.

4.5 Other Techniques

Several other techniques have been described for introducing mRNA into cells but are less frequently used than those discussed above. Mandl et al. [105] used a GeneGun to introduce in vitro-synthesized infectious flavivirus RNA, coated onto gold microcarrier particles, into mice, and this induced a protective immunity. Lipid nanoparticles were developed for delivery of siRNAs, miRNAs, and other noncoding RNAs but have been adapted to deliver self-amplifying RNA vaccines [106, 107]. Finally, a self-assembling polyplex nanomicelle composed of a polyethylene glycol-polyamino acid block copolymer was used to administer luciferase-expressing mRNA with nucleoside modification into the CNS by intrathecal injection into the cisterna magna of mice [108]. Sustained protein expression was observed in the cerebrospinal fluid for one week. Immune responses were suppressed compared with naked mRNA introduction.

5 Exogenous mRNA and the Innate Immunity Response

With the ability to synthesize long, capped mRNAs in vitro, researcher began investigating the expression of specific proteins in cells. This uncovered a new problem: incoming mRNAs are viewed by the cell as viruses and trigger the innate immunity response, leading to accelerated mRNA decay and a shut-off of translation, among other responses [12, 109]. In retrospect, this should have come as no surprise since it had been known since the early 1970s that the dsRNA associated with viral infection shuts down protein synthesis due to activation of PKR, a kinase that phosphorylates initiation factor eIF2α [110], and also induces an endonuclease that degrades mRNA [111]. Ultimately it was discovered that these, as well as other consequences of activating the innate immune response, are due to recognition of both dsRNA [112] and ssRNA [113, 114] by Toll-like receptors.

Shutdown of protein synthesis and degradation of mRNA present a major obstacle to mRNA-based therapeutic approaches. The beginnings of a solution to this problem came from the observation by Karikó et al. [115] that methylated CpG motifs in DNA do not activate the Toll-like receptors. This led them to discover that methylated or otherwise modified ribonucleoside residues such as 5-methyluridine, 5-methylcytidine, N^6-methyladenosine, pseudouridine, and 2-thiouridine also fail to activate the Toll-like receptors. DCs exposed to such modified RNAs expressed significantly less cytokines and activation markers than those treated with unmodified RNA. Hornung et al. [116] then found that RNA containing a 5′-triphosphate is the ligand for retinoic acid-inducible protein I (RIG-I), another receptor that is essential for controlling viral infection. This is particularly relevant for synthetic mRNA that is incompletely capped and therefore retains a 5′-triphosphate (see above). Karikó et al. [117] applied their findings with modified nucleoside residues to the issue of RNA-based therapeutics and found that pseudouridine-containing mRNAs had a higher translational capacity than unmodified mRNAs when tested in mammalian cells and lysates or administered intravenously into mice. Furthermore the substituted mRNAs were more stable. These findings reflect the fact that PKR and endonucleases are not being induced by activation of the innate immunity system. Experiments looking directly at PKR confirmed that the pseudouridine-for-uridine substitution spares translation by failing to activate PKR [118]. A further improvement in translational efficiency of synthetic mRNA after introduction into cells was obtained by high performance liquid chromatography (HPLC), which removes contaminants like dsRNA in nucleoside-modified mRNA that are responsible for activation of the innate immune response [119]. The purified mRNAs are translated 10- to 1000-fold better in primary cells. The chapter by Boros et al. in this volume utilizes HPLC to purify synthetic mRNA.

Nucleoside-modified mRNAs are now in wide use by those exploring mRNA-based therapeutics. For instance, the production of erythropoietin and surfactant protein B in mice, mentioned in Subheading 4.2, employed mRNA that contained 2-thiouridine, 5-methylcytidine, pseudouridine, and N^6-methyladenosine [99]. In some situations, however, the mRNA is left unmodified intentionally because a vigorous immune response is desired, e.g., in developing vaccines against influenza virus infection [95]. Three protocols in the current volume utilize nucleoside-modified mRNA: Boros et al., Mahiny and Karikó, and Lu et al.

It has long been recognized that, not surprisingly, viruses attempt to thwart host defense mechanism [120]. Recently, Hyde et al. [121] discovered that secondary structures in the 5′-UTRs of many alphaviruses can antagonize host innate immune responses that inhibit translation of nonself mRNAs. This finding potentially could be exploited for therapeutic mRNA. The chapter by Gardner et al. in this volume presents a protocol for synthesizing and transfecting mRNA containing these alphavirus elements as a tool to study the innate immune system.

6 Applications for Synthetic mRNA

Synthetic mRNAs with high translational efficiency, high stability, and low immunogenicity have been used for a number of applications. The following are some broad topics currently being addressed with mRNA-based approaches and some representative studies.

6.1 Protein Replacement to Correct Diseases

Many diseases are caused by the partial or complete absence of a single protein. Alternatively, a protein may be present but have diminished catalytic or regulatory functions. In these cases, delivery of mRNA encoding that protein may be preferable to protein-based therapy. Unlike mRNA-based vaccines, it is disadvantageous in this type of therapy to activate the innate immune system since it shuts down protein synthesis and accelerates mRNA decay. Hence, the use of nucleoside-modified mRNA to suppress this is critical.

Arginine vasopressin was the first protein made with synthetic mRNA in vivo to treat a disease, as noted above [7]. The

Brattleboro rat strain suffers from chronic diabetes insipidus. Injection of a naked, synthetic mRNA encoding vasopressin into the hypothalamus led to selective uptake, retrograde transport, and expression of vasopressin in magnocellular neurons. Reversal of diabetes insipidus was observed for up to 5 days beginning within hours of the injection.

Kormann et al. [99] corrected a mouse model of a lethal congenital lung disease caused by a lack of surfactant protein B, as noted above. They used an aerosol of mRNA encoding surfactant protein B that contained 2-thiouridine and 5-methylcytidine. Twice weekly treatment protected mice from respiratory failure and prolonged their life spans.

Erythropoietin is a practical choice for protein replacement because it is effective at low doses and causes a readily detected increase in hematocrit. Two studies have described injection of nucleoside-modified, HPLC-purified mRNA encoding erythropoietin in mice and primates [99, 122]. The chapter by Mahiny et al. in this volume gives a detailed protocol for measuring hematocrit in mice.

Mays et al. [123] created a disease model of asthma in mice by sensitizing them to ovalbumin. They then delivered a nucleoside-modified mRNA encoding the regulatory T cell transcription factor FOXP3 intratracheally by aerosol. The result was that modified FOXP3 mRNA rebalanced pulmonary T helper cell responses and protected from allergen-induced tissue inflammation, airway hyper-responsiveness, and goblet cell metaplasia. The chapter by Devi and Nath in this volume describes the use of FOXP3-specific cytotoxic T-lymphocytes to treat inflammatory breast cancer cells.

Vascular endothelial growth factor A was expressed from a synthetic mRNA containing pseudouridine and 5-methylcytidine by intramyocardial injection in a mouse model of myocardial infarction [124, 125]. Overexpression of this factor promoted endothelial specification as well as engraftment, proliferation, and survival of human $Isl1^+$ progenitors. This markedly improved heart function and enhanced long-term survival of animals due to mobilization of epicardial progenitor cells and redirection of their differentiation toward cardiovascular cell types.

Finally, expression of CPD-photolyase in human keratinocytes was shown to increase the repair of DNA damage [126]. The chapter by Boros et al. in the current volume presents a protocol for this.

6.2 Vaccines Against Cancer

In 1909, Paul Ehrlich suggested that the immune system may suppress tumor development. Today his prediction is coming true—one of the most exciting and promising applications for synthetic mRNA is immunotherapy for cancer. Because of its great potential, this field has attracted many talented researchers and led to many scholarly publications. A very comprehensive review of the field has been published by Sahin et al. [9].

Furthermore, an entire volume of *Immunological Reviews* containing 17 review articles has been published on this topic, with an introductory chapter by Vonderheide and June [127]. The introductory chapter of the current volume will not attempt to cover this rapidly evolving field but rather will mention only topics that are relevant to the protocols presented here.

6.2.1 Ex Vivo Introduction of Synthetic mRNA into DCs

The idea of using mRNA to program DCs to present tumor antigens for cancer immunotherapy began with the pioneering work of Boczkowski et al. [128], who introduced ovalbumin mRNA into DCs with cationic lipids and then showed that mice vaccinated with these DCs were protected against a challenge with ovalbumin-expressing tumor cells. Seven of the chapters in this volume give protocols relating to introduction of synthetic mRNA into DCs.

Benteyn et al. present methodology to increase the translational efficiency of WT1 mRNA, and Coosemans et al. describe methodology for electroporation of mRNAs encoding WT1, survivin, and TriMix. The Wilms' tumor 1 antigen (WT1) was ranked by the National Cancer Institute as the most relevant for immunotherapeutic targeting [129]. Survivin is an inhibitor of apoptosis that is highly expressed in most cancers and associated with resistance to chemotherapy, tumor recurrence, and shorter patient survival [130]. TriMix is described in Subheading 4.1.

Devi and Nath show how to introduce FOXP3 mRNA into human DCs, generate mature DCs and FOXP3-specific cytotoxic T lymphocytes, and use these T-lymphocytes against inflammatory breast cancer cells. Immunization of mice with DCs transfected with the mRNA for FOXP3, a member of the forkhead winged helix family of transcriptional regulators, stimulates FOXP3-specific cytotoxic T lymphocytes that target the IBC cells of an aggressive subtype of breast cancer [131].

Derdelinckx et al. present a standardized and reproducible method for the manufacturing of GMP-grade mRNA-transfected DCs for clinical use. Selmeczi et al. describe the instrumentation and methods needed for the efficient transfection by electroporation of millions of DCs in one continuous flow process. Borch et al. present a method for performing immune monitoring using peripheral blood mononuclear cells and autologous DCs transfected with tumor antigen-encoding mRNA.

Kreiter et al. [92] discovered that the potency of the immunization can be enhanced if the FLT3 ligand is co-administered with the tumor-associated antigen. They present methods for analysis of FLT3 ligand effects as an adjuvant in combination with antitumor immunization using internodally injected naked mRNA.

6.2.2 Electroporation of T Cells with Synthetic mRNA

Two of the chapters in this volume present protocols in which synthetic mRNA is introduced into T cells by electroporation. Idorn et al. address the problem that only a small fraction of adoptively transferred T cells reach the tumor site by transfecting tumor infiltrating lymphocytes with mRNA encoding the chemokine receptor CXCR2. Koh et al. seek to redirect T cell specificity towards tumors expressing peptides from hepatitis B virus in hepatocellular carcinoma. They do this by electroporating primary human T lymphocytes with synthetic mRNA encoding a T cell receptors specific for hepatitis B viral antigens expressed on hepatocellular carcinoma cells.

6.3 Vaccines Against Infectious Diseases

The advantages of using synthetic mRNA to immunize against infectious diseases are the same as for cancer—only a transient exposure to antigen-encoding mRNA is needed to prime the immune system, and there is no possibility of integration into the host genome as with DNA. Several successes have been reported for mRNA-based immunotherapy against infectious diseases.

6.3.1 Influenza

The first report of an mRNA-based vaccine was for the influenza A virus [8]. Martinon et al. injected mice subcutaneously with liposome-entrapped synthetic mRNA encoding the nucleoprotein of influenza A virus. The cytotoxic T lymphocytes obtained were indistinguishable from those obtained in vivo with infectious virus in terms of specificity and lysed both peptide-sensitized and virus-infected targets. Petsch et al. [95] made synthetic mRNAs (without modified nucleosides) encoding the neuraminidase and hemagglutinin antigens of influence A virus and administered them by intradermal injection in mice, ferrets, and domestic pigs. They observed B and T cell-dependent protection against multiple influenza antigens. The protective effects were similar to those of a licensed influenza vaccine in pigs. Hekele et al. [107] developed a self-amplifying mRNA encoding the hemagglutinin antigen of influenza A virus. This was complexed with synthetic lipid nanoparticles and used to immunize mice by intramuscular injection. Two weeks after the second immunization, all mice had hemagglutinin inhibition titers that were considered protective.

6.3.2 HIV

An immunotherapy was devised to induce antigen-specific $CD8^+$ and $CD4^+$ T cell responses against HIV-1 proteins [132]. DCs were electroporated with mRNA encoding HIV-1 antigens (Gag, Vpr, Rev, and Nef) without modified nucleosides. The DCs were then injected intradermally into patients every four weeks for four treatments. Full or partial HIV-specific proliferative immune responses occurred in seven of nine subjects. In a second study, patients similarly received four vaccinations with autologous DCs electroporated with mRNA encoding Tat, Rev, and Nef [133]. The patients were then taken off antiretroviral therapy and remained so for 96 weeks. As before, enhanced $CD4^+$ and $CD8^+$ T cell responses specific for the immunogens were observed in most of the patients. In a third study [134], autologous DCs were electroporated with synthetic mRNA without modified nucleosides encoding Gag or a chimeric Tat-Rev-Nef protein. The DCs were administered to patients four times every four weeks. This caused an increase in the magnitude of the HIV-1-specific IFN-γ response and of T-cell proliferation. The antiviral response to HIV-1 was correlated with the magnitude of the Gag-specific IFN-γ response.

6.4 Engineering the Genome

There are two broad technologies that permit genome editing, which might be considered to be another form of gene therapy: (1) transposases [135] and (2) site-specific nucleases, which are further subdivided into zinc finger nucleases (ZFNs) and truncated transcription activator-like effector nucleases (TALENs) [136]. Both of these technologies can be implemented through introduction of either DNA or synthetic mRNA into cells. Since the transposases and nucleases are only required for a short duration for their action in genome editing, their transient expression from synthetic mRNA serves to reduce off-target consequences as well as unintended genome integration that can occur with DNA vectors. In one example, injecting ZFN-encoding ARCA-capped mRNA without nucleoside modifications into one-cell zebrafish (*Danio rerio*) embryos yielded a high percentage of animals carrying distinct mutations at the ZFN-specified position and exhibiting the expected loss-of-function phenotypes [137]. In another example, injection of either DNA or synthetic mRNA (ARCA-capped, posttranscriptionally polyadenylated) encoding ZFNs into the one-cell rat embryo produced animals carrying 25–100 % disruption at the target locus [138]. This has also been done for mRNA encoding Cas9 [139]. For the transposon termed *Sleeping Beauty*, injection of transposon vectors along with an ARCA-capped mRNA encoding the transposase (without nucleoside modification) into one-cell mouse embryos produced transposition, germ-line transmission, and expression from transposed elements [140]. The chapter by Bire et al. in this volume presents a protocol for producing mRNA encoding the transposase for the *piggyBac* transposon-based system and introducing it into both HeLa and stromal mesenchymal cells.

6.5 Generation of iPSCs

One of the most important biomedical advances of this decade is the ability to convert somatic cells to stem cells by the introduction of DNA encoding a small set of transcription factors [141]. This achievement by Shinya Yamanaka and coworkers, along with much earlier work by John Gurdon on replacing the nucleus of a fertilized *Xenopus* egg with the nucleus of a somatic cell, was recognized by award of the 2012 Nobel Prize in Physiology or Medicine "for the discovery that mature cells can be reprogrammed to become pluripotent". The resulting iPSCs have enormous potential for hematopoietic cell-based therapies and regenerative medicine approaches for the treatment of diabetes, liver disease, neurologic and retinal diseases, muscular dystrophies, and heart disease [142, 143].

Although the initial work of Yamanaka utilized DNA vectors, it was subsequently learned that this could be achieved with synthetic mRNA [13, 144, 145].

Warren et al. [146] prepared synthetic ARCA-capped mRNA with modified nucleosides encoding the four canonical Yamanaka factors, KLF4, c-MYC, OCT4, and SOX2. A poly(A) tract was added with a PCR reaction using a T_{120}-heeled reverse primer. mRNAs were introduced with cationic lipids into murine embryonic fibroblasts, human epidermal keratinocytes, and other cell types. This resulted in reprogramming of cells to pluripotency with efficiencies that greatly surpassed established protocols. The same technology was used to direct the differentiation of these iPSCs into terminally differentiated myogenic cells.

Yakubov et al. [147] prepared synthetic ARCA-capped mRNA encoding the same four transcription factors with a poly(A) tract being transcribed from the plasmid. After five consecutive transfections of human foreskin fibroblasts using a cationic lipid (Lipofectamine 2000), these authors observed formation of iPSC colonies that expressed alkaline phosphatase and several embryonic stem cell markers.

Plews et al. [148] synthesized mRNAs encoding OCT4, SOX2, cMYC, KLF4, and SV40 large T antigen with the AmpliCap-Max™ system, which employs a symmetrical m^7Gpppm^7G cap and provides for poly(A) tract transcription from the plasmid. They electroporated human fibroblast cells with the mixture of mRNAs. After 30 days of culture in human embryonic stem cell medium, they observed small aggregates positive for alkaline phosphatase activity and OCT4 protein.

Subsequent publications have developed and improved the methodology for iPSC generation by synthetic mRNA [149–151]. Lu et al. present a protocol in this volume for introducing synthetic mRNA into human pancreatic progenitor cells as a first step toward their differentiation for transplantation.

7 Concluding Remarks

Paradigm-breaking discoveries and the development of new protocols over the past decade have established synthetic mRNA as a powerful tool in such varied fields as proteins production for disease mitigation, induction of specific immune responses to tumor antigens or infectious agents, editing the genome, and creation of iPSCs for regenerative medicine. Methods have been developed to synthesize mRNAs with increased stability, increased translational efficiency, and decreased immunogenicity for these medical applications as well as for such investigative purposes as photo-cross-linking to macromolecules, biophysical measurements using fluorescence, and understanding the mechanisms involved in mRNA synthesis, translation, and degradation. The 20 chapters of this volume presenting protocols cover a range of topics in the broad areas of mRNAs synthesis, introduction into cells, and measurement of the resulting physiological changes. The future looks bright for fulfillment of the initial promise of gene therapy, albeit through RNA-mediated rather than DNA-mediated technologies.

Acknowledgements

The author thanks all those who generously contributed their time and talent to write protocols for this volume as well as Dr. Anren Song, University of Texas Medical School, and Dr. Dixie Jones, Louisiana State University Health Sciences Center in Shreveport, for assistance with the scientific literature.

References

1. Nirenberg M, Caskey T, Marshall R, Brimacombe R, Kellogg D, Doctor B, Hatfield D, Levin J, Rottman F, Pestka S, Wilcox M, Anderson F (1966) The RNA code and protein synthesis. Cold Spring Harb Symp Quant Biol 31:11–24
2. Anderson WF, Fletcher JC (1980) Sounding boards. Gene therapy in human beings: when is it ethical to begin? N Engl J Med 303:1293–1297
3. Anderson WF (1992) Human gene therapy. Science 256:808–813
4. Waldrop MM (1992) Molecular Biology—Finding RNA makes proteins gives 'RNA World' a big boost. Science 256:1396–1397
5. Tan PH, Janes SE, Handa AI, Friend PJ (2008) More trouble ahead; is gene therapy coming of age? Expert Opin Biol Ther 8:561–567
6. Wolff JA, Malone RW, Williams P, Chong W, Acsadi G, Jani A, Felgner PL (1990) Direct gene transfer into mouse muscle *in vivo*. Science 247:1465–1468
7. Jirikowski GF, Sanna PP, Maciejewski-Lenoir D, Bloom FE (1992) Reversal of diabetes insipidus in Brattleboro rats: intrahypothalamic injection of vasopressin mRNA. Science 255:996–998
8. Martinon F, Krishnan S, Lenzen G, Magne R, Gomard E, Guillet JG, Levy JP, Meulien P (1993) Induction of virus-specific cytotoxic T lymphocytes in vivo by liposome-entrapped mRNA. Eur J Immunol 23:1719–1722
9. Sahin U, Karikó K, Tureci O (2014) mRNA-based therapeutics—developing a new class of drugs. Nat Rev Drug Discov 13 (10):759–780
10. Kallen KJ, Thess A (2014) A development that may evolve into a revolution in medicine: mRNA as the basis for novel, nucleotide-based vaccines and drugs. Ther Adv Vaccines 2:10–31
11. Quabius ES, Krupp G (2015) Synthetic mRNAs for manipulating cellular phenotypes: an overview. N Biotechnol 32:229–235
12. Weissman D (2015) mRNA transcript therapy. Expert Rev Vaccines 14:265–281
13. Bernal JA (2013) RNA-based tools for nuclear reprogramming and lineage-conversion: towards clinical applications. J Cardiovasc Transl Res 6:956–968
14. Caruthers MH (2013) The chemical synthesis of DNA/RNA: our gift to science. J Biol Chem 288:1420–1427
15. Studier FW, Moffatt BA (1986) Use of bacteriophage T7 RNA polymerase to direct selective high-level expression of cloned genes. J Mol Biol 189:113–130
16. Krieg PA, Melton DA (1984) Functional messenger RNAs are produced by SP6 in vitro transcription of cloned cDNAs. Nucleic Acids Res 12:7057–7070
17. Furuichi Y, Shatkin AJ (2000) Viral and cellular mRNA capping: past and prospects. Adv Virus Res 55:135–184
18. Martin SA, Moss B (1975) Modification of RNA by mRNA guanylyltransferase and mRNA (guanine-7-)methyltransferase from vaccinia virions. J Biol Chem 250:9330–9335

19. Contreras R, Cheroutre H, Degrave W, Fiers W (1982) Simple, efficient *in vitro* synthesis of capped RNA useful for direct expression of cloned eukaryotic genes. Nucleic Acids Res 10:6353–6362
20. Konarska MM, Padgett RA, Sharp PA (1984) Recognition of a cap structure in splicing in vitro of mRNA precursors. Cell 38:731–736
21. Yisraeli JK, Melton DA (1989) Synthesis of long, capped transcripts in vitro by SP6 and T7 RNA polymerases. Methods Enzymol 180:42–50
22. Gallie DR, Tanguay R (1994) Poly(A) binds to initiation factors and increases cap-dependent translation *in vitro*. J Biol Chem 269:17166–17173
23. Decker CJ, Parker R (1993) A turnover pathway for both stable and unstable mRNAs in yeast: evidence for a requirement for deadenylation. Genes Dev 7:1632–1643
24. Martin G, Keller W (1998) Tailing and 3′-end labeling of RNA with yeast poly(A) polymerase and various nucleotides. RNA 4:226–230
25. Palmiter RD, Carey NH (1974) Rapid inactivation of ovalbumin messenger ribonucleic acid after acute withdrawal of estrogen. Proc Natl Acad Sci U S A 71:2357–2361
26. Lodish HF, Small B (1976) Different lifetimes of reticulocyte messenger RNA. Cell 7:59–65
27. Guyette WA, Matusik RJ, Rosen JM (1979) Prolactin-mediated transcriptional and post-transcriptional control of casein gene expression. Cell 17:1013–1023
28. Sittman DB, Graves RA, Marzluff WF (1983) Histone mRNA concentrations are regulated at the level of transcription and mRNA degradation. Proc Natl Acad Sci U S A 80:1849–1853
29. Gallie DR (1991) The cap and poly(A) function synergistically to regulate mRNA translation efficiency. Genes Dev 5:2108–2116
30. Preiss T, Hentze MW (1998) Dual function of the messenger RNA cap structure in poly (A)-tail-promoted translation in yeast. Nature 392:516–520
31. Michel YM, Poncet D, Piron M, Kean KM, Borman AM (2000) Cap-poly(A) synergy in mammalian cell-free extracts. J Biol Chem 275:32268–32276
32. Lall S, Friedman CC, Jankowska-Anyszka M, Stepinski J, Darzynkiewicz E, Davis RE (2004) Contribution of *trans*-splicing, 5′ - leader length, cap-poly(A) synergism, and initiation factors to nematode translation in an *Ascaris suum* embryo cell-free system. J Biol Chem 279:45573–45585
33. Mockey M, Goncalves C, Dupuy F, Lemoine F, Pichon C, Midoux P (2006) mRNA transfection of dendritic cells: synergistic effect of ARCA mRNA capping with Poly(A) chains in *cis* and in *trans* for a high protein expression level. Biochem Biophys Res Commun 340:1062–1068
34. Pasquinelli AE, Dahlberg JE, Lund E (1995) Reverse 5′ caps in RNAs made *in vitro* by phage RNA polymerases. RNA 1:957–967
35. Marcotrigiano J, Gingras A-C, Sonenberg N, Burley SK (1997) Cocrystal structure of the messenger RNA 5′ cap-binding protein (eIF4E) bound to 7-methyl-GDP. Cell 89:951–961
36. Matsuo H, Li H, McGuire AM, Fletcher CM, Gingras A-C, Sonenberg N, Wagner G (1997) Structure of translation factor eIF4E bound to m^7GDP and interaction with 4E-binding protein. Nat Struct Biol 4:717–724
37. Deshmukh MV, Jones BN, Quang-Dang DU, Flinders J, Floor SN, Kim C, Jemielity J, Kalek M, Darzynkiewicz E, Gross JD (2008) mRNA decapping is promoted by an RNA-binding channel in Dcp2. Mol Cell 29:324–336
38. Stepinski J, Waddell C, Stolarski R, Darzynkiewicz E, Rhoads RE (2001) Synthesis and properties of mRNAs containing the novel "anti-reverse" cap analogues 7-methyl(3′-O-methyl)GpppG and 7-methyl(3′-deoxy) GpppG. RNA 7:1486–1495
39. Jemielity J, Fowler T, Zuberek J, Stepinski J, Lewdorowicz M, Niedzwiecka A, Stolarski R, Darzynkiewicz E, Rhoads RE (2003) Novel "anti-reverse" cap analogues with superior translational properties. RNA 9:1108–1122
40. Peng Z-H, Sharma V, Singleton SF, Gershon PD (2002) Synthesis and application of a chain-terminating dinucleotide mRNA cap analog. Org Lett 4:161–164
41. Kore AR, Shanmugasundaram M, Charles I, Cheng AM, Barta TJ (2007) Synthesis and application of 2′-fluoro-substituted cap analogs. Bioorg Med Chem Lett 17:5295–5299
42. Kore AR, Shanmugasundaram M, Vlassov AV (2008) Synthesis and application of a new 2′,3′-isopropylidene guanosine substituted cap analog. Bioorg Med Chem Lett 18:4828–4832
43. Kore AR, Shanmugasundaram M, Charles I, Vlassov AV, Barta TJ (2009) Locked nucleic acid (LNA)-modified dinucleotide mRNA cap analogue: synthesis, enzymatic incorporation, and utilization. J Am Chem Soc 131:6364–6365
44. Grudzien E, Kalek M, Jemielity J, Darzynkiewicz E, Rhoads RE (2006) Differential

inhibition of mRNA degradation pathways by novel cap analogs. J Biol Chem 281:1857–1867

45. Rabinovich P, Komarovskaya M, Ye Z, Imai C, Campana D, Bahceci E, Weissman S (2006) Synthetic messenger RNA as a tool for gene therapy. Hum Gene Ther 17:1027–1035
46. Li Y, Kiledjian M (2010) Regulation of mRNA decapping. Wiley Interdiscip Rev RNA 1:253–265
47. Chen CY, Shyu AB (2011) Mechanisms of deadenylation-dependent decay. Wiley Interdiscip Rev RNA 2:167–183
48. Liu H, Rodgers ND, Jiao X, Kiledjian M (2002) The scavenger mRNA decapping enzyme DcpS is a member of the HIT family of pyrophosphatases. EMBO J 21:4699–4708
49. Wang Z, Jiao X, Carr-Schmid A, Kiledjian M (2002) The hDcp2 protein is a mammalian mRNA decapping enzyme. Proc Natl Acad Sci U S A 99:12663–12668
50. Lykke-Andersen J (2002) Identification of a human decapping complex associated with hUpf proteins in nonsense-mediated decay. Mol Cell Biol 22:8114–8121
51. van Dijk E, Cougot N, Meyer S, Babajko S, Wahle E, Seraphin B (2002) Human Dcp2: a catalytically active mRNA decapping enzyme located in specific cytoplasmic structures. EMBO J 21:6915–6924
52. Piccirillo C, Khanna R, Kiledjian M (2003) Functional characterization of the mammalian mRNA decapping enzyme hDcp2. RNA 9:1138–1147
53. She M, Decker CJ, Chen N, Tumati S, Parker R, Song H (2006) Crystal structure and functional analysis of Dcp2p from *Schizosaccharomyces pombe*. Nat Struct Mol Biol 13:63–70
54. Kalek M, Jemielity J, Grudzien E, Zuberek J, Bojarska E, Cohen LS, Stepinski J, Stolarski R, Davis RE, Rhoads RE, Darzynkiewicz E (2005) Synthesis and biochemical properties of novel mRNA 5′ cap analogs resistant to enzymatic hydrolysis. Nucleosides Nucleotides Nucleic Acids 24:615–621
55. Grudzien-Nogalska E, Jemielity J, Kowalska J, Darzynkiewicz E, Rhoads RE (2007) Phosphorothioate cap analogs stabilize mRNA and increase translational efficiency in mammalian cells. RNA 13:1745–1755
56. Kowalska J, Lewdorowicz M, Zuberek J, Grudzien-Nogalska E, Bojarska E, Stepinski J, Rhoads RE, Darzynkiewicz E, Davis RE, Jemielity J (2008) Synthesis and characterization of mRNA cap analogs containing phosphorothioate substitutions that bind tightly to eIF4E and are resistant to the decapping pyrophosphatase DcpS. RNA 14:1119–1131
57. Kuhn AN, Diken M, Kreiter S, Selmi A, Kowalska J, Jemielity J, Darzynkiewicz E, Huber C, Tureci O, Sahin U (2010) Phosphorothioate cap analogs increase stability and translational efficiency of RNA vaccines in immature dendritic cells and induce superior immune responses *in vivo*. Gene Ther 17:961–971
58. Su W, Slepenkov S, Grudzien-Nogalska E, Kowalska J, Kulis M, Zuberek J, Lukaszewicz M, Darzynkiewicz E, Jemielity J, Rhoads RE (2011) Translation, stability, and resistance to decapping of mRNAs containing caps substituted in the triphosphate chain with BH_3, Se, and NH. RNA 17:978–988
59. Ziemniak M, Szabelski M, Lukaszewicz M, Nowicka A, Darzynkiewicz E, Rhoads RE, Wieczorek Z, Jemielity J (2013) Synthesis and evaluation of fluorescent cap analogues for mRNA labelling. RSC Adv 3:20943–20958
60. Gunawardana D, Domashevskiy AV, Gayler KR, Goss DJ (2015) Efficient preparation and properties of mRNAs containing a fluorescent cap analog: Anthraniloyl-m^7GpppG. Translation 3, e988538
61. Baranowski MR, Nowicka A, Rydzik AM, Warminski M, Kasprzyk R, Wojtczak BA, Wojcik J, Claridge TD, Kowalska J, Jemielity J (2015) Synthesis of fluorophosphate nucleotide analogues and their characterization as tools for ^{19}F NMR studies. J Org Chem 80:3982–3997
62. Paredes E, Das SR (2011) Click chemistry for rapid labeling and ligation of RNA. Chembiochem 12:125–131
63. Holstein JM, Stummer D, Rentmeister A (2015) Enzymatic modification of 5′-capped RNA with a 4-vinylbenzyl group provides a platform for photoclick and inverse electron-demand Diels–Alder reaction. Chem Sci 6:1362–1369
64. Nowakowska M, Kowalska J, Martin F, d'Orchymont A, Zuberek J, Lukaszewicz M, Darzynkiewicz E, Jemielity J (2014) Cap analogs containing 6-thioguanosine—reagents for the synthesis of mRNAs selectively photo-crosslinkable with cap-binding biomolecules. Org Biomol Chem 12:4841–4847
65. Pisareva VP, Pisarev AV, Komar AA, Hellen CU, Pestova TV (2008) Translation initiation on mammalian mRNAs with structured 5′UTRs requires DExH-box protein DHX29. Cell 135:1237–1250

66. Olivares E, Landry DM, Caceres CJ, Pino K, Rossi F, Navarrete C, Huidobro-Toro JP, Thompson SR, Lopez-Lastra M (2014) The 5′ untranslated region of the human T-cell lymphotropic virus type 1 mRNA enables cap-independent translation initiation. J Virol 88:5936–5955

67. Bugaut A, Balasubramanian S (2012) 5′-UTR RNA G-quadruplexes: translation regulation and targeting. Nucleic Acids Res 40:4727–4741

68. Kuhn AN, Beibetaert T, Simon P, Vallazza B, Buck J, Davies BP, Tureci O, Sahin U (2012) mRNA as a versatile tool for exogenous protein expression. Curr Gene Ther 12:347–361

69. Karikó K, Kuo A, Barnathan E (1999) Overexpression of urokinase receptor in mammalian cells following administration of the in vitro transcribed encoding mRNA. Gene Ther 6:1092–1100

70. Holtkamp S, Kreiter S, Selmi A, Simon P, Koslowski M, Huber C, Tureci O, Sahin U (2006) Modification of antigen-encoding RNA increases stability, translational efficacy, and T-cell stimulatory capacity of dendritic cells. Blood 108:4009–4017

71. Zinckgraf JW, Silbart LK (2003) Modulating gene expression using DNA vaccines with different 3′-UTRs influences antibody titer, seroconversion and cytokine profiles. Vaccine 21:1640–1649

72. Gustafsson C, Govindarajan S, Minshull J (2004) Codon bias and heterologous protein expression. Trends Biotechnol 22:346–353

73. Bossi L, Ruth JR (1980) The influence of codon context on genetic code translation. Nature 286:123–127

74. Kimchi-Sarfaty C, Oh JM, Kim IW, Sauna ZE, Calcagno AM, Ambudkar SV, Gottesman MM (2007) A "silent" polymorphism in the MDR1 gene changes substrate specificity. Science 315:525–528

75. Mauro VP, Chappell SA (2014) A critical analysis of codon optimization in human therapeutics. Trends Mol Med 20:604–613

76. Tarun SZ Jr, Wells SE, Deardorff JA, Sachs AB (1997) Translation initiation factor eIF4G mediates in vitro poly(A) tail-dependent translation. Proc Natl Acad Sci U S A 94:9046–9051

77. Preiss T, Muckenthaler M, Hentze MW (1998) Poly(A)-tail-promoted translation in yeast: implications for translational control. RNA 4:1321–1331

78. Wakiyama M, Imataka H, Sonenberg N (2000) Interaction of eIF4G with poly(A)-binding protein stimulates translation and is critical for *Xenopus* oocyte maturation. Curr Biol 10:1147–1150

79. Amrani N, Ghosh S, Mangus DA, Jacobson A (2008) Translation factors promote the formation of two states of the closed-loop mRNP. Nature 453:1276–1280

80. Tomek W, Wollenhaupt K (2012) The "closed loop model" in controlling mRNA translation during development. Anim Reprod Sci 134:2–8

81. Archer SK, Shirokikh NE, Hallwirth CV, Beilharz TH, Preiss T (2015) Probing the closed-loop model of mRNA translation in living cells. RNA Biol 12:248–254

82. Richter JD (2008) Breaking the code of polyadenylation-induced translation. Cell 132:335–337

83. Eckmann CR, Rammelt C, Wahle E (2011) Control of poly(A) tail length. Wiley Interdiscip Rev RNA 2:348–361

84. Brown JA, Valenstein ML, Yario TA, Tycowski KT, Steitz JA (2012) Formation of triple-helical structures by the 3′-end sequences of MALAT1 and MENβ noncoding RNAs. Proc Natl Acad Sci U S A 109:19202–19207

85. Marzluff WF, Wagner EJ, Duronio RJ (2008) Metabolism and regulation of canonical histone mRNAs: life without a poly(A) tail. Nat Rev Genet 9:843–854

86. Su W, Slepenkov SV, Slevin MK, Lyons SM, Ziemniak M, Kowalska J, Darzynkiewicz E, Jemielity J, Marzluff WF, Rhoads RE (2013) mRNAs containing the histone 3′ stem-loop are degraded primarily by decapping mediated by oligouridylation of the 3′ end. RNA 19:1–16

87. Mullen TE, Marzluff WF (2008) Degradation of histone mRNA requires oligouridylation followed by decapping and simultaneous degradation of the mRNA both 5′ to 3′ and 3′ to 5′. Genes Dev 22:50–65

88. Pascolo S (2008) Vaccination with messenger RNA (mRNA). Handb Exp Pharmacol 221–235

89. Wang W, Li W, Ma N, Steinhoff G (2013) Non-viral gene delivery methods. Curr Pharm Biotechnol 14:46–60

90. Diken M, Kreiter S, Selmi A, Britten CM, Huber C, Tureci O, Sahin U (2011) Selective uptake of naked vaccine RNA by dendritic cells is driven by macropinocytosis and abrogated upon DC maturation. Gene Ther 18:702–708

91. Kreiter S, Selmi A, Diken M, Koslowski M, Britten CM, Huber C, Tureci O, Sahin U (2010) Intranodal vaccination with naked

antigen-encoding RNA elicits potent prophylactic and therapeutic antitumoral immunity. Cancer Res 70:9031–9040

92. Kreiter S, Diken M, Selmi A, Diekmann J, Attig S, Husemann Y, Koslowski M, Huber C, Tureci O, Sahin U (2011) FLT3 ligand enhances the cancer therapeutic potency of naked RNA vaccines. Cancer Res 71:6132–6142
93. Van Lint S, Goyvaerts C, Maenhout S, Goethals L, Disy A, Benteyn D, Pen J, Bonehill A, Heirman C, Breckpot K, Thielemans K (2012) Preclinical evaluation of TriMix and antigen mRNA-based antitumor therapy. Cancer Res 72:1661–1671
94. Rittig SM, Haentschel M, Weimer KJ, Heine A, Muller MR, Brugger W, Horger MS, Maksimovic O, Stenzl A, Hoerr I, Rammensee HG, Holderried TA, Kanz L, Pascolo S, Brossart P (2011) Intradermal vaccinations with RNA coding for TAA generate $CD8^+$ and $CD4^+$ immune responses and induce clinical benefit in vaccinated patients. Mol Ther 19:990–999
95. Petsch B, Schnee M, Vogel AB, Lange E, Hoffmann B, Voss D, Schlake T, Thess A, Kallen KJ, Stitz L, Kramps T (2012) Protective efficacy of in vitro synthesized, specific mRNA vaccines against influenza A virus infection. Nat Biotechnol 30:1210–1216
96. Malone RW, Felgner PL, Verma IM (1989) Cationic liposome-mediated RNA transfection. Proc Natl Acad Sci U S A 86:6077–6081
97. Conry RM, LoBuglio AF, Wright M, Sumerel L, Pike MJ, Johanning F, Benjamin R, Lu D, Curiel DT (1995) Characterization of a messenger RNA polynucleotide vaccine vector. Cancer Res 55:1397–1400
98. Granstein RD, Ding W, Ozawa H (2000) Induction of anti-tumor immunity with epidermal cells pulsed with tumor-derived RNA or intradermal administration of RNA. J Invest Dermatol 114:632–636
99. Kormann MS, Hasenpusch G, Aneja MK, Nica G, Flemmer AW, Herber-Jonat S, Huppmann M, Mays LE, Illenyi M, Schams A, Griese M, Bittmann I, Handgretinger R, Hartl D, Rosenecker J, Rudolph C (2011) Expression of therapeutic proteins after delivery of chemically modified mRNA in mice. Nat Biotechnol 29:154–157
100. Hoerr I, Obst R, Rammensee HG, Jung G (2000) *In vivo* application of RNA leads to induction of specific cytotoxic T lymphocytes and antibodies. Eur J Immunol 30:1–7
101. Weide B, Pascolo S, Scheel B, Derhovanessian E, Pflugfelder A, Eigentler TK, Pawelec G, Hoerr I, Rammensee HG, Garbe C (2009) Direct injection of protamine-protected mRNA: results of a phase 1/2 vaccination trial in metastatic melanoma patients. J Immunother 32:498–507
102. Su Z, Dannull J, Yang BK, Dahm P, Coleman D, Yancey D, Sichi S, Niedzwiecki D, Boczkowski D, Gilboa E, Vieweg J (2005) Telomerase mRNA-transfected dendritic cells stimulate antigen-specific CD8+ and CD4+ T cell responses in patients with metastatic prostate cancer. J Immunol 174:3798–3807
103. Selmeczi D, Hansen TS, Met O, Svane IM, Larsen NB (2011) Efficient large volume electroporation of dendritic cells through micrometer scale manipulation of flow in a disposable polymer chip. Biomed Microdevices 13:383–392
104. Lenz P, Bacot SM, Frazier-Jessen MR, Feldman GM (2003) Nucleoporation of dendritic cells: efficient gene transfer by electroporation into human monocyte-derived dendritic cells. FEBS Lett 538:149–154
105. Mandl CW, Aberle JH, Aberle SW, Holzmann H, Allison SL, Heinz FX (1998) In vitro-synthesized infectious RNA as an attenuated live vaccine in a flavivirus model. Nat Med 4:1438–1440
106. Geall AJ, Verma A, Otten GR, Shaw CA, Hekele A, Banerjee K, Cu Y, Beard CW, Brito LA, Krucker T, O'Hagan DT, Singh M, Mason PW, Valiante NM, Dormitzer PR, Barnett SW, Rappuoli R, Ulmer JB, Mandl CW (2012) Nonviral delivery of self-amplifying RNA vaccines. Proc Natl Acad Sci U S A 109:14604–14609
107. Hekele A, Bertholet S, Archer J, Gibson DG, Palladino G, Brito LA, Otten GR, Brazzoli M, Buccato S, Bonci A, Casini D, Maione D, Qi ZQ, Gill JE, Caiazza NC, Urano J, Hubby B, Gao GF, Shu Y, De Gregorio E, Mandl CW, Mason PW, Settembre EC, Ulmer JB, Craig Venter J, Dormitzer PR, Rappuoli R, Geall AJ (2013) Rapidly produced SAM® vaccine against H7N9 influenza is immunogenic in mice. Emerg Microbes Infect 2, e52
108. Uchida S, Itaka K, Uchida H, Hayakawa K, Ogata T, Ishii T, Fukushima S, Osada K, Kataoka K (2013) In vivo messenger RNA introduction into the central nervous system using polyplex nanomicelle. PLoS One 8, e56220
109. Nallagatla SR, Toroney R, Bevilacqua PC (2011) Regulation of innate immunity through RNA structure and the protein kinase PKR. Curr Opin Struct Biol 21:119–127
110. Mathews MB, Korner A (1970) The inhibitory action of a mammalian viral RNA on the

initiation of protein synthesis in a reticulocyte cell-free system. Eur J Biochem 17:339–343

111. Graziadei WD 3rd, Weideli H, Lengyel P (1973) Endonuclease of high specific activity in a purified mouse interferon preparation. Biochem Biophys Res Commun 54:40–46
112. Alexopoulou L, Holt AC, Medzhitov R, Flavell RA (2001) Recognition of double-stranded RNA and activation of NF-kappaB by Toll-like receptor 3. Nature 413:732–738
113. Diebold SS, Kaisho T, Hemmi H, Akira S, Reis e Sousa C (2004) Innate antiviral responses by means of TLR7-mediated recognition of single-stranded RNA. Science 303:1529–1531
114. Heil F, Hemmi H, Hochrein H, Ampenberger F, Kirschning C, Akira S, Lipford G, Wagner H, Bauer S (2004) Species-specific recognition of single-stranded RNA via toll-like receptor 7 and 8. Science 303:1526–1529
115. Karikó K, Buckstein M, Ni H, Weissman D (2005) Suppression of RNA recognition by Toll-like receptors: the impact of nucleoside modification and the evolutionary origin of RNA. Immunity 23:165–175
116. Hornung V, Ellegast J, Kim S, Brzozka K, Jung A, Kato H, Poeck H, Akira S, Conzelmann KK, Schlee M, Endres S, Hartmann G (2006) 5′-Triphosphate RNA is the ligand for RIG-I. Science 314:994–997
117. Karikó K, Muramatsu H, Welsh FA, Ludwig J, Kato H, Akira S, Weissman D (2008) Incorporation of pseudouridine into mRNA yields superior nonimmunogenic vector with increased translational capacity and biological stability. Mol Ther 16:1833–1840
118. Anderson BR, Muramatsu H, Nallagatla SR, Bevilacqua PC, Sansing LH, Weissman D, Karikó K (2010) Incorporation of pseudouridine into mRNA enhances translation by diminishing PKR activation. Nucleic Acids Res 38:5884–5892
119. Karikó K, Muramatsu H, Ludwig J, Weissman D (2011) Generating the optimal mRNA for therapy: HPLC purification eliminates immune activation and improves translation of nucleoside-modified, protein-encoding mRNA. Nucleic Acids Res 39, e142
120. Katze MG (1992) The war against the interferon-induced dsRNA-activated protein kinase—Can viruses win? J Interferon Res 12:241–248
121. Hyde JL, Gardner CL, Kimura T, White JP, Liu G, Trobaugh DW, Huang C, Tonelli M, Paessler S, Takeda K, Klimstra WB, Amarasinghe GK, Diamond MS (2014) A viral RNA structural element alters host recognition of nonself RNA. Science 343:783–787
122. Karikó K, Muramatsu H, Keller JM, Weissman D (2012) Increased erythropoiesis in mice injected with submicrogram quantities of pseudouridine-containing mRNA encoding erythropoietin. Mol Ther 20:948–953
123. Mays LE, Ammon-Treiber S, Mothes B, Alkhaled M, Rottenberger J, Muller-Hermelink ES, Grimm M, Mezger M, Beer-Hammer S, von Stebut E, Rieber N, Nurnberg B, Schwab M, Handgretinger R, Idzko M, Hartl D, Kormann MS (2013) Modified Foxp3 mRNA protects against asthma through an IL-10-dependent mechanism. J Clin Invest 123:1216–1228
124. Lui KO, Zangi L, Silva EA, Bu L, Sahara M, Li RA, Mooney DJ, Chien KR (2013) Driving vascular endothelial cell fate of human multipotent Isl1+ heart progenitors with VEGF modified mRNA. Cell Res 23:1172–1186
125. Zangi L, Lui KO, von Gise A, Ma Q, Ebina W, Ptaszek LM, Spater D, Xu H, Tabebordbar M, Gorbatov R, Sena B, Nahrendorf M, Briscoe DM, Li RA, Wagers AJ, Rossi DJ, Pu WT, Chien KR (2013) Modified mRNA directs the fate of heart progenitor cells and induces vascular regeneration after myocardial infarction. Nat Biotechnol 31:898–907
126. Boros G, Miko E, Muramatsu H, Weissman D, Emri E, Rozsa D, Nagy G, Juhasz A, Juhasz I, van der Horst G, Horkay I, Remenyik E, Karikó K, Emri G (2013) Transfection of pseudouridine-modified mRNA encoding CPD-photolyase leads to repair of DNA damage in human keratinocytes: a new approach with future therapeutic potential. J Photochem Photobiol B 129:93–99
127. Vonderheide RH, June CH (2014) Engineering T cells for cancer: our synthetic future. Immunol Rev 257:7–13
128. Boczkowski D, Nair SK, Snyder D, Gilboa E (1996) Dendritic cells pulsed with RNA are potent antigen-presenting cells in vitro and in vivo. J Exp Med 184:465–472
129. Cheever MA, Allison JP, Ferris AS, Finn OJ, Hastings BM, Hecht TT, Mellman I, Prindiville SA, Viner JL, Weiner LM, Matrisian LM (2009) The prioritization of cancer antigens: a national cancer institute pilot project for the acceleration of translational research. Clin Cancer Res 15:5323–5337
130. Fukuda S, Pelus LM (2006) Survivin, a cancer target with an emerging role in normal adult tissues. Mol Cancer Ther 5:1087–1098

131. Nair S, Aldrich AJ, McDonnell E, Cheng Q, Aggarwal A, Patel P, Williams MM, Boczkowski D, Lyerly HK, Morse MA, Devi GR (2013) Immunologic targeting of FOXP3 in inflammatory breast cancer cells. PLoS One 8, e53150
132. Routy JP, Boulassel MR, Yassine-Diab B, Nicolette C, Healey D, Jain R, Landry C, Yegorov O, Tcherepanova I, Monesmith T, Finke L, Sekaly RP (2010) Immunologic activity and safety of autologous HIV RNA-electroporated dendritic cells in HIV-1 infected patients receiving antiretroviral therapy. Clin Immunol 134:140–147
133. Allard SD, De Keersmaecker B, de Goede AL, Verschuren EJ, Koetsveld J, Reedijk ML, Wylock C, De Bel AV, Vandeloo J, Pistoor F, Heirman C, Beyer WE, Eilers PH, Corthals J, Padmos I, Thielemans K, Osterhaus AD, Lacor P, van der Ende ME, Aerts JL, van Baalen CA, Gruters RA (2012) A phase I/IIa immunotherapy trial of HIV-1-infected patients with Tat, Rev and Nef expressing dendritic cells followed by treatment interruption. Clin Immunol 142:252–268
134. Van Gulck E, Vlieghe E, Vekemans M, Van Tendeloo VF, Van De Velde A, Smits E, Anguille S, Cools N, Goossens H, Mertens L, De Haes W, Wong J, Florence E, Vanham G, Berneman ZN (2012) mRNA-based dendritic cell vaccination induces potent antiviral T-cell responses in HIV-1-infected patients. AIDS 26:F1–F12
135. Wilson MH, Coates CJ, George AL Jr (2007) *PiggyBac* transposon-mediated gene transfer in human cells. Mol Ther 15:139–145
136. Wood AJ, Lo TW, Zeitler B, Pickle CS, Ralston EJ, Lee AH, Amora R, Miller JC, Leung E, Meng X, Zhang L, Rebar EJ, Gregory PD, Urnov FD, Meyer BJ (2011) Targeted genome editing across species using ZFNs and TALENs. Science 333:307
137. Doyon Y, McCammon JM, Miller JC, Faraji F, Ngo C, Katibah GE, Amora R, Hocking TD, Zhang L, Rebar EJ, Gregory PD, Urnov FD, Amacher SL (2008) Heritable targeted gene disruption in zebrafish using designed zinc-finger nucleases. Nat Biotechnol 26:702–708
138. Geurts AM, Cost GJ, Freyvert Y, Zeitler B, Miller JC, Choi VM, Jenkins SS, Wood A, Cui X, Meng X, Vincent A, Lam S, Michalkiewicz M, Schilling R, Foeckler J, Kalloway S, Weiler H, Menoret S, Anegon I, Davis GD, Zhang L, Rebar EJ, Gregory PD, Urnov FD, Jacob HJ, Buelow R (2009) Knockout rats via embryo microinjection of zinc-finger nucleases. Science 325:433
139. Wang H, Yang H, Shivalila CS, Dawlaty MM, Cheng AW, Zhang F, Jaenisch R (2013) One-step generation of mice carrying mutations in multiple genes by CRISPR/Cas-mediated genome engineering. Cell 153:910–918
140. Dupuy AJ, Clark K, Carlson CM, Fritz S, Davidson AE, Markley KM, Finley K, Fletcher CF, Ekker SC, Hackett PB, Horn S, Largaespada DA (2002) Mammalian germ-line transgenesis by transposition. Proc Natl Acad Sci U S A 99:4495–4499
141. Takahashi K, Tanabe K, Ohnuki M, Narita M, Ichisaka T, Tomoda K, Yamanaka S (2007) Induction of pluripotent stem cells from adult human fibroblasts by defined factors. Cell 131:861–872
142. Okano H, Nakamura M, Yoshida K, Okada Y, Tsuji O, Nori S, Ikeda E, Yamanaka S, Miura K (2013) Steps toward safe cell therapy using induced pluripotent stem cells. Circ Res 112:523–533
143. Fox IJ, Daley GQ, Goldman SA, Huard J, Kamp TJ, Trucco M (2014) Stem cell therapy. Use of differentiated pluripotent stem cells as replacement therapy for treating disease. Science 345:1247391
144. Li M, Sancho-Martinez I, Izpisua Belmonte JC (2011) Cell fate conversion by mRNA. Stem Cell Res Ther 2:5
145. Schlaeger TM, Daheron L, Brickler TR, Entwisle S, Chan K, Cianci A, DeVine A, Ettenger A, Fitzgerald K, Godfrey M, Gupta D, McPherson J, Malwadkar P, Gupta M, Bell B, Doi A, Jung N, Li X, Lynes MS, Brookes E, Cherry AB, Demirbas D, Tsankov AM, Zon LI, Rubin LL, Feinberg AP, Meissner A, Cowan CA, Daley GQ (2015) A comparison of non-integrating reprogramming methods. Nat Biotechnol 33:58–63
146. Warren L, Manos PD, Ahfeldt T, Loh YH, Li H, Lau F, Ebina W, Mandal PK, Smith ZD, Meissner A, Daley GQ, Brack AS, Collins JJ, Cowan C, Schlaeger TM, Rossi DJ (2010) Highly efficient reprogramming to pluripotency and directed differentiation of human cells with synthetic modified mRNA. Cell Stem Cell 7:618–630
147. Yakubov E, Rechavi G, Rozenblatt S, Givol D (2010) Reprogramming of human fibroblasts to pluripotent stem cells using mRNA of four transcription factors. Biochem Biophys Res Commun 394:189–193
148. Plews JR, Li J, Jones M, Moore HD, Mason C, Andrews PW, Na J (2010) Activation of pluripotency genes in human fibroblast cells by a novel mRNA based approach. PLoS One 5, e14397

149. Warren L, Ni Y, Wang J, Guo X (2012) Feeder-free derivation of human induced pluripotent stem cells with messenger RNA. Sci Rep 2:657
150. Mandal PK, Rossi DJ (2013) Reprogramming human fibroblasts to pluripotency using modified mRNA. Nat Protoc 8:568–582
151. Yoshioka N, Gros E, Li HR, Kumar S, Deacon DC, Maron C, Muotri AR, Chi NC, Fu XD, Yu BD, Dowdy SF (2013) Efficient generation of human iPSCs by a synthetic self-replicative RNA. Cell Stem Cell 13:246–254

Part II

Synthesis of mRNA

Chapter 2

Synthetic Capped mRNAs for Cap-Specific Photo-Cross-Linking Experiments

Joanna Kowalska, Franck Martin, and Jacek Jemielity

Abstract

The 7-methylguanosine triphosphate cap present at the 5′ ends of eukaryotic mRNAs plays numerous roles in mRNA expression and metabolism. The identification and studies on cap-binding partners can be significantly advanced using tailored chemical tools such as synthetic cap analogues or RNAs carrying modified cap structures. Here we provide protocols for the production of mRNAs specifically labeled within the 5′ cap with a nucleoside capable of being photo-activated, either 6-thioguanosine or 7-methyl-6-thioguanosine, which can be used in photo-cross-linking experiments to identify or characterize cap-binding biomolecules. We also describe a protocol for the cross-linking experiments with capped RNAs to map histone H4 cap-binding pocket.

Key words 5′ cap analogs, 6-thioguanosine, mRNA, Transcription, Photo-cross-linking, Cap binding, Histone mRNA

1 Introduction

The 7-methylguanosine triphosphate cap present at the 5′ ends of eukaryotic mRNAs serves as a specific recognition site for several specialized biomolecules involved in mRNA expression and metabolism [1–3]. 6-Thioguanosine (^{6S}G) and other thionucleosides (Fig. 1) absorb light at wavelengths longer than 300 nm (Fig. 2); thus, nucleic acid modified with such nucleosides can be selectively photo-activated in the presence of natural nucleic acids and proteins (which absorb UV light at shorter wavelengths). In the presence of a binding partner a selective covalent cross-link can be formed upon irradiation. Therefore, RNAs capped with ^{6S}G-containing cap should be particularly useful to cross-linking experiments targeting 5′ cap binding biomolecules.

The synthesis of 5′ capped mRNA can be achieved by two principle approaches, both based on in vitro transcription. In the

Robert E. Rhoads (ed.), *Synthetic mRNA: Production, Introduction Into Cells, and Physiological Consequences*, Methods in Molecular Biology, vol. 1428, DOI 10.1007/978-1-4939-3625-0_2,

first approach, RNA synthesis is carried out in the presence of all four NTPs and a cap dinucleotide such as m^7GpppG [4]. The DNA template is designed so that G is the first transcribed nucleotide. The polymerase can initiate the transcription from the nucleoside moiety of GTP or m^7GpppG, thereby incorporating one of the nucleotides at the 5′-end of the nascent RNA. To increase the percentage of cap analog incorporation (capping efficiency), the concentration of GTP is lowered relative to the other NTPs and the concentration of cap dinucleotide is elevated (from four- to tenfold excess relative to GTP). The capping efficiency may reach up to 90 % [5]. To avoid so called reverse incorporation of cap dinucleotide resulting in non-functional $Gpppm^7G$-RNA transcript, a group of anti-reverse cap analogs (ARCAs) has been developed [6]. ARCAs contain a chemical modification (usually an *O*-methyl group) at either the 2′ or 3′ positions of the m^7Guo moiety to ensure only correct orientation [6–9]. An additional benefit of the ARCA modification is that it ensures precise location of additional modifications introduced into the cap with respect to the mRNA body. ARCA-type nucleotides can be used to introduce into the 5′ RNA cap various modifications such as bulky ribose substituents, [10, 11] phosphate-stabilizing modifications, [12–15] biotin [16] and fluorescent labels [17]. In the second approach for the synthesis of capped RNA, known as post-transcriptional capping, a polynucleotide chain is synthesized in vitro by an RNA polymerase, and then the m^7G cap is incorporated at the 5′ end by the capping enzyme from vaccinia virus which uses GTP as a substrate with concomitant release of P_i (from pppRNA) and PP_i (from GTP), followed by 7-methylation of the 5′-terminal G with S-adenosylmethionine [18]. This method is advantageous because it yields higher percentage of capping, but the repertoire of cap modifications is restricted by the capping enzyme's specificity towards GTP analogs [19]. A mixed approach for the synthesis of capped RNA combining chemical synthesis of Gppp-RNA followed by enzymatic methylation with human methyltransferase has also been reported [20].

We have recently shown that the co-transcriptional RNA capping method can be used to site-specifically incorporate a photo-reactive 6-thioguanosine (^{6S}G) moiety at the +1 position of capped mRNA (Fig. 3) [21]. This was achieved by the use of a properly designed novel ARCA containing ^{6S}G as the second nucleoside ($m_2{}^{7,2'\text{-}O}Gppp^{6S}G$, Fig. 1). The 2′-*O*-methyl group ensures only correctly oriented (forward) incorporation of the cap analog into RNA (i.e., to make the analog an ARCA). We have shown that the ^{6S}G moiety in $m_2{}^{7,2'\text{-}O}Gppp^{6S}G$ does not disturb the interaction with translation initiation factor 4E, which is a critical factor for efficient translation of capped mRNA. Luciferase mRNA capped with $m_2{}^{7,2'\text{-}O}Gppp^{6S}G$ was translated with lower efficiency than luciferase mRNA capped with $m_2{}^{7,3'\text{-}O}GpppG$, but it

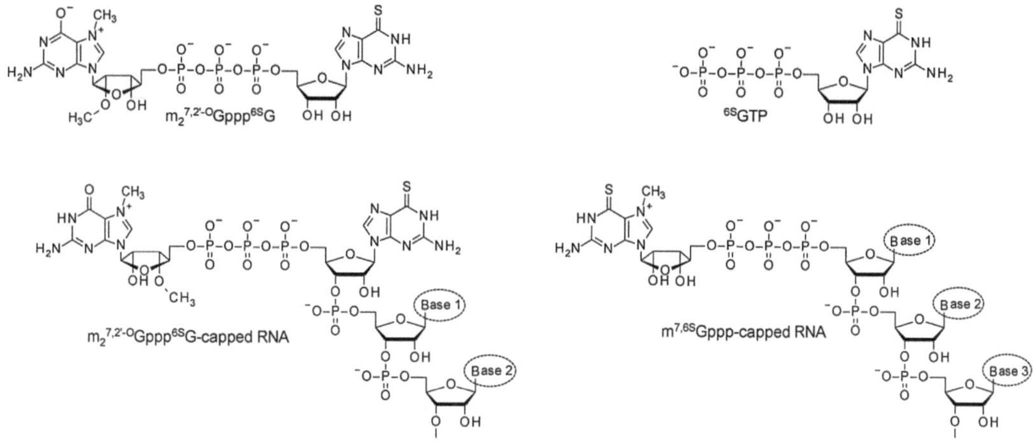

Fig. 1 Structures of 5′ capped RNAs containing 6-thioguanosine and 7-methyl-6-thioguanosine together with substrates for their preparation $m_2^{7,2'-O}Gppp^{6S}G$ and ^{6S}GTP, respectively

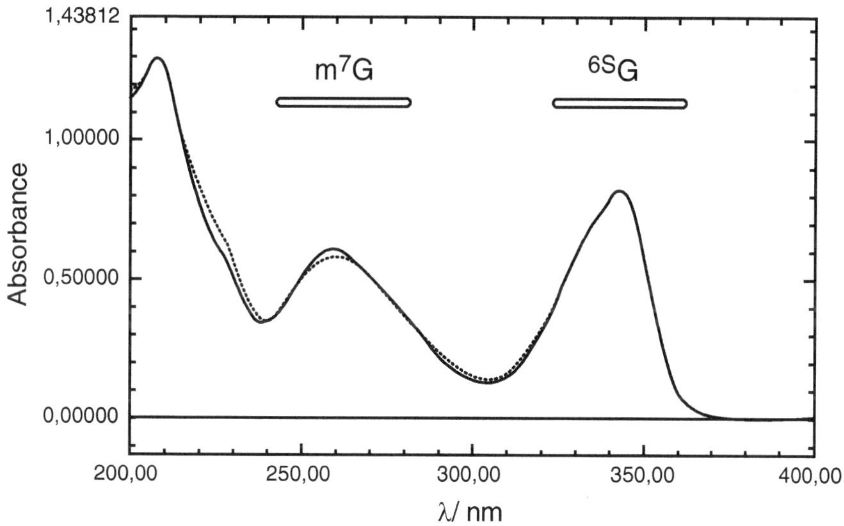

Fig. 2 UV spectrum of $m_2^{7,2'-O}Gppp^{6S}G$ in 0.1 M phosphate buffer pH 7 (*continuous line*) and pH6 (*dotted line*)

was still chiefly in a cap dependent manner [21]. In the UV spectrum of $m_2^{7,2'-O}Gppp^{6S}G$ the m^7G and ^{6S}G absorption bands are separated by ~60 nm (Fig. 2) confirming possibility of selective ^{6S}G activation. The post-transcriptional capping method, on the other hand, gives access to capped mRNAs carrying a photo-cross-linkable 7-methyl-6-thioguanosine moiety ($m^{7,6S}G$, Fig. 1). This can be achieved by replacing GTP with ^{6S}GTP (Fig. 3) in the capping reaction catalysed by commercially available VCE capping enzyme, with capping efficiencies reaching 95–100 % [22].

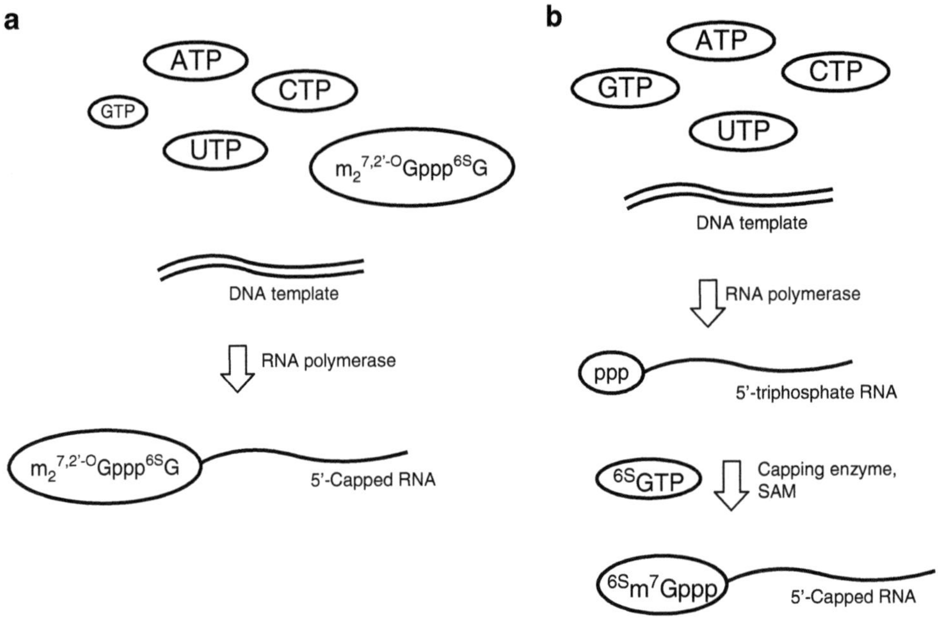

Fig. 3 Two strategies for site-specific incorporation of ^{6S}G into mRNA cap by transcription in vitro. (**a**) Co-transcriptional capping with $m_2^{7,2'-O}Gppp^{6S}G$ (cap analog **1**) (**b**) post-transcriptional capping using ^{6S}GTP and S-adenosyl methionine (SAM)-dependent capping enzyme

Interestingly, not only proteins but also nucleic acids can specifically recognize the cap. Histone H4 mRNA forms a pocket within the ORF region capable of sequestering its 5′ cap, which contributes to an atypical translation initiation regulatory mechanism [22]. We have shown that both $m_2^{7,2'-O}Gppp^{6S}G$- and $m^{7,6S}$GpppG-capped histone H4 mRNAs are useful for mapping the cap binding pocket using photo-cross-linking experiments (Fig. 4) [21, 22].

In this chapter, we provide protocols for: (1) the synthesis of the cross-linkable cap-analog $m_2^{7,2'-O}Gppp^{6S}G$, (2) construction of a plasmid template for the synthesis of a capped histone H4 mRNA by in vitro T7 RNA polymerase run-off transcription, (3) in vitro synthesis of $m_2^{7,2'-O}Gppp^{6S}G$-RNA-H4 and $m^{7,6S}$Gppp-RNA-H4 mRNA transcripts using co-transcriptional and posttranscriptional capping method, respectively, (4) cross-linking experiments with the capped RNAs to map histone H4 cap binding pocket, (5) analysis of cross-linking products by reverse-transcription and polyacrylamide gel electrophoresis.

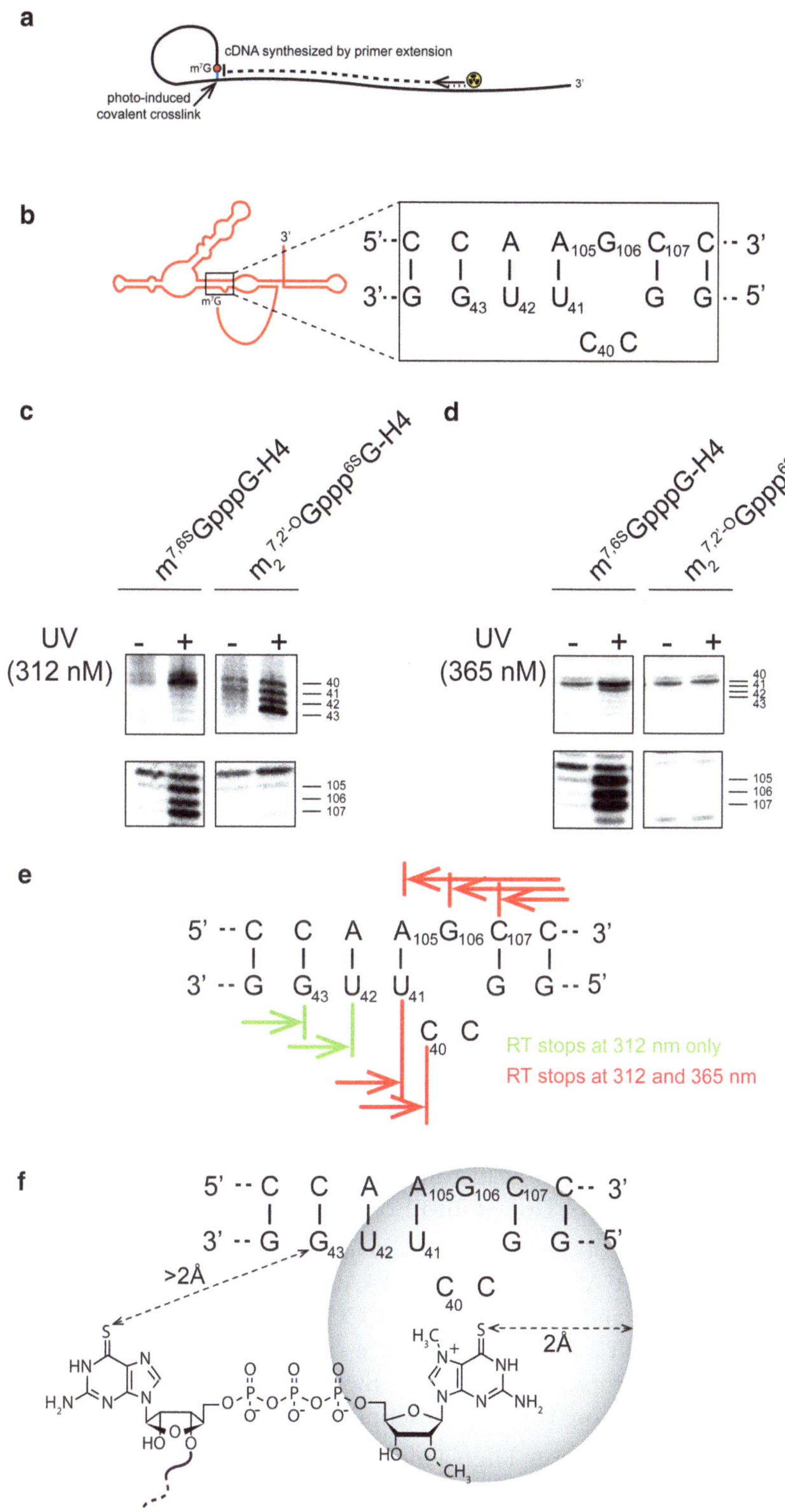

Fig. 4 The UV-cross-linking experiments on capped histone H4-RNAs. $m_2^{7,2'\text{-}O}Gppp^{6S}G$-RNA was obtained by co-transcriptional capping with $m_2^{7,2'\text{-}O}Gppp^{6S}G$, and $m^{7,6S}GpppG$-RNA by post-transcriptional capping with ^{6S}GTP. (**a**) The overview of the experiment. (**b**) the secondary structure of the cap-binding pocket (**c**) results of the cross-linking experiments at 312 nm and (**d**) 365 nm revealed by reverse-transcription of the cross-linked mRNA (**e**) Cross-linking sites detected after irradiation at 312 (*green*) and 365 nm (*red*) (365 nm reveals so-called "zero-length" cross-links). (**f**) Nucleotides responsible for cap binding identified by cross-linking

2 Materials

2.1 Synthesis of $m_2^{7,2'-O}Gppp^{6S}G$

1. 6-Thioguanosine. This is the commercially available staring material for the synthesis of $m_2^{7,2'-O}Gppp^{6S}G$.
2. $m_2^{7,2'-O}GDP$. This is the second nucleotide starting material and can be synthesized as described previously [6, 9].
3. Freshly distilled $POCl_3$.
4. Solid $NaHCO_3$.
5. 1.2 M triethylammonium bicarbonate buffer.
6. Imidazole.
7. 2,2′-dithiodipyridine.
8. Triphenylphosphine.
9. Triethylamine.
10. Sodium perchlorate.
11. Lithium perchlorate.
12. Anhydrous zinc chloride.
13. Disodium EDTA dihydrate.
14. Dry trimethylphosphate (dried over 4 Å molecular sieves).
15. DMF.
16. Acetone.
17. Deionized water.
18. Water/ice bath.
19. 1.2 M triethylammonium bicarbonate buffer (*see* **Note 1**).
20. DEAE Sephadex A-25 resin in HCO_3^- form (*see* **Note 2**).
21. 0.1 M phosphate buffer pH 7.0.

2.2 Template for the Synthesis of H4-12 mRNA

1. pUC19-H4 plasmid containing the coding region and UTRs of the murine histone H4-12 gene amplified from mouse genomic DNA cloned into EcoRI site [22, 23] (10 ng/μL).
2. 100 μM of the following primers: -5′-ATATTAATACGACT-CACTATAGGTCATAACCATGTCTGGACG-3′ (forward) and 5′-TGGGTGGCCCTGAAAAGGGC-3′ (reverse). The T7 promoter sequence (underlined) has been inserted in the forward primer and the reverse primer was designed to promote run-off transcription at the desired position.
3. Phusion High-Fidelity DNA polymerase (2 U/μL), Thermo Fisher Scientific Inc.
4. 5× Buffer High Fidelity (Thermo Fisher Scientific Inc), 25 mM of each dNTP.
5. DMSO 100 %.

6. Phenol.
7. Phenol–chloroform.
8. 4 M NaCl.
9. 100 % ethanol.
10. 80 % ethanol.

2.3 In Vitro Synthesis of Histone H4-12 mRNA

1. Autoclaved Milli-Q® water.
2. 5× TMSDT Transcription buffer: 200 mM Tris–HCl, pH 8.1, 110 mM $MgCl_2$, 5 mM spermidine, 25 mM DTT, 0.05 % (v/v) Triton X-100 (*see* **Note 2**).
3. RNasin® Ribonuclease Inhibitor (40 U/μL, Promega).
4. 50 mM ATP, 50 mM UTP, 50 mM CTP, and 50 mM GTP.
5. 10 mM cap dinucleotide (the substrate for co-transcriptional capping, *see* Subheading 2.1) or ^{6S}GTP (the substrate for post-transcriptional capping).
6. Vaccinia Capping Enzyme (10 U/μL) (ScriptCap™ capping system, CELLSCRIPT™) (only for reactions with ^{6S}GTP).
7. 2 mM S-adenosyl methionine (SAM) (only for reactions with ^{6S}GTP).
8. 1 μg/μL DNA template (PCR products).
9. Recombinant -His-tagged T7 RNA polymerase (0.2 mg/mL).
10. Pyrophosphatase (1 μg/μL).
11. RNase-free DNase I (2000 U/mL).
12. Illustra™ Sephadex™ G-25 purification columns.
13. 100 % ethanol.
14. 80 % ethanol.
15. 4 M NaCl.
16. Formamide dye.
17. 95 % formamide.
18. 0.025 % bromophenol blue.
19. 0.025 % xylene cyanol.

2.4 Photo-cross-linking Experiment

1. 1 μM solution of Capped RNA produced as described in Subheading 3.3.
2. Autoclaved Milli-Q® water and concentrated buffer for achieving the final composition.
3. AMV -Reverse transcriptase 15 U/μL.
4. 50,000 cpm/μL of 5′ ^{32}P-labeled H4-specific reverse primers 5′- GCAGGCTTGGTGATGCCCTG -3′ and 5′- TGAGGCC GGAGATGCGCTTC -3′

3 Methods

3.1 Synthesis of $m_2^{7,2\text{-}O}Gppp^{6S}G$

3.1.1 Synthesis of ^{6S}GMP-Im

1. Mix 1.24 g of 6-thioguanosine (3.11 mmol) in 19 mL of trimethyl phosphate in a round bottom flask: equipped with magnetic stir bar, close with a stopper.
2. Cool the mixture to 0 °C on ice/water bath.
3. Add phosphorus oxychloride (782 μL, 7.78 mmol).
4. Keep at 0 °C under vigorous stirring for 48 h.
5. Quench the reaction by the addition of 5 % aqueous $NaHCO_3$ solution.
6. Purify by DEAE-Sephadex using a linear gradient from 0 to 0.7 M TEAB (*see* **Note 3**).
7. Analyze fractions at 340 nm, collect the product (^{6S}GMP) and concentrate the eluate under reduced pressure until white solid residue is obtained (*see* **Note 4**). The expected yield is around 1 g (2.12 mmol, 65–70 %) of ^{6S}GMP.
8. To the flask containing solid ^{6S}GMP (around 1 g, 2.12 mmol), add 1.44 g (21.2 mmol) of imidazole.
9. Add 1.40 g (6.35 mmol) 2,2′-dithiodipyridine.
10. Add 10 mL of DMF.
11. Add 800 μL of triethylamine.
12. Stir the mixture for few min to obtain suspension of all reagents in the solvent.
13. Add 1.66 g (6.35 mmol) of triphenylphosphine.
14. Stir the mixture for 6–8 h. A clear solution should be obtained with reaction progress.
15. Dissolve anhydrous $LiClO_4$ (1.04 g, 8.47 mmol) in dry acetone (100 mL)
16. Precipitate the product (^{6S}GMP-Im) from the reaction mixture with the solution prepared in p. 15.
17. Cool the mixture at 4 °C for 2 h.
18. Separate the precipitate from the supernatant by filtration or centrifugation. Discard the supernatant. Wash the precipitate three times with 50 mL of acetone.
19. Dry the product under in a vacuum desiccator over P_4O_{10}. The expected yields is around 0.9 g of ^{6S}GMP-Im.

3.1.2 Coupling ^{6S}GMP-Im and $m_2^{7,2\text{-}O}$GDP

1. Mix 52 mg (0.11 mmol) of $m_2^{7,2\text{-}O}$GDP and 60 mg (0.13 mmol) of ^{6S}GMP in a round bottom flask.
2. Suspended in 1.85 mL of DMF and stir for 5 min.

3. Add anhydrous 179 mg (1.33 mmol) of zinc chloride and continue stirring. Rapid dissolution of the reagents should be observed.
4. Continue stirring at room temperature for 24 h.
5. Prepare a solution of solid $NaHCO_3$ (246 mg, 2.93 mmol) and EDTA (493 mg, 1.33 mmol) in 18.5 mL of water.
6. Add the solution of -EDTA and $NaHCO_3$ to the reaction mixture.
7. Purify the product on a DEAE-Sephadex using a linear gradient from 0 to 1.2 M TEAB (*see* **Note 3**).
8. Analyze fractions at 340 nm, collect the product ($m_2^{7,2'\text{-}O}Gppp^{6S}G$), and concentrate the eluate to solid residue under reduced pressure (*see* **Note 4**).
9. Prepare a solution of $NaClO_4$ (76 mg, 0.624 mmol) in acetone (20 mL).
10. Dissolve solid $m_2^{7,2'\text{-}O}Gppp^{6S}G$ in 0.5–1.0 mL of water and add the $NaClO_4$ solution.
11. Cool the mixture at 4 °C for 2 h.
12. Separate the precipitate (the sodium salt of $m_2^{7,2'\text{-}O}Gppp^{6S}G$) from the supernatant by filtration or centrifugation. Discard the supernatant. Wash the precipitate three times with 20 mL of acetone. Dry in a vacuum desiccator over P_2O_5.
13. Redissolve the precipitate in water and freeze-dry. The expected yield is around 36 mg (0.043 mmol, 40 %) of $m_2^{7,2'\text{-}O}Gppp^{6S}G$, trisodium salt.
14. Prepare ~15 mM solution of $m_2^{7,2'\text{-}O}Gppp^{6S}G$ by dissolving ~1.8 mg in 200 μl deionized water. Determine the exact concentration spectrophotometrically in 0.1 M phosphate buffer pH 7 at 340 nm using ε_{340} 24800 cm^{-1} M^{-1}. Dilute an aliquot of the ~15 mM solution to obtain 10 mM solution of the cap.

3.2 Amplification of a PCR Template for H4-12 mRNA Transcript

1. Prepare 100 μL PCR amplification mixture in a 0.5 mL centrifuge tube by adding the following reagents on ice in the order listed: 20 μL 5× Buffer HF, 1 μL of dNTP, 2 μL of each primer (forward and reverse), 2 μL of plasmid pUC19-H4 -template, 3 μL of DMSO, and 1 μL of Phusion High-Fidelity DNA polymcrasc.
2. Incubate in a thermocycler for 32 cycles of 15 s at 95 °C, 15 s at 55 °C, and 1 min 30 s at 72 °C.
3. Transfer the reaction mixture in a 1.5 mL centrifuge tube and add 100 μL of H_2O.
4. Treat the reaction mixture with 200 μL of phenol, vortex, briefly centrifuge.

5. Discard the organic phase and transfer the aqueous phase in a 1.5 mL centrifuge tube.
6. Treat the aqueous phase with 200 μL of phenol–chloroform, vortex, and briefly centrifuge.
7. Discard the organic phase and transfer the aqueous phase in a 1.5 mL centrifuge tube.
8. Precipitate the PCR products by adding 12.5 μL of NaCl and 600 μL of ethanol 100 %, incubate at −20 °C.
9. Centrifuge, discard the supernatant and wash the DNA pellet with ethanol 80 %, dry the DNA pellet, and dissolve the DNA pellet in 20 μL of H_2O.

3.3 In Vitro Synthesis of Histone H4-12 mRNA Capped with $m_2^{7,2'-O}Gppp^{6S}G$

1. Prepare 100 μL transcription reaction mixture in a 1.5 mL centrifuge tube by adding the following reagents at room temperature in the order listed: 20 μL of DNA template (Subheading 3.3), 20 μL 5× TMSDT transcription buffer; 1 μL RNasin Inhibitor; 10 μL ATP; 10 μL UTP; 10 μL CTP; 5 μL of cap analog; 5 μL of T7 RNA polymerase.
2. Incubate the reaction at 37 °C.
3. Add 2 μL of GTP after 10 min, then another 2 μL after 20, 30, 40, and 50 min.
4. Treat the reaction mixture with 2 μL of Pyrophosphatase and incubate 30 min at 37 °C to degrade the excess of pyrophosphate.
5. Treat the reaction mixture with 2 μL of RNase-free DNase I and incubate 30 min at 37 °C to degrade the DNA template.
6. The unincorporated nucleotides are trapped on a 1 mL-Sephadex G-25 column and the transcripts are phenol-extracted and precipitated with ethanol 100 %.

3.4 In Vitro Synthesis of Histone H4-12 mRNA Capped with $m^{7,6S}Gppp$

1. Prepare 100 μL transcription reaction mixture in a 1.5 mL centrifuge tube by adding the following reagents at room temperature in the order listed: 20 μL of DNA template (Subheading 3.3), 20 μL 5× TMSDT transcription buffer; 1 μL RNasin Inhibitor; 10 μL ATP; 10 μL UTP; 10 μL CTP; 10 μL GTP; 5 μL of T7 RNA polymerase.
2. Incubate the reaction at 37 °C for 1 h.
3. Treat the reaction mixture with 2 μL of Pyrophosphatase and incubate 30 min at 37 °C to degrade the excess of pyrophosphate.
4. Treat the reaction mixture with 2 μL of RNase-free DNase I and incubate 30 min at 37 °C to degrade the DNA template.
5. Purify the transcripts by denaturing 4 % urea-polyacrylamide gel electrophoresis and electro-elute from gel slices with a Biotrap apparatus.

6. Precipitate the eluted transcripts with ethanol 100 % at −20 °C, centrifuge, wash with ethanol 80 % and dry the RNA pellet.
7. Dissolve the RNA -transcript in 20 μL of H_2O and determine the RNA concentration by absorbance at 260 nm. 60 μg of pure RNA transcript is used to cap with ^{6S}GTP using the ScriptCap™ kit.
8. Prepare 100 μL capping reaction mixture in a 1.5 mL centrifuge tube by adding the following reagents on ice in the order listed: 60 μg of pure RNA transcript, 10 μL 10× Buffer ScriptCap™, 5 μL SAM, 10 μL ^{6S}GTP, 2,5 μL of ScriptCap™ inhibitor, 4 μL of ScriptCap™ Vaccine Capping Enzyme. Incubate for 1 h at 37 °C.
9. Clean the capped-RNA by phenol-extraction and precipitate with ethanol 100 % at −20 °C.

3.5 Photo-cross-linking Experiments

1. Irradiate 1 μM solutions of $m_2^{7,2'-O}$Gppp^{6S}G-RNA-H4 or $^{6S}m^7$GpppG-RNA-H4 with 312 and 365 nm for 30 min on ice (e.g., using a Bio-Link BLX 312).
2. Perform a reverse transcription reaction containing 1 μM of cross-linked RNA, with 50,000 cpm of ^{32}P-labelled H4-specific reverse primers and 10 U of AMV reverse transcriptase.
3. Incubate at 37 °C for 30 min.
4. Add NaCl and ethanol 100 % to precipitate the synthesized cDNA overnight at −20 °C.
5. Centrifuge then wash the pellet with ethanol 80 % and dry the cDNA pellet.
6. Dissolve the cDNA pellet in 6 μL of Formamide Dye.
7. Separate the synthesized cDNA by denaturing polyacrylamide gel electrophoresis 8 % with 8 M urea and reveal by autoradiography.

4 Notes

1. 1.2 M TEAB buffer can either be prepared by saturating a mixture of 500 mL of triethylamine and 2500 mL of deionized water with gaseous CO_2 or purchased commercially.
2. The resin is available commercially in the Cl^- (chloride) form and before use should be converted into HCO_3^- form by swelling in 1 M $NaHCO_3$ followed by excessive washing with 1 M $NaHCO_3$ and then with deionized water.
3. TEAB is unstable at room temperature, therefore, the chromatography should be performed at 4–8 °C.

4. The temperature during evaporation should not exceed 30–35 °C. Repeated additions of ethanol during evaporation are recommended to accelerate decomposition of TEAB buffer.

Acknowledgements

This research was supported by grant UMO-2012/05/E/ST5/03893 from the National Science Centre (to J.J.) and grants ANR-06-BLAN-0206-01 and ANR-2011-SVSE8-025-01 from CNRS (Centre National de la Recherche Scientifique) and Agence Nationale pour la Recherche (to F.M.).

References

1. Furuichi Y, Shatkin AJ (2000) Advances in virus research, vol 55. Academic Press Inc, San Diego, pp 135–184
2. Cougot N, van Dijk E, Babajko S, Séraphin B (2004) Cap-tabolism. Trends Biochem Sci 29:436–444
3. Rhoads RE (2009) eIF4E: new family members, new binding partners, new roles. J Biol Chem 284:16711–16715
4. Contreras R, Cheroutre H, Degrave W, Fiers W (1982) Simple, efficient in vitro synthesis of capped RNA useful for direct expression of cloned eukaryotic genes. Nucleic Acids Res 10:6353–6362
5. Grudzien-Nogalska E, Kowalska J, Su W, Kuhn A, Slepenkov S, Darzynkiewicz E, Sahin U, Jemielity J, Rhoads R (2013) In: Rabinovich PM (ed) Synthetic messenger RNA and cell metabolism modulation, vol 969. Humana Press, Totowa, pp 55–72
6. Grudzien-Nogalska E, Stepinski J, Jemielity J, Zuberek J, Stolarski R, Rhoads RE, Darzynkiewicz E (2007) Synthesis of anti-reverse cap analogs (ARCAs) and their applications in mRNA translation and stability. Method Enzymol 431:203–227
7. Stepinski J, Waddell C, Stolarski R, Darzynkiewicz E, Rhoads RE (2001) Synthesis and properties of mRNAs containing the novel "anti-reverse" cap analogs 7-methyl(3′-O-methyl) GpppG and 7-methyl(3′-deoxy)GpppG. RNA 7:1486–1495
8. Peng ZH, Sharma V, Singleton SF, Gershon PD (2002) Synthesis and application of a chain-terminating dinucleotide mRNA cap analog. Org Lett 4:161–164
9. Jemielity J, Fowler T, Zuberek J, Stepinski J, Lewdorowicz M, Niedzwiecka A, Stolarski R, Darzynkiewicz E, Rhoads RE (2003) Novel "anti-reverse" cap analogs with superior translational properties. RNA 9:1108–1122
10. Kore AR, Shanmugasundaram M, Charles I, Vlassov AV, Barta TJ (2009) Locked nucleic acid (LNA)-modified dinucleotide mRNA Cap analogue: synthesis, enzymatic incorporation, and utilization. J Am Chem Soc 131 (18):6364–6365
11. Kore AR, Charles I (2010) Synthesis and evaluation of 2′-O-allyl substituted dinucleotide cap analog for mRNA translation. Bioorg Med Chem 18:8061–8065
12. Kalek M, Jemielity J, Darzynkiewicz ZM, Bojarska E, Stepinski J, Stolarski R, Davis RE, Darzynkiewicz E (2006) Enzymatically stable 5′ mRNA cap analogs: Synthesis and binding studies with human DcpS decapping enzyme. Bioorg Med Chem 14:3223–3230
13. Jemielity J, Kowalska J, Rydzik AM, Darzynkiewicz E (2010) Synthetic mRNA cap analogs with a modified triphosphate bridge—synthesis, applications and prospects. New J Chem 34:829–844
14. Rydzik AM, Kulis M, Lukaszewicz M, Kowalska J, Zuberek J, Darzynkiewicz ZM, Darzynkiewicz E, Jemielity J (2012) Synthesis and properties of mRNA cap analogs containing imidodiphosphate moiety-fairly mimicking natural cap structure, yet resistant to enzymatic hydrolysis. Bioorg Med Chem 20 (5):1699–1710
15. Kowalska J, Wypijewska del Nogal A, Darzynkiewicz ZM, Buck J, Nicola C, Kuhn AN, Lukaszewicz M, Zuberek J, Strenkowska M, Ziemniak M et al (2014) Synthesis, properties, and biological activity of boranophosphate analogs of the mRNA cap: versatile tools for manipulation of therapeutically relevant

cap-dependent processes. Nucleic Acids Res 42:10245–10264

16. Jemielity J, Lukaszewicz M, Kowalska J, Czarnecki J, Zuberek J, Darzynkiewicz E (2012) Synthesis of biotin labelled cap analogue—incorporable into mRNA transcripts and promoting cap-dependent translation. Org Biomol Chem 10:8570–8574
17. Ziemniak M, Szabelski M, Lukaszewicz M, Nowicka A, Darzynkiewicz E, Rhoads RE, Wieczorek Z, Jemielity J (2013) Synthesis and evaluation of fluorescent cap analogues for mRNA labelling. RSC Adv 3:20943–20958
18. Martin SA, Paoletti E, Moss B (1975) Purification of messenger-Rna guanylyltransferase and messenger-RNA(guanine-7-)-methyltransferase from vaccinia virions. J Biol Chem 250:9322–9329
19. Issur M, Bougie I, Despins S, Bisaillon M (2013) Enzymatic synthesis of RNAs capped with nucleotide analogues reveals the molecular basis for substrate selectivity of RNA capping enzyme: impacts on RNA metabolism. PLoS One 8, e75310
20. Thillier Y, Decroly E, Morvan F, Canard B, Vasseur J-J, Debart F (2012) Synthesis of 5′ cap-0 and cap-1 RNAs using solid-phase chemistry coupled with enzymatic methylation by human (guanine-N7)-methyl transferase. RNA 18:856–868
21. Nowakowska M, Kowalska J, Martin F, d'Orchymont A, Zuberek J, Lukaszewicz M, Darzynkiewicz E, Jemielity J (2014) Cap analogs containing 6-thioguanosine—reagents for the synthesis of mRNAs selectively photo-crosslinkable with cap-binding biomolecules. Org Biomol Chem 12:4841–4847
22. Martin F, Barends S, Jaeger S, Schaeffer L, Prongidi-Fix L, Eriani G (2011) Cap-assisted internal initiation of translation of histone H4. Mol Cell 41:197–209
23. Meier VS, Böhni R, Schümperli D (1989) Nucleotide sequence of two mouse histone H4 genes. Nucleic Acids Res 17:795

Chapter 3

Enzymatic Modification of 5′-Capped RNA and Subsequent Labeling by Click Chemistry

Josephin M. Holstein, Daniela Stummer, and Andrea Rentmeister

Abstract

The combination of enzymatic modification and bioorthogonal click chemistry provides a powerful approach for site-specific labeling of different classes of biomolecules in vitro and even in cellular environments. Herein, we describe a chemoenzymatic method to site specifically label 5′-capped model mRNAs independent of their sequence. A trimethylguanosine synthase was engineered to introduce alkyne, azido, or 4-vinylbenzyl moieties to the 5′-cap. These functional groups were then used for labeling using typical click reactions, such as the azide-alkyne cycloaddition or the tetrazine ligation.

Key words RNA, Cap, Chemoenzymatic, Click chemistry, AdoMet analog, RNA modification, RNA labeling

1 Introduction

Localization of eukaryotic mRNAs to distinct regions within the cytoplasm provides a basis for developmental processes such as axial patterning [1–3], asymmetric cell division [4], cell migration [5,6], and neuronal maturation [7,8]. The ability to acquire complete spatiotemporal profiles of RNA transport and dynamics is therefore of tremendous importance to understand cell function and behavior in conditions of health and disease and in response to external stimuli. To study mRNA localization and dynamics, various imaging techniques have been developed ranging from hybridization-based probes [9–11] to genetically encoded protein- or RNA-based reporters [12–16]. Recently, so-called chemoenzymatic approaches have emerged as flexible tools to label RNAs in complex surroundings [17–20]. In contrast to approaches based on chemical synthesis [21–23], a reactive moiety is firstly introduced by enzymes that either accept nonnatural nucleotides [17,24,25] or that modify RNA site specifically. In a second step, the reactive handle is

Robert E. Rhoads (ed.), *Synthetic mRNA: Production, Introduction Into Cells, and Physiological Consequences*, Methods in Molecular Biology, vol. 1428, DOI 10.1007/978-1-4939-3625-0_3,

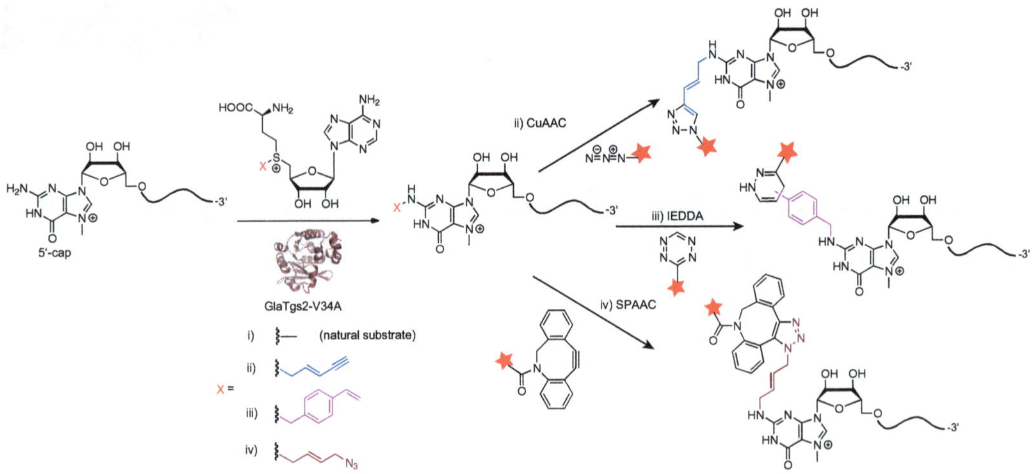

Fig. 1 Strategy for chemo-enzymatic labeling of 5′-capped RNAs. A trimethylguanosine synthase variant (GlaTgs2-V34A) can transfer (*i*) methyl, (*ii*) pentenynyl, (*iii*) 4-vinylbenzyl, or (*iv*) 4-azidobut-2-enyl residues to the position N^2 of the 5′-cap using AdoMet or respective analogs. The introduced functional groups can subsequently be labeled with a tag (shown as *red star*) by Cu(I)-catalyzed azide-alkyne cycloaddition (CuAAC), inverse electron-demand Diels-Alder (IEDDA) reaction, or strain-promoted azide-alkyne cycloaddition (SPAAC), respectively

covalently coupled to a reporter molecule using click chemistry [18,19,26]. From the repertoire of existing click reactions Cu(I)-catalyzed azide-alkyne cycloaddition (CuAAC) [18,19], Staudinger ligation [22], strain-promoted azide-alkyne cycloaddition (SPAAC) [27,28], photoclick reaction [29,30], and the tetrazine ligation [23,24] have successfully been applied to RNA.

In this chapter, we describe a chemoenzymatic approach for site-specific introduction of a reactive moiety to the 5′-cap typical of eukaryotic mRNAs and subsequent derivatization using bioorthogonal click chemistry (Fig. 1). A variant of the *Giardia lamblia* trimethylguanosine synthase 2 (GlaTgs2), which catalyzes the methyl transfer from *S*-adenosyl-L-methionine (AdoMet) to the N^2 atom of RNA 5′-caps [31], was used to install alkyne [18], azido [28], or 4-vinylbenzyl groups [29] on a eukaryotic model mRNA. In a next step, terminal alkyne groups are reacted in the CuAAC to label in vitro produced and capped RNA with a fluorophore of choice [18]. In a similar way, bioorthogonal labeling via SPAAC was achieved by reacting the azido moiety with a reporter molecule [28]. The 4-vinylbenzyl-modified 5′-cap could be converted in a tetrazine ligation based on the inverse electron-demand Diels-Alder (IEDDA) reaction [29]. This approach (1) enables site-specific installation of small fluorescent reporters on endogenous (i.e., non-tagged) mRNA, (2) is compatible with cellular functionalities, and (3) not toxic to living cells when bioorthogonal reactions like SPAAC or IEDDA are used. This strategy bears potential for cap-dependent isolation and labeling of eukaryotic mRNAs in living cells that in combination with modern sensitive live cell imaging techniques could open up new possibilities for investigation of transport mechanisms, dynamics, and misregulation of these mRNAs.

2 Materials

Prepare all solutions using ultrapure, RNase-free water, and analytical grade reagents. Handle and store all reagents at room temperature (unless indicated otherwise). Final compound concentrations are given in parenthesis.

2.1 Synthesis of AdoMet Analogs

1. *S*-Adenosyl-L-homocysteine.
2. 1,4-Dibromobut-2-ene.
3. 4-Vinylbenzyl chloride.
4. Trifluoroacetic acid (TFA).
5. Silver perchlorate.
6. Mesyl chloride.
7. Sodium azide.
8. (*E*)-2-Penten-4-yn-1-ol (Alfa Aesar).
9. Diethyl ether, petroleum ether, ethyl acetate, dichloromethane, and DMF should be used in technical and acetonitrile in HPLC grade.
10. Saturated sodium hydrogencarbonate solution.
11. Formic and acetic acid solution (1:1), freshly prepared.
12. Preparative HPLC column (Nucleodur® C_{18} Pyramid (5 μm, 125 × 10 mm, 10 mm ID), Macherey Nagel).
13. Thin layer chromatography (TLC): silica gel plates with fluorescence indicator on aluminum sheets.
14. Proton nuclear magnetic resonance spectra (^{1}H NMR) can be recorded on an Agilent DD2 600 MHz or a Bruker 300 MHz instrument.
15. ESI-TOF mass spectra can be recorded on an Agilent 6224 instrument and ESI-Orbitrap-MS on a LTQ Orbitrap XL™ Hybrid Ion Trap-Orbitrap Mass Spectrometer from Thermo Scientific.

2.2 Recombinant Production and Purification of Enzymes

1. Plasmid pRSET-A-GlaTgs2-V34A was generated by cloning GlaTgs2-V34A into the expression vector pRSET-A (Invitrogen) via *BamHI* and *XhoI* sites.
2. The plasmids pET29a-LuxS and pProEx-MTAN were a gift from Prof. Birgit Dräger (University of Halle, Germany) (*see* Acknowledgments).

3. *E. coli* Tuner(DE3)pLacI cells (Novagen).
4. 2YT medium: 1.6 % (w/v) tryptone, 125 mM sodium chloride, 1 % (w/v) yeast extract.
5. Ampicillin: 100 mg/mL in 70 % ethanol, sterile-filtrated. Store at −20 °C.
6. Kanamycin: 25 mg/mL in water, sterile-filtrated. Store at −20 °C.
7. Glucose: 33 wt% in water, sterile-filtrated.
8. IPTG: 0.16 M in water, sterile-filtrated. Store at −20 °C.
9. Absolute ethanol (molecular biology grade).
10. Protease inhibitor: 10 mM PMSF in 2-propanol. Store at −20 °C.
11. GlaTgs2-V34A lysis buffer: 50 mM Tris–HCl, 200 mM NaCl, 10 % glycerol (v/v), pH 8.
12. GlaTgs2-V34A elution buffer: lysis buffer plus 500 mM imidazole.
13. GlaTgs2-V34A storage buffer: 50 mM Tris–HCl, 100 mM NaCl, 1 mM EDTA, 10 % glycerol (v/v), 2 mM DTT, pH 8 (*see* **Note 1**).
14. 5′-methylthioadenosine nucleosidase (MTAN) lysis buffer: 50 mM sodium phosphate, pH 7.5.
15. MTAN elution buffer: lysis buffer plus 250 mM imidazole.
16. MTAN storage buffer: lysis buffer plus 50 mM HEPES and 10 % glycerol.
17. *S*-Ribosylhomocysteine lyase (LuxS) lysis buffer: 30 mM sodium phosphate, pH 7.2, 500 mM NaCl.
18. LuxS elution buffer: lysis buffer plus 500 mM imidazole.
19. LuxS storage buffer: lysis buffer plus 50 mM HEPES, 10 % glycerol.
20. Filtropur S (0.25 μm).
21. ÄKTA purifier™ protein purification system.
22. HisTrap™ FF 1 mL columns (GE Healthcare).
23. Vivaspin 500 (Sartorius) protein concentration system.

2.3 T7 In Vitro Transcription

1. DNA template (~100 nt, *see* Acknowledgments).
2. 2 U/μL Phusion® DNA polymerase.
3. NaOAc solution: 3 M in water, pH 5.5.
4. dNTPs (each 25 mM). Store at −20 °C.
5. 3× transcription buffer: 120 mM Tris–HCl, pH 8.1, 15 mM DTT, 6 mM spermidine, 0.03 % (v/v) Triton-X-100, 4.5 % (w/v) PEG 6000, sterile-filtrated. Store at −20 °C.

6. $Mg(OAc)_2$ solution: 500 mM in water.
7. NTPs (each 25 mM). Store at −20 °C.
8. 50 U/μL T7 RNA polymerase.
9. 1 U/μL DNase I.
10. 20 U/μL inorganic pyrophosphatase.
11. Phenol: Roti®-Aqua-Phenol, pH 4.5–5.
12. Chloroform (molecular biology grade).

2.4 Capping of RNA

1. Vaccinia capping system (NEB).
2. 40 U/μL RiboLock RNase Inhibitor.
3. Tip of a spatula of cation exchange P11 cellulose phosphate resuspended in 50 μL 1× PBS.

2.5 Labeling of RNA

1. Dibenzylcyclooctyne-Sulforhodamine B (DBCO-SRB, Click Chemistry Tools or Jena Bioscience) was dissolved in DMSO to obtain a 20 mM stock solution.
2. Tetrazine-5-TAMRA (Jena Bioscience) was dissolved in DMSO to obtain a 20 mM stock solution.
3. 40 U/μL RiboLock RNase Inhibitor.
4. 1× PBS.
5. PUS buffer: 100 mM Tris-HCl, 10 mM $MgCl_2$, 100 mM NH_4OAc, 0.1 mM EDTA, 0.5 mM DTT, pH 8.0.
6. Click Solution: DMSO:*t*BuOH 3:1.
7. Eterneon™-Azide (480/653) (Jena Bioscience): 2.5 mM in Click Solution.
8. Tris[(1-benzyl-1*H*-1,2,3-triazol-4-yl)methyl]amin (TBTA, Jena Bioscience): 111 mM in Click Solution.
9. Low-range RNA Ladder.
10. 2× RNA loading buffer: 95 % (v/v) formamide, 87 μM SDS, 0.5 mM EDTA.
11. Four drops of 1 % ethidium bromide solution in 500 mL 1× TAE buffer.
12. Fluorescence imager: The ChemoCam Imager ECL TypeHR 16-3200 from Intas or the Imager Versa Doc™ MP 4000 from Biorad can be used.

3 Methods

3.1 Synthesis of AdoViBe

1. To synthesize 5′-[(*R/S*)-[(3*S*)-3-amino-3-carboxypropyl]4-vinylbenzylsulfonio]-5′-deoxyadenosine (AdoViBe) dissolve *S*-adenosyl-L-homocysteine (10 mg, 26 mmol) in 1.5 mL of a

mixture of formic and acetic acid (1:1) and cool the solution to 0 °C in an ice bath. To protect the reaction mixture from light, wrap the flask into aluminum foil.

2. Slowly add 4-vinylbenzyl chloride (400 μL, 2.8 mmol) to the stirred solution. Allow the solution to warm up to room temperature by removing the ice bath and continue stirring for 4 days (*see* **Notes 2** and **3**).
3. Stop the reaction after 4 days by gently transferring the reaction mixture to cool water (15 mL).
4. Extract the aqueous phase 3× with 5 mL diethyl ether and keep the aqueous phase.
5. Remove the solvent from the crude product by lyophilization. Do this by flash freezing the solution with liquid nitrogen and connecting the flask to a freeze dryer. Make sure that the sample stays frozen during lyophilization (*see* **Note 4**).
6. Redissolve the obtained white powder in 1.5 mL water containing 0.01 % TFA.
7. Purify the crude product by preparative reversed-phase HPLC. Adenine-containing components can be detected at 260 nm. Perform the elution at a flowrate of 5 mL/min using water containing 0.01 % TFA as eluent A and a gradient of acetonitrile containing 0.01 % TFA from 0 to 7 % within 15 min, followed by a steep gradient to 70 % acetonitrile in the following 5 min. The diastereomeric mixture elutes with a retention time of 14 min (*see* **Note 5**).
8. Pool and freeze the product-containing fractions with liquid nitrogen and freeze dry them (*see* **Note 4**).
9. Redissolve the lyophilisate in water (40 μL) and determine the yield by UV absorption at 260 nm (extinction coefficient $\varepsilon_{260\ nm}$: 15,400 L/(mol·cm) for the adenine chromophore [32]).
10. Validate formation of AdoViBe by mass spectrometry and NMR.
11. Purified AdoViBe can be stored at −20 °C for at least 1 year (*see* **Note 4**).
12. Analytical data: ESI-Orbitrap-MS $[M]^+$ calculated for AdoViBe ($C_{23}H_{29}N_6O_5S^+$): 501.19202; found 501.19110. **^{1}H NMR** (600 MHz, D_2O): $\delta = 2.09$ (q, $^3J = 8.0$ Hz, 2H, Hβ), 3.26–3.30 (m, 1H, Hγ), 3.50–3.40 (m, 1H, Hγ), 3.53 (t, $^3J = 6$ Hz, 1H, Hα), 3.65–3.66 (m, 2H, H5′), 4.16–4.18 (m, 1H, H4′), 4.3 (t, $^3J = 5.8$ Hz, 1H, H3′), 4.33 (dd, $^4J = 12.0$ Hz, 1H, H1″), 4.37–4.39 (m, 1H, H2′), 4.46 (dd, $^4J = 12.0$ Hz, 1H, H1″), 5.01 (d, $^3J = 10.5$ Hz, 1H, vinyl), 5.47 (d, $^3J = 18.2$ Hz, 1H, vinyl), 4.69 (d, $^3J = 3.4$ Hz, 1H, H1′), 6.26–6.31 (dd, $^3J = 10.5$ and 18.2 Hz, 1H, vinyl), 6.82–6.87 (m, 4H, arom. H), 7.83 (s, 1H, arom. H), 7.99 (s, 1H, arom. H).

3.2 Synthesis of Ab-SAM

Synthesize 5′-[(*R/S*)-[(3*S*)-3-amino-3-carboxypropyl]-4-azidobut-2-enylsulfonio]-5′-deoxy-adenosine (Ab-SAM) via the steps given below, which followed the procedure originally described by Islam et al. [33].

3.2.1 (E)-1-Azido-4-Bromobut-2-Ene

1. Dissolve (*E*)-1,4-dibromobut-2-ene (2.0 g, 9.4 mmol) in 8 mL DMF and add sodium azide (0.60 g, 9.4 mmol) to the solution.
2. Stir the reaction mixture for 15 h at 50 °C.
3. Dilute the reaction with ethyl acetate (60 mL) and wash the organic layer with water (15 mL) and a saturated sodium hydrogencarbonate solution (15 mL).
4. Dry the organic phase over anhydrous sodium sulfate and evaporate it under reduced pressure.
5. Purify the crude product by silica gel column chromatography using 100 % petroleum ether as mobile phase. The product as well as side products can be monitored by TLC.
6. Analytical data: **^{1}H NMR** (400 MHz, $CDCl_3$): $\delta = 6.03$–5.97 (m, 1H, =CH), 5.86–5.82 (m, 1H, =CH), 4.0 (d, 2H, $^3J =$ 4.0 Hz), 3.82 (d, 2H).

3.2.2 Synthesis and Characterization of Ab-SAM

1. Dissolve *S*-adenosyl-L-homocysteine (10 mg, 26 mmol) in 1.0 mL of a mixture of formic and acetic acid (1:1) and place it in an ice bath. To protect the reaction mixture from light, wrap the flask into aluminum foil.
2. Add (*E*)-1-azido-4-bromobut-2-ene (269 mg, 1.53 mmol) and a tip of a spatula $AgClO_4$ to the stirred solution (*see* **Notes 2** and **3**).
3. After stirring for 5 min remove the ice bath and allow the solution to warm up to room temperature.
4. After 5 h reaction time, add again (*E*)-1-azido-4-bromobut-2-ene (269 mg, 1.53 mmol) and a tip of a spatula $AgClO_4$ to drive the reaction to completion.
5. After 5 h, stop the reaction by gently adding the reaction mixture to cool water (20 mL) containing 0.01 % TFA.
6. Extract the aqueous phase 3× with 15 mL diethyl ether.
7. Freeze the aqueous phase with liquid nitrogen and concentrate it by lyophilization (*see* **Note 4**).
8. Redissolve the obtained product in 1.5 mL water containing 0.01 % TFA.
9. Purify the crude product by preparative reversed-phase HPLC. Adenine-containing components can be detected at 260 nm. Perform the elution at a flowrate of 5 mL/min using water containing 0.01 % TFA as eluent A and a gradient of

acetonitrile containing 0.01 % TFA from 0 to 7 % within 15 min. The diastereomeric mixture elutes with a retention time of 5.2 min (*see* **Note 5**).

10. Concentrate the diastereomeric mixture by lyophilization. Make sure that the sample stays frozen during lyophilization (*see* **Note 4**).
11. Redissolve the obtained product in water (40 μL) and determine the concentration by UV absorption at 260 nm (extinction coefficient $\varepsilon_{260\ \mathrm{nm}}$: 15,400 L/(mol·cm) for the adenine chromophore [32]).
12. Validate formation of Ab-SAM by mass spectrometry.
13. Purified Ab-SAM can be stored at −20 °C for approximately 1 year (*see* **Note 4**).
14. Analytical data: ESI-MS $[M]^+$ calculated for Ab-SAM ($C_{18}H_{26}N_9O_5S^+$): 480.1778; found 480.1785.

3.3 Synthesis of AdoEnYn

Synthesize 5′-[(*R*/*S*)-[(3*S*)-3-amino-3-carboxypropyl](*E*)-pent-2-en-4-ynylsulfonio]-5′-deoxy-adenosine (AdoEnYn) via the steps given below, which follow the procedure originally described by Peters et al. [34].

1. Resuspend sodium hydroxide (480 mg, 12 mmol) in dichloromethane (12 mL), add mesyl chloride (1.26 g, 11 mmol) and cool the suspension to 0 °C in an ice bath.
2. Add (*E*)-pent-2-en-4-yn-1-ol (821 mg, 10 mmol) and stir the reaction mixture at ambient temperature overnight.
3. Wash the reaction mixture with saturated sodium hydrogencarbonate solution until the aqueous phase remains basic.
4. Evaporation of the organic solvent should yield a yellow oil.
5. Dissolve the activated alcohol directly in 1 mL of formic and acetic acid (1:1).
6. Add the solution to *S*-adenosyl-L-homocysteine (10 mg, 26 μmol) and stir the reaction mixture at ambient temperature for 14 h.
7. Stop the reaction by gently transferring it to cool water (30 mL).
8. Extract the aqueous phase 3× with 60 mL diethyl ether.
9. Freeze the aqueous phase with liquid nitrogen and concentrate it by lyophilization and redissolve the lyophilisate in 1.5 mL water containing 0.01 % TFA (*see* **Note 4**).
10. Purify the crude product by preparative reversed-phase HPLC. Adenine-containing components can be detected at 260 nm. Perform the elution at a flowrate of 5 mL/min using water containing 0.01 % TFA as eluent A and a gradient of acetonitrile containing 0.01 % TFA from 0 to 7 % within 15 min.

The diastereomeric mixture elutes with a retention time of 3.9 min (*see* **Note 5**).

11. Concentrate the diastereomeric mixture by lyophilization (*see* **Note 4**).
12. Redissolve the obtained product in water (40 μL) and determine the concentration by UV absorption at 260 nm (extinction coefficient $\varepsilon_{260\ \mathrm{nm}}$: 15,400 L/(mol·cm) for the adenine chromophore [32]).
13. Validate generation of AdoEnYn by mass spectrometry.
14. Purified AdoEnYn can be stored at −20 °C for approximately 1 year (*see* **Note 4**).
15. Analytical data: ESI-MS $[M]^+$ calculated for AdoEnYn ($C_{19}H_{25}N_6O_5S^+$): 449.1602; found 449.1600.

3.4 Recombinant Production and Purification of Enzymes

3.4.1 GlaTgs2-V34A

1. Produce GlaTgs2-V34A by transforming *E. coli* Tuner(DE3) pLacI cells with pRSET-A-GlaTgs2-V34A (*see* Fig. 2 for sequence data).
2. Grow overnight cultures in 2YT medium in presence of ampicillin (100 μg/mL) and 2 % (v/v) glucose solution (*see* **Note 6**).

```
atgagcacctggctgctggatagcaaatgtgttgaacgtatgaaatggctgtttagcgat
 M  S  T  W  L  L  D  S  K  C  V  E  R  M  K  W  L  F  S  D
ctgccggaagaaaaacgtgtgatgatcaaaatgaatgaagcggcctttttagcgttaca
 L  P  E  E  K  R  V  M  I  K  M  N  E  A  A  F  F  S  V  T
ccggcagtttatgcagatgaagttgcacgtatgatgcgtaccgttctggcactgctgggt
 P  A  V  Y  A  D  E  V  A  R  M  M  R  T  V  L  A  L  L  G
aaaccgccttatgcagttattgatggcaccgcatgtgttggtggtgatacccgtctgctg
 K  P  P  Y  A  V  I  D  G  T  A  C  V  G  G  D  T  R  L  L
gcaaaacattttgatatgaccgttgccattgaacgtgatccggaaacctatgcactgctg
 A  K  H  F  D  M  T  V  A  I  E  R  D  P  E  T  Y  A  L  L
caggataatctgaccacctggggtgttgatgcaaaaaccattagcggtgataccgcagca
 Q  D  N  L  T  T  W  G  V  D  A  K  T  I  S  G  D  T  A  A
ctgattccgcagttttggaccctgattggtgcagttgcaacctttagcctgtatctggac
 L  I  P  Q  F  W  T  L  I  G  A  V  A  T  F  S  L  Y  L  D
cctccttggggtggtgttgattatcgtagccagaccgatattcagctgaccctgggtagc
 P  P  W  G  G  V  D  Y  R  S  Q  T  D  I  Q  L  T  L  G  S
ctggcagttgaagatgttgttaatcgtgcatttgaagcacatctgagcatgaaactggca
 L  A  V  E  D  V  V  N  R  A  F  E  A  H  L  S  M  K  L  A
gttctgaaactgcctcgcaactataattgcggttacctgtttcgcaaactgggtaaacat
 V  L  K  L  P  R  N  Y  N  C  G  Y  L  F  R  K  L  G  K  H
gaagtgtttcgtattacccagggcaatttttttgtgtttttttgttgcacgtcgtggtagc
 E  V  F  R  I  T  Q  G  N  F  F  V  F  F  V  A  R  R  G  S
cgtgttaaagaacatggtcgtaccgcaatgctgcagctgcgtaaagcacgtgaagaagca
 R  V  K  E  H  G  R  T  A  M  L  Q  L  R  K  A  R  E  E  A
aaagcacgtagcgaagaaaccaaagaagatggcgaaacacgcggtagcggtgaa
 K  A  R  S  E  E  T  K  E  D  G  E  T  R  G  S  G  E
```

Fig. 2 Nucleic and amino acid sequence of GlaTgs2-V34A

3. Inoculate the main culture supplemented as described above with 2.5 % (v/v) overnight culture.
4. Cultivate the cells for 4 h at 37 °C and constant shaking.
5. Cool to room temperature and induce protein production by adding IPTG (0.32 mM) and 2 % (v/v) ethanol.
6. Cultivate the cells for 20 h at 17 °C.
7. Harvest cells and store the pellets at −20 °C.

3.4.2 MTAN

1. Use a glycerol stock of *E. coli* BL21(DE3) cells containing the plasmid pProEx-MTAN to inoculate an overnight culture.
2. Use 5 mL of the overnight culture to inoculate 400 mL 2YT medium supplemented with ampicillin (100 μg/mL).
3. Cultivate the cells at 37 °C and constant shaking till an OD_{600} of 0.7 is reached.
4. Induce protein production by adding IPTG (1 mM) and cultivate the cells for another 3 h at 37 °C.
5. Harvest the cells by centrifugation and store the pellets at −20 °C.

3.4.3 LuxS

1. Use a glycerol stock of *E. coli* BL21(DE3) cells containing pET29-LuxS to inoculate an overnight culture.
2. Use the overnight culture to inoculate 400 mL 2YT medium supplemented with kanamycin (25 μg/mL) to give an OD_{600} of 0.1.
3. Cultivate the cells at 37 °C and constant shaking till an OD_{600} of 1 is reached.
4. Induce protein production by adding IPTG (0.8 mM) and cultivate the cells for another 6 h at 37 °C.
5. Harvest the cells by centrifugation and store the pellets at −20 °C.

3.4.4 Purification

1. Resuspend cell pellets obtained from a 200 mL culture in 10 mL of the corresponding lysis buffer (in case of GlaTgs2-V34A also add PMSF (15 μM)).
2. Sonicate 3× at an amplitude of 30 % for 3 min while cooling the cells in an ice bath.
3. After centrifugation, filter the cell lysate using Filtropur S.
4. Use an ÄKTA purifier™ protein purification system in combination with HisTrap™ FF 1 mL columns for immobilized metal ion affinity chromatography (IMAC).
5. Perform the elution at a flowrate of 1 mL/min applying eight column volumes 10 % elution buffer followed by a gradient to 100 % elution buffer within 50 column volumes.

6. To obtain RNase-free enzyme preparations, fractions can be concentrated individually in storage buffer using Vivaspin 500 (*see* **Notes 7–9**).
7. Store aliquots at −80 °C (*see* **Note 10**).

3.5 In Vitro T7 Transcription

1. To amplify the DNA template, perform a PCR reaction using 5 μg DNA-template (*see* Acknowledgments) or 15 μL from a former PCR reaction, add 10 μL dNTPs (0.5 mM), 100 μL 5× HF-Puffer, 5 μL DMSO, 2.5 μL Phusion® DNA polymerase (5 U), 1 nmol of each primer and adjust the volume to 500 μL. Perform the PCR in 100 μL aliquots for 25 cycles.
2. Precipitate DNA from the PCR reaction using one volume 2-propanol and 1/10 volume 3 M NaOAc [35].
3. Redissolve the resulting pellet in 256 μL ddH_2O and add 166.5 μL 3× transcription buffer, 50 μL NTPs (2.5 mM), 15 μL $Mg(OAc)_2$ (15 mM) as well as 12.5 μL T7 RNA-polymerase (1.25 U/μL). Incubate the reaction for 180–240 min at 300 rpm and 37 °C.
4. Add 1 μL DNase (1 U) and 1 μL inorganic pyrophosphatase (20 U) 30 min before finishing the incubation period.
5. Perform a phenol-chloroform extraction and precipitate the RNA with one volume 2-propanol and 1/10 volume 3 M NaOAc [35].
6. Redissolve the obtained pellet in 50 μL 1× PBS by incubating the samples for up to 3 h at room temperature.
7. Verify success of the in vitro transcription by gel electrophoresis followed by ethidium bromide staining and detection under UV light.

3.6 Capping of In Vitro Produced RNA

1. Prepare a solution of 0.9 nmol RNA (*see* Subheading 3.5), 1 μL GTP (0.5 mM), 2 μL vaccinia capping enzyme (20 U), with or without 1 μL AdoMet (0.1 mM) in a final volume of 20 μL (*see* **Note 11**). Incubate the reaction mixture for 1 h at 37 °C.
2. Heat the samples for 15 min at 65 °C to stop the reaction and to degrade nonreacted AdoMet.
3. Incubate the solution for 15 min at 4 °C with 5 μL of freshly prepared cation exchange P11 cellulose phosphate solution.
4. Precipitate the RNA using one volume of 2-propanol and 1/10 volume 3 M NaOAc.

3.7 Labeling of Azido-Modified RNA by SPAAC

1. Resuspend precipitated, capped RNA (*see* Subheading 3.6) in GlaTgs2-V34A (110 μM), MTAN (1.2 μM), LuxS (0.6 μM), and Ab-SAM (430 μM). Adjust the reaction volume to 5 μL with 1× PBS and add 0.5 μL RiboLock RNase Inhibitor (*see* **Notes 1** and **12**). Incubate the samples for 90 min at 37 °C.

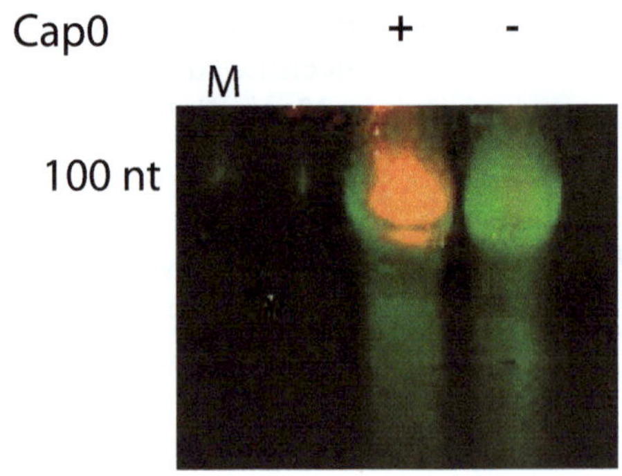

Fig. 3 Labeling of enzymatically modified RNA by SPAAC. Chemo-enzymatic labeling of in vitro produced and capped 100 nt RNA by SPAAC using DBCO-SRB^{TM}568/585 (+). An overlay of SRB (*red*) and ethidium bromide fluorescence (*green*) is shown. The negative control (–) contained RNA of the same length that is capped with an unmethylated guanosine

2. Adjust the volume to 25 μL with ddH_2O and precipitate the modified RNA by adding one volume 2-propanol and 1/10 volume 3 M NaOAc.
3. Redissolve the pellet in 5 μL 1× PBS and add DBCO-SRB (210 μM). Incubate the samples for 1 h at 37 °C protected from light.
4. Add 2× RNA Loading Dye and analyze the samples on 10 % denaturing PAA gels by fluorescence imaging.
5. Stain RNA by ethidium bromide, visualize it by UV transillumination and prepare an overlay of the fluorescence and stained RNA images (*see* **Note 13** and Fig. 3).

3.8 Chemoenzymatic Labeling of Capped RNA by IEDDA

1. Resuspend precipitated RNA (*see* Subheading 3.6) in GlaTgs2-V34A (100 μM), MTAN (1.2 μM), LuxS (0.6 μM), and AdoViBe (860 μM). Adjust the volume to 5 μL with 1× PBS and add 0.5 μL RiboLock RNase Inhibitor. Incubate the reaction mixture for 90 min at 37 °C (*see* **Note 12**).
2. Precipitate modified RNA by adjusting the volume to 25 μL with water and adding an equal volume of 2-propanol and 1/10 volume 3 M NaOAc.
3. Dissolve precipitated RNA in 5 μL 1× PBS and add tetrazine-5-TAMRA (400 μM) to the solution. Incubate the reaction mixture for 3 h at 37 °C protected from light.
4. Add 2× RNA loading buffer to the samples and analyze them on 10 % denaturing PAA gels by fluorescence imaging.
5. Stain the RNA by ethidium bromide, visualize it by UV transillumination and prepare an overlay of the fluorescence and stained RNA images (*see* **Note 13**).

3.9 Chemoenzymatic Labeling of Capped RNA by CuAAC

1. Resuspend precipitated, capped RNA (*see* Subheading 3.6) in GlaTgs2-V34A (50 μM), MTAN (1.2 μM), LuxS (0.6 μM), AdoEnYn (1.7 mM), and 0.25 μL RiboLock RNase Inhibitor. Adjust the reaction volume to 5 μL with 1× PUS buffer. Incubate the samples for 90 min at 37 °C and 300 rpm (*see* **Note 12**).
2. Precipitate modified RNA by adjusting the volume to 25 μL with water and addition of one volume 2-propanol and 1/10 volume 3 M NaOAc.
3. Resuspend the precipitated RNA in 4 μL PUS buffer (including 0.5 mM DTT) by incubating it at room temperature for 15 min and add 3.2 μL Click Solution.
4. Freshly prepare 314 mM CuBr solution (1 mg per 23 μL Click Solution), immediately dilute it 1:10 with a TBTA solution and transfer 1.2 μL to the reaction tube (*see* **Note 14**). Add 0.8 μL Eterneon™-Azide (480/653) and incubate the samples for 1 h at 37 °C protected from light.
5. Add 2× RNA Loading Dye and analyze the samples on 10 % denaturing PAA gels by fluorescence imaging.
6. Stain RNA by ethidium bromide and visualize by UV transillumination.

4 Notes

1. Due to oxidation, DTT should be added to the buffers just before use. However, in the case of the azido modification, buffers and solutions should be prepared without DTT due to its azide reducing properties.
2. During the synthesis of the AdoMet analogs, product formation is accompanied by slow product decomposition [32,36]. Thus, the ideal reaction time leading to a maximal yield needs to be accurately determined.
3. The progress of the product formation can be analyzed by reversed-phase HPLC. Adenine-containing components can be detected at 260 nm.
4. Keep the AdoMet analog solutions always cooled (0 °C or frozen if possible) and under slightly acidic conditions to reduce product decomposition [37].
5. Synthesis of the AdoMet analogs results in a diastereomeric mixture that can only partially be separated by HPLC purification. However, only the (*S*,*S*)-AdoMet analog (where the designations refer to the sulfur and the α-carbon, respectively) is converted by GlaTgs2-V34A and other methyltransferases [36].
6. For recombinant production of GlaTgs2-V34A the use of plastic flasks was observed to increase protein production.

7. Ensure that all solutions and protein preparations are RNase-free.
8. To obtain RNase-free enzyme preparations, gel filtration after IMAC is recommended by, e.g., using Superdex 200 Increase 10/300 GL (GE Healthcare) column.
9. Purified enzymes should be tested regarding their activity before performing chemoenzymatic labeling of RNA [18,38,39].
10. Protein stocks of MTAN and LuxS can be frozen and thawed several times, but in the case of GlaTgs2-V34A it is recommended to use aliquots only once.
11. The sample without AdoMet should be used as negative control, containing RNA with a 5′–5′ linked guanosine cap, but without the methyl group at the *N7* atom.
12. During enzymatic transfer reactions the presence of MTAN and LuxS is important to degrade the coproduct *S*-adenosylhomocysteine from the reaction because it is known to inhibit GlaTgs2-V34A and other methyltransferases [39].
13. To create an overlay of fluorescence and ethidium bromide stained RNA images, it is necessary that the gel is located at the same position during the two detection procedures.
14. Freshly prepare CuBr solution for each experiment and after solving CuBr in Click Solution immediately transfer the needed amount into TBTA.

Acknowledgments

A. R. gratefully acknowledges financial support from the Emmy Noether-Programme of the Deutsche Forschungsgemeinschaft (RE 2796/2-1) and the Fonds der Chemischen Industrie. This work was partly supported by the Deutsche Forschungsgemeinschaft, DFG EXC 1003 Cells in Motion—Cluster of Excellence, Münster, Germany. We thank Prof. Hahn (University of Hamburg) for providing the DNA template for transcription. We would like to thank Prof. Birgit Dräger (University of Halle, Germany) for plasmids encoding LuxS and MTAN. J. M. H. thanks the Fonds der Chemischen Industrie for a doctoral fellowship.

References

1. Januschke J, Gervais L, Dass S et al (2002) Polar transport in the Drosophilia oocyte requires Dynein and Kinesin I cooperation. Curr Biol 12:1971–1981
2. King ML, Messitt TJ, Mowry KL (2005) Putting RNAs in the right place at the right time: RNA localization in the frog oocyte. Biol Cell 97:19–33

3. Melton DA (1987) Translocation of a localized maternal mRNA to the vegetal pole of Xenopus oocytes. Nature 328:80–82
4. Bertrand E, Chartrand P, Schaefer M et al (1998) Localization of ASH1 mRNA particles in living yeast. Mol Cell 2:437–445
5. Condeelis J, Singer RH (2005) How and why does beta-actin mRNA target? Biol Cell 97:97–110
6. Mili S, Moissoglu K, Macara IG (2008) Genome-wide screen reveals APC-associated RNAs enriched in cell protrusions. Nature 453:115–119
7. Sutton MA, Schuman EM (2006) Dendritic protein synthesis, synaptic plasticity, and memory. Cell 127:49–58
8. Lin AC, Holt CE (2007) Local translation and directional steering in axons. EMBO J 26:3729–3736
9. Politz JC, Browne ES, Wolf DE et al (1998) Intranuclear diffusion and hybridization state of oligonucleotides measured by fluorescence correlation spectroscopy in living cells. Proc Natl Acad Sci U S A 95:6043–6048
10. Rinne JS, Kaminski TP, Kubitscheck U et al (2013) Light-inducible molecular beacons for spatio-temporally highly defined activation. Chem Commun 49:5375–5377
11. Kummer S, Knoll A, Socher E et al (2012) PNA FIT-probes for the dual color imaging of two viral mRNA targets in influenza H1N1 infected live cells. Bioconjug Chem 23:2051–2060
12. Strack RL, Disney MD, Jaffrey SR (2013) A superfolding Spinach2 reveals the dynamic nature of trinucleotide repeat-containing RNA. Nat Methods 10:1219–1224
13. Yoshimura H, Inaguma A, Yamada T et al (2012) Fluorescent probes for imaging endogenous beta-actin mRNA in living cells using fluorescent protein-tagged pumilio. ACS Chem Biol 7:999–1005
14. Ozawa T, Natori Y, Sato M et al (2007) Imaging dynamics of endogenous mitochondrial RNA in single living cells. Nat Methods 4:413–419
15. Kellermann SJ, Rath AK, Rentmeister A (2013) Tetramolecular fluorescence complementation for detection of specific RNAs in vitro. Chembiochem 14:200–204
16. Lionnet T, Czaplinski K, Darzacq X et al (2011) A transgenic mouse for in vivo detection of endogenous labeled mRNA. Nat Methods 8:165–170
17. Jao CY, Salic A (2008) Exploring RNA transcription and turnover in vivo by using click chemistry. Proc Natl Acad Sci U S A 105:15779–15784
18. Schulz D, Holstein JM, Rentmeister A (2013) A chemo-enzymatic approach for site-specific modification of the RNA cap. Angew Chem Int Ed Engl 52:7874–7878
19. Motorin Y, Burhenne J, Teimer R et al (2011) Expanding the chemical scope of RNA: methyltransferases to site-specific alkynylation of RNA for click labeling. Nucleic Acids Res 39:1943–1952
20. Tomkuviene M, Clouet-D'orval B, Cerniauskas I et al (2012) Programmable sequence-specific click-labeling of RNA using archaeal box C/D RNP methyltransferases. Nucleic Acids Res 40:6765–6773
21. Seidu-Larry S, Krieg B, Hirsch M et al (2012) A modified guanosine phosphoramidite for click functionalization of RNA on the sugar edge. Chem Commun 48:11014–11016
22. Aigner M, Hartl M, Fauster K et al (2011) Chemical synthesis of site-specifically 2′-azido-modified RNA and potential applications for bioconjugation and RNA interference. Chembiochem 12:47–51
23. Pyka AM, Domnick C, Braun F et al (2014) Diels-Alder cycloadditions on synthetic RNA in mammalian cells. Bioconjug Chem 25:1438–1443
24. Schoch J, Ameta S, Jaschke A (2011) Inverse electron-demand Diels-Alder reactions for the selective and efficient labeling of RNA. Chem Commun 47:12536–12537
25. Grammel M, Hang H, Conrad NK (2012) Chemical reporters for monitoring RNA synthesis and poly(A) tail dynamics. Chembiochem 13:1112–1115
26. Li F, Dong J, Hu X et al (2015) A covalent approach for site-specific RNA labeling in mammalian cells. Angew Chem Int Ed Engl 54:4597–4602
27. Winz ML, Samanta A, Benzinger D et al (2012) Site-specific terminal and internal labeling of RNA by poly(A) polymerase tailing and copper-catalyzed or copper-free strain-promoted click chemistry. Nucleic Acids Res 40:e78
28. Holstein JM, Schulz D, Rentmeister A (2014) Bioorthogonal site-specific labeling of the 5′-cap structure in eukaryotic mRNAs. Chem Commun 50:4478–4481
29. Holstein JM, Stummer D, Rentmeister A (2015) Enzymatic modification of 5′-capped RNA with a 4-vinylbenzyl group provides a platform for photoclick and inverse electron-demand Diels–Alder reaction. Chem Sci 6:1362–1369

30. Stummer D, Herrmann C, Rentmeister A (2015) Quantum chemical calculations and experimental validation of the photoclick reaction for fluorescent labeling of the 5′ cap of eukaryotic mRNAs. Chem Open 4:295–301
31. Hausmann S, Shuman S (2005) Giardia lamblia RNA cap guanine-N2 methyltransferase (Tgs2). J Biol Chem 280:32101–32106
32. Dalhoff C, Lukinavicius G, Klimasauakas S et al (2006) Synthesis of S-adenosyl-L-methionine analogs and their use for sequence-specific transalkylation of DNA by methyltransferases. Nat Protoc 1:1879–1886
33. Islam K, Bothwell I, Chen Y et al (2012) Bioorthogonal profiling of protein methylation using azido derivative of S-adenosyl-L-methionine. J Am Chem Soc 134:5909–5915
34. Peters W, Willnow S, Duisken M et al (2010) Enzymatic site-specific functionalization of protein methyltransferase substrates with alkynes for click labeling. Angew Chem Int Ed 49:5170–5173
35. Sambrook J (2001) Molecular cloning: a laboratory manual/Joseph Sambrook, David W. Russell. Cold Spring Harbor Laboratory, Cold Spring Harbor, NY
36. Hoffman JL (1986) Chromatographic analysis of the chiral and covalent instability of S-adenosyl-L-methionine. Biochemistry 25:4444–4449
37. Iwig DF, Booker SJ (2004) Insight into the polar reactivity of the onium chalcogen analogues of S-adenosyl-L-methionine. Biochemistry 43:13496–13509
38. Schulz D, Rentmeister A (2012) An enzyme-coupled high-throughput assay for screening RNA methyltransferase activity in E. coli cell lysate. RNA Biol 9:577–586
39. Hendricks CL, Ross JR, Pichersky E et al (2004) An enzyme-coupled colorimetric assay for S-adenosylmethionine-dependent methyltransferases. Anal Biochem 326: 100–105

Chapter 4

Preparation of Functional, Fluorescently Labeled mRNA Capped with Anthraniloyl-m^{7}GpppG

Artem V. Domashevskiy, David J. Rodriguez, Dilantha Gunawardana, and Dixie J. Goss

Abstract

Fluorescent mRNA molecules offer a wide range of applications for studying capping/decapping reactions, translation, and other biophysical studies. Furthermore, fluorescent tags prove invaluable for tracking RNA molecules in cells. Here, we describe an efficient synthesis of a fluorescent cap analog, anthranioyl-GTP, its purification, and in vitro cap labeling of transcribed mRNA catalyzed by the recombinant vaccinia capping enzyme to produce anthranioyl-m^{7}GpppG-capped RNA.

Key words Fluorescence, Anthranioyl-m^{7}G, Ant-GTP, Chromatography, Fluorescently labeled RNA, Capped RNA, RNA labeling

1 Introduction

The 5′-terminal 7-methylguanosine (m^{7}GpppN, where N is any nucleotide) cap structure (Cap 0) [1] of eukaryotic mRNAs is involved in RNA stabilization [2], transport [3], translation [4], and is recognized by many cellular proteins involved in pre-mRNA splicing, RNA export, translation initiation, and RNA turnover [5]. Enzymatic production of capped RNA is an easy way to improve the stability and translational competence of RNA used for in vitro translation, transfection, and microinjection. Fluorescence spectroscopy is widely used to study these protein–RNA interactions [6–9]. The presence of a fluorescent cap derivative incorporated into the 5′ end of RNA provides an alternative method to study these cap-binding interactions. Fluorescent derivatives have a greater intensity than intrinsic protein and nucleic acid fluorescence. Furthermore, absorbance and emission maxima of the extrinsic fluorophore are far removed from native protein fluorescence, allowing interpretation of the interactions involving several proteins.

Robert E. Rhoads (ed.), *Synthetic mRNA: Production, Introduction Into Cells, and Physiological Consequences*, Methods in Molecular Biology, vol. 1428, DOI 10.1007/978-1-4939-3625-0_4,

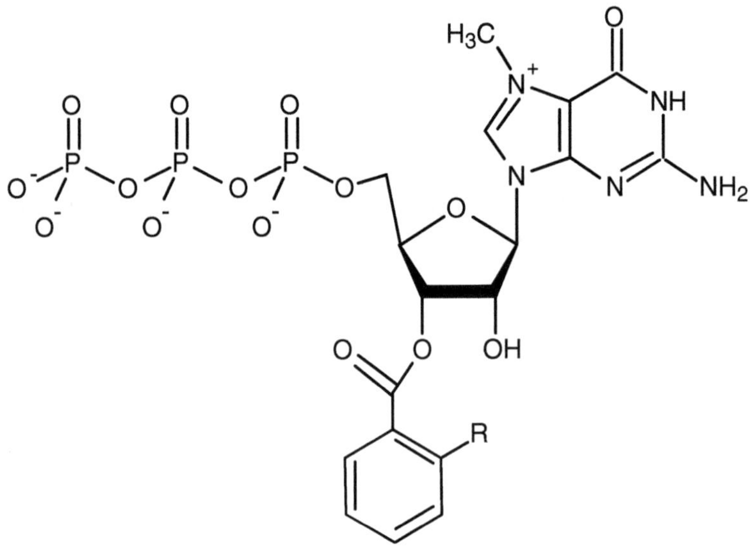

Fig. 1 Structure of the fluorescent anthraniloyl (Ant) derivative of m^{7}GTP ($R = -NH_2$)

Here, we describe a novel synthesis of a fluorescent anthraniloyl-GTP (Ant-GTP) cap analog [10, 11] and its purification by Sephadex LH-20 resin. The fluorescent cap analog is further methylated to produce Ant-m^{7}GTP (Fig. 1) and incorporated into mRNA using a vaccinia virus capping enzyme and in vitro transcription. We also spectroscopically quantify this fluorescent mRNA, and demonstrate its efficient translation in a wheat germ in vitro translation system [10] and decapping efficiency by Malachite Green phosphatase assay. Vaccinia virus capping enzyme efficiently executes all three steps in m^{7}GpppRNA synthesis and has been widely used as a reagent for capping and cap-labeling RNAs in vitro [12–15]. This enzyme carries of the following steps in m^{7}GpppRNA synthesis [10]:

$$\text{ppp5}'\text{-}r(n)\text{-3}' \rightarrow \text{pp5}'\text{-}r(n)\text{-3}' + \text{P}_1\,(\text{triphosphatase})$$

$$\text{pp5}'\text{-}r(n)\text{-3}' + \text{GTP} \rightarrow \text{Gppp5}'\text{-}r(n)\text{-3}' + \text{PP}_1\,(\text{guanylyltransferase})$$

$$\text{Gppp5}'\text{-}r(n)\text{-3}' + \text{AdoMet} \rightarrow \text{7-Methyl-Gppp5}'\text{-}r(n)\text{-3}'\ (\text{methyltransferase})$$

The Ant-m^{7}GTP-capped RNA serves as an effective substrate for the DCP2 decapping enzyme that hydrolyzes the cap structure prior to the degradation of mRNA in the $5' \rightarrow 3'$ direction [16–18]. Additionally, Ant-m^{7}GTP cap analog and labeled RNA have been used for binding studies with eukaryotic translation initiation factors and other cap-binding proteins [e.g., eIF4E and eIF*iso*4E, eIF*iso*4F, pokeweed antiviral protein (PAP), etc.] [11, 19–21]. This fluorescently labeled RNA enables one to carry out a variety of biophysical studies on cap-binding protein interactions, capping/decapping reactions, individual steps in translation, and other processes involving mRNA.

2 Materials

All solutions and buffers are prepared using ultrapure deionized water (18 MΩ cm at 25 °C) and biochemical or molecular biology grade reagents. All solutions and reagents are prepared at room temperature, unless otherwise indicated.

2.1 Thin-Layer Chromatography (TLC) Development Solvents

1. *System A*: silica gel plate developed in 1-propanol/NH_4OH/water, 6:3:1, v/v, containing 0.5 g/L EDTA. To prepare 100 mL, mix 60 mL of 1-propanol with 30 mL NH_4OH, conc. Add 10 mL of ultrapure deionized water. Weigh out 50 mg of EDTA (free acid) and transfer into the mixture. Mix well to dissolve.
2. *System B*: cellulose plate developed in 2-propanol/water/HCl, 65:18.4:16.6, v/v. To prepare 100 mL, mix 65 mL of 2-propanol with 18.4 mL of ultrapure deionized water. Add 16.6 mL of 12.1 *N* HCl, conc. Mix well to dissolve.

2.2 Solutions and Buffers for the In Vitro Transcription of RNA

1. Transcription 5× Buffer: 400 mM HEPES-KOH, pH 7.5, 160 mM $MgCl_2$ for SP6 (or 120 mM $MgCl_2$ for T7), 10 mM spermidine, 200 mM DTT. To prepare 10 mL, dissolve 0.953 g HEPES in ultrapure Nuclease-free water and adjust pH to 7.5 with 1 *N* KOH. Add 0.152 g $MgCl_2$ for SP6 (or 0.114 g $MgCl_2$ for T7), 15.7 μL spermidine, and 0.308 g DTT. Mix well to dissolve. Aliquot into 1 mL and store at −20 °C.
2. TE Buffer: 10 mM Tris–HCl (pH 8.0), 1 mM EDTA. To prepare 100 mL, dissolve 0.121 g Trizma base in ultrapure Nuclease-Free water and adjust pH to 8.0 with 1 *N* HCl. Add 0.0292 g EDTA (free acid) and adjust volume to 100 mL.
3. TE-saturated phenol:chloroform:*iso*-amyl alcohol (25:24:1, v/v) (pH 8.0). Mix equal parts of the TE buffer and phenol and allow the phases to separate. Then, mix 1 part of the lower phenol phase with 1 part of the chloroform:*iso*-amyl alcohol (24:1, v/v).
4. MOPS Buffer (5×): 0.2 M MOPS (pH 7.0), 50 mM sodium acetate, 5 mM EDTA (pH 8.0). To prepare 10 mL, dissolve 0.418 g MOPS in ultrapure water and adjust pH to 7.0. Add 0.041 g of anhydrous sodium acetate, 100 μL 0.5 M EDTA (pH 8.0), and water up to 10 mL.
5. RNA Loading Buffer: 50 % glycerol, 1 mM EDTA, 0.4 % bromophenol blue, 1 mg/mL ethidium bromide. To prepare

10 mL, dissolve 40 mg bromophenol blue in few milliliters of nuclease-free water, add 20 μL EDTA (pH 8.0), 10 mg ethidium bromide, 5 mL glycerol (use a high-grade glycerol; lower grades of glycerol contain ribonuclease activity). Add water up to 10 mL. Aliquot RNA loading buffer and store at −20 °C.

6. RNA Sample Buffer: Mix 10.0 mL deionized formamide with 3.5 mL 37 % formaldehyde and 2.0 mL 5× MOPS buffer (final concentration 7 %). Dispense into aliquots and store at −20 °C for up to 6 months. Do not freeze-thaw more than twice.

2.3 Buffers for RNA Capping and Tailing

1. RNA Capping Buffer (1×): 50 mM Tris–HCl (pH 7.9), 6 mM KCl, 1.25 mM $MgCl_2$, 2.5 mM DTT, 0.1 mg/mL BSA, 40 U RNaseOUT ribonuclease inhibitor, 0.5 mM SAM (*S*-Adenosyl Methionine), up to 176 μg of uncapped RNA, 90–150 μM Ant-GTP analog, and 10 U of vaccinia capping enzyme (New England BioLabs).
2. *E. coli* Poly(A) Polymerase Reaction Buffer (1×): 250 mM NaCl, 50 mM Tris–HCl, pH 7.9, 10 mM $MgCl_2$.

2.4 Reagents for Determination of Capping Efficiency

1. Malachite Green Reagent: To prepare the Malachite Green solution, add one volume of 4.2 % (w/v) ammonium molybdate in 4 M HCl to 3 volumes of 0.045 % (w/v) Malachite Green. Stir the solution for a minimum of 30 min. Filter the solution through a 0.22 μm-pore diameter filter (Millipore) equipped syringe. Store the filtered solution for up to 6 months at 4 °C. Add Tween 20 (0.01 %, v/v) to an aliquot of filtered Malachite Green solution prior to use.
2. Phosphatase Activity Buffer E: 100 mM Tris–HCl, pH 8.0, 5 mM 2-mercaptoethanol. To prepare 100 mL, dissolve 1.211 g of Trizma base in ultrapure water and adjust pH to 8.0. Add 35.1 μL of 2-mercaptoethanol.
3. Decapping CE Buffer (10×): 200 mM Tris–HCl, pH 8.0, 1 M KCl, 10 mM $MgCl_2$, 10 mM DTT, 0.5 mg/mL BSA.

3 Methods

Carry out all procedures at room temperature, unless otherwise specified.

3.1 Synthesis of a Fluorescent Cap Analog: Ant-GTP

1. Weigh out 1 mmol of GTP (*see* **Note 1**) [10, 11, 22] and dissolve it in a minimum amount of ultrapure deionized water (15 mL) at 38 °C. Adjust the pH of the resulting solution to 9.6 with 2 *N* NaOH solution. To this solution with continuous stirring, add 1.5 mmol of crystalline isatoic anhydride. Maintain the pH of the reaction mixture at 9.6 with 2 *N* NaOH during the 2 h reaction.

Table 1
Isolated yield and R_f values of Ant-GTP and Ant-m^{7}GTP analogs

		R_f in system	
Compound	**Yield (%)**	**A**	**B**
Anthraniloyl-GTP (Ant-GTP)	37	0.09	0.22
Anthraniloyl-m^{7}GTP (Ant-m^{7}GTP)	20	0.10	0.30
Anthranilic acid	–	0.63	0.81

2. After the reaction is complete, adjust the pH of the reaction mixture to 7.0 with 1 *N* HCl solution.
3. Assemble a Sephadex LH-20 column (2.4 × 56 cm) packaged in ultrapure deionized water (*see* **Note 2**). Load the reaction products directly on the column. Elute the column with ultrapure water at a flow rate of about 6 mL/h (*see* **Note 3**). Collect 1 mL elution fractions.
4. Assay the fractions by spotting 0.2–0.5 μL aliquots of each fraction on a silica gel TLC plates. Develop the dried TLC plates in either *System A* or *B* developing solvent (*see* **Note 4**) (for R_f values, *see* Table 1).
5. Pool the peak fractions of the fluorescent analog and combine them together. Lyophilize the combined fractions in vacuo at the temperature of liquid nitrogen to prevent degradation.
6. Dissolve the resulting residue in a minimum amount of ultrapure water (~0.5 mL), and add an excess of cold anhydrous ethanol to induce the precipitation of the compound. Leave the ethanolic fractions at −20 °C for 30 min to ensure the completeness of the precipitation. Collect the precipitate by centrifugation at 12,000 × *g*.
7. Dry the fluorescent cap analog in vacuo over phosphoric anhydride at 4 °C to obtain an amorphous powder.
8. Analyze the purity of the product by TLC plates (chromatographically pure sample shows single brilliant blue spot) (Fig. 2).
9. Measure the absorption spectrum of the product at two maxima at 254 and 332 nm (Fig. 3). Unreacted GTP fractions show one maximum at 254 nm and one shoulder at 280 nm. The product fractions show two maxima as described above, and one fluorescent spot on TLC plates. The unreacted isatoic anhydride and byproduct, anthraniloyl acid, fractions show two maxima at 210 and 300 nm and a fluorescent spot on TLC plates which runs much faster than the product (Table 1).
10. Measure the concentration of the product by UV absorbance at 332 nm ($\varepsilon = 4600/\mathrm{M/cm}$). Divided into aliquots and store at −20 °C.

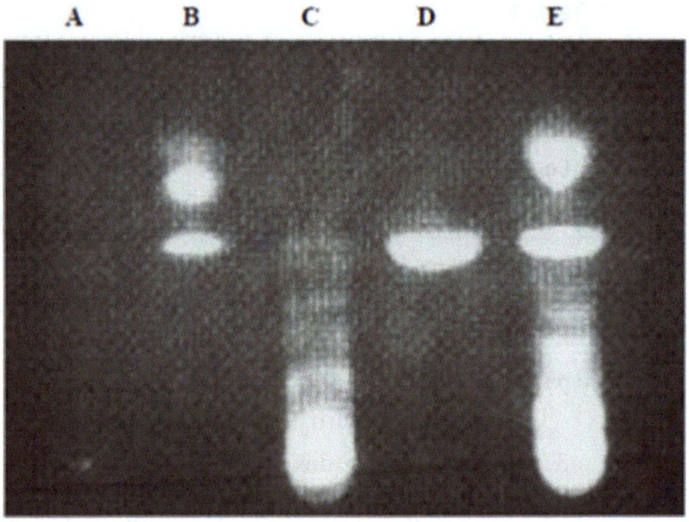

Fig. 2 Silica gel TLC of (**a**) m^7GTP, (**b**) isatoic anhydride, (**c**) Ant-m^7GTP, (**d**) anthranilic acid, and (**e**) reaction mixture, as monitored by UV absorbance at 366 nm. The solvent was *n*-propyl alcohol:ammonia:water, 6:3:1, v/v containing 0.5 g/L EDTA

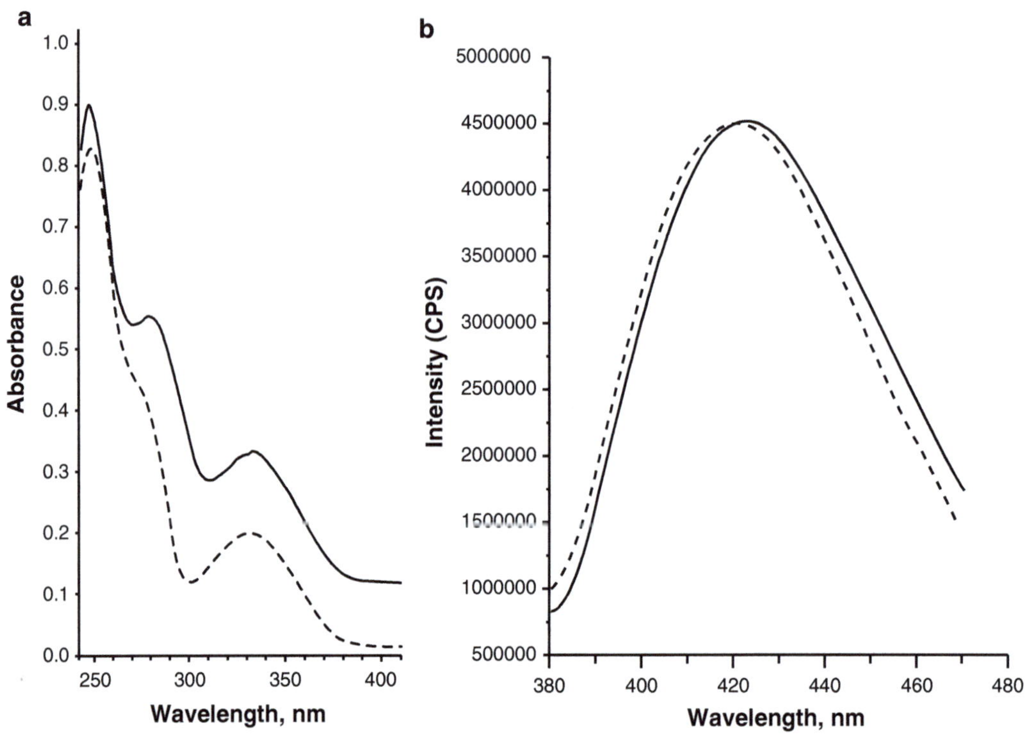

Fig. 3 Absorption (**a**) and fluorescence emission (**b**) spectra of Ant-GTP (*dashed line*) and Ant-m^7GTP (*solid line*). The spectra were measured in 20 mM HEPES-KOH, pH 7.6, 22 °C

3.2 Synthesis of RNA Transcripts: In Vitro Transcription

In vitro transcription reactions can be used to synthesize microgram amounts of RNA probes from recombinant DNA templates, and for other applications that require large amounts of biologically active RNA, including in vitro translation and for synthesis of tRNA, rRNA, other small functional RNAs, RNA virus genomes, and ribozymes. Large-scale RNA preparation is also useful for production of substrates for studies of RNA splicing, RNA secondary structure, antisense RNA, and RNA–protein interactions. Promega® RiboMAX™ Large Scale RNA Production Systems (SP6 or T7) is a quick and efficient method to produce milligram amounts of viable and biologically active RNA [23–25].

1. Setup the appropriate transcription reaction (100 μL final volume) (*see* **Note 5**) for either SP6 or T7 RNA Polymerase at room temperature. Add the reaction components in the following order (*see* **Note 6**) [26]: 20 μL SP6 (or T7) Transcription 5× buffer, 20 μL rNTPs (25 mM ATP, CTP, GTP, UTP) (*see* **Note 7**), 50 μL linear DNA template (5–10 μg total plus nuclease-free water) (*see* **Notes 8** and **9**) [27], 10 μL SP6 (or T7) enzyme mix (RNA polymerase, recombinant RNasin® Ribonuclease Inhibitor, and recombinant inorganic phyrophosphatase). Make sure to dissolve the DNA template in water before adding it to the reaction.
2. Gently pipet the reaction to mix and incubate at 37 °C for 2–4 h (*see* **Note 10**).
3. Removal of the DNA Template Following Transcription (if DNase treatment is not performed, proceed to **step 4**): Add RQ1 RNase-Free DNase to 1 U/μg of DNA template. Incubate for 15 min at 37 °C.
4. Extract with 1 volume of citrate-saturated phenol (pH 4.7): chloroform:*iso*-amyl alcohol (125:24:1, v/v). Vortex for 1 min and spin at top speed in a microcentrifuge for 2 min.
5. Transfer the upper layer, aqueous phase to a fresh tube, and add 1 volume of chloroform:*iso*-amyl alcohol (24:1, v/v). Vortex for 1 min and centrifuge as described above. At this point, unincorporated nucleotides may be removed (*see* **Note 11**), or the RNA may be precipitated directly (**step 6**).
6. Transfer the upper layer, aqueous phase to a fresh tube. Any transferred chloroform can be removed by performing a quick spin (10 s) in a microcentrifuge, followed by removal of the bottom phase with a micropipette. Add 0.1 volume of 3 M sodium acetate (pH 5.2) and 1 volume of *iso*-propanol (or 2.5 volumes of 95 % ethanol). Mix and place on ice for 2–5 min. Spin at top speed in a microcentrifuge for 10 min.

7. Carefully pour off or aspirate the supernatant and wash the pellet with 1 mL of 70 % ethanol. Dry the pellet in vacuo and resuspend the RNA sample in TE buffer or nuclease-free water to a volume identical to that of the transcription reaction. Store at −70 °C.

3.3 Incorporation of Ant-GTP Analog into mRNA: Capping of mRNA

The protocol is designed to cap up to 100 μg of RNA (100 nt or larger) in a 200-μL reaction. Reaction size can be scaled up as needed.

1. Combine RNA (*see* **Note 12**) and nuclease-free water in a 1.5 mL Eppendorf tube to a final volume of 150 μL. Heat at 65 °C for 5 min. Place tube on ice for 5 min.
2. Add the following components in the following order: Denatured RNA (from above) 150 μL, 10× Capping Buffer 20 μL, Ant-GTP (10 mM) 10 μL, SAM (2 mM, dilute 32 mM stock to 2 mM) 10 μL, vaccinia capping enzyme 10 μL. Total reaction volume is 200 μL. Incubate at 37 °C for 4–6 h and precipitate the total RNA with 3.25 M lithium chloride. Resuspend the precipitated RNA in 200 μL of RNase-free water and purify further, if desired. If the RNA requires a poly(A) tail, poly(A) polymerase can be used (**step 3**).
3. Poly(A) tailing of RNA using *E. coli* Poly(A) polymerase (*see* **Note 13**): To tail up to 10 μg of RNA (*see* **Note 14**) in a 20-μL reaction, mix 1–10 μg RNA in 15 μL nuclease-free water, 2 μL 10× *E. coli* poly(A) polymerase Reaction Buffer, 2 μL 10 mM ATP, 1 μL of *E. coli* poly(A) polymerase. Incubate reaction at 37 °C for 30 min. Stop the reaction by adding EDTA to a final concentration of 10 mM or directly proceeding to the cleanup step.

3.4 Measurement of RNA Quantity and Determination of Capping Efficiency

The traditional method for assessing RNA concentration and purity is UV spectroscopy (*see* **Note 15**).

3.4.1 RNA Quantification

1. Measure the absorbance of a diluted RNA sample at 260 and 280 nm (*see* **Note 16**).
2. Calculate the nucleic acid concentration using Beer-Lambert law, which predicts a linear change in absorbance with concentration: $A = \varepsilon\, c\, l$ (where A = absorbance at a particular wavelength, c = concentration of nucleic acid, l = path length of the spectrophotometer cuvette, and ε = the extinction coefficient, for RNA ε is 0.025/(mg/mL)/cm). Using this equation, an A_{260} reading of 1.0 is equivalent to ~40 μg/mL single-stranded RNA (*see* **Note 17**).
3. The A_{260}/A_{280} ratio is used to assess RNA purity. An A_{260}/A_{280} ratio of 1.8–2.1 is indicative of highly purified RNA (*see* **Note 18**) [28].

Absorption and emission spectra of Ant-GTP, Ant-m^7GTP, Ant-GpppG-capped TEV RNA, and Ant-m^7GpppG-capped TEV RNA show no shift in absorption/emission maxima (Fig. 4). The label does not perturb properties of the RNA: K_d for PAP-m^7GTP interactions is 43.3 ± 0.1 nM [29] and PAP-Ant-m^7GTP is 41.7 ± 2 nM at 25 °C [21].

3.4.2 Determination of Capping Efficiency

Capping efficiency may be determined using the Malachite Green phosphatase assay [10, 30, 31] on half-volume 96-well plates.

1. Treat both capped and uncapped RNAs with calf alkaline phosphatase (Promega, Co., USA) in Buffer E. Incubate at 37 °C for 1 h (25 μL, final volume) to hydrolyze the exposed phosphate groups prior to the assay with Malachite Green reagent.
2. Terminate the reactions by the addition of 50 μL of Malachite Green solution.
3. Incubate the plates for 15 min at room temperature before measuring absorbance at 620 nm. Report the capping efficiency based on the percentage of capped transcripts in a total pool of RNA species that has undergone capping, assuming that alkaline phosphatase treatment releases all three 5′ terminal phosphate groups on the uncapped RNA. Capping efficiency of 73 % is observed for Ant-m^7GTP with TEV RNA to produce Ant-m^7GpppG-capped TEV RNA transcript. Fluorescence emission spectrum of Ant-m^7GpppG-capped 5′-leader TEV RNA shows a maximum at 420 nm (Fig. 4).

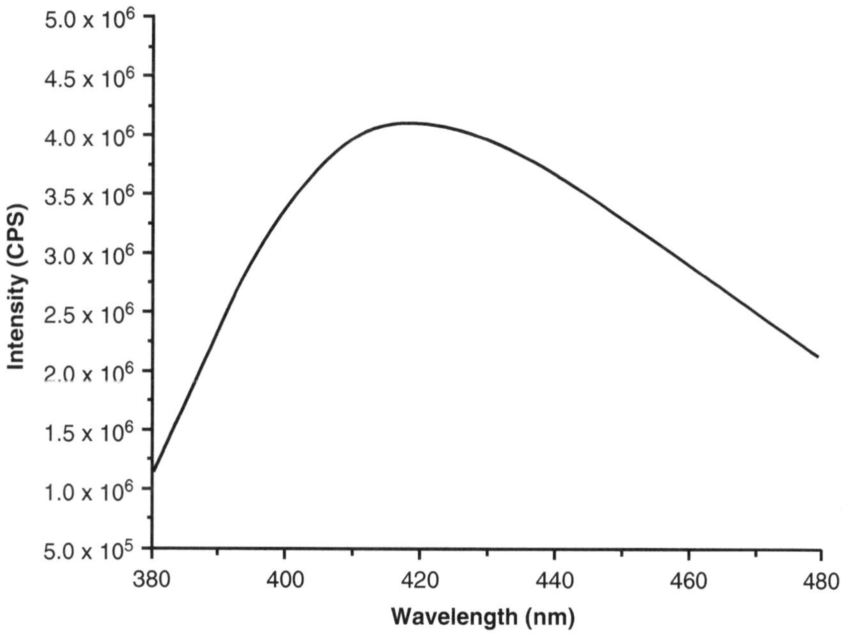

Fig. 4 Fluorescence emission spectra of Ant-m^7G-capped 5′-leader TEV RNA. The spectra were measured in 20 mM HEPES-KOH, pH 7.6, 22 °C

3.4.3 Decapping Reaction

To assess the presence of the Ant-labeled cap at the 5′ end of the RNA, the cap structure may be hydrolyzed using a recombinant enzyme (AtDCP2 decapping protein) specific for the hydrolysis of the cap structure (*see* **Note 19**) [10, 17, 32–34]. Alternatively tobacco acid pyrophosphatase (TAP) hydrolyzes various pyrophosphate bonds [35], cleaving the pyrophosphate bond of the 5′-terminal methylated guanine nucleotide "cap" of eukaryotic mRNAs [36]. Similarly, human NUDT16 is a RNA-binding and decapping enzyme that catalyzes the cleavage of the cap structure of snoRNAs [37, 38] and mRNAs in a metal-dependent manner and a bacterial RNA 5′ pyrophosphohydrolase (RppH) which removes pyrophosphates from the 5′ terminus of triphosphorylated RNA and release a 5′ monophosphate RNA [39] may be used for the decapping reactions.

1. Treat up to 20 μg of Ant-labeled RNA with 10 μg of the AtDCP2 decapping protein in the 1× CE decapping buffer, containing $MnCl_2$ (metal ion cofactor) in a final volume of 200 μL. Run the decapping reactions overnight.
2. Precipitate the RNA with 3.25 M lithium chloride.
3. Dissolve the RNA in 50 μL of nuclease-free water (*see* **Note 20**).
4. Assay the effect of decapping on translation in any of the standard available cell-free expression systems (e.g., rabbit reticulocyte lysate or wheat germ extract).

3.4.4 In Vitro Translation Assay

Cell-free extracts of wheat germ and rabbit reticulocyte lysate support the in vitro translation of a wide variety of viral, prokaryotic, and eukaryotic mRNAs [10, 40, 41]. The importance of m^7G for the translation of mRNA is strongly influenced by K^+ concentration in both the wheat germ and the reticulocyte cell-free systems [42].

1. For the 20 μL reaction, combine wheat germ extract (Promega, Co., USA) in 100 mM potassium acetate, 1 mM amino acid mixture (minus methionine), 40 U RNaseOUT ribonuclease inhibitor (Invitrogen, Co.), [^{35}S]-methionine (1175 Ci/mmol) with 0.560 ng of an in vitro transcribed RNA.
2. Incubate the reactions for 2.5 h at 25 °C, and analyze the protein products by 16 % Tris-tricine (and/or 15 % Tris-glycine) PAGE.
3. Dry the gels and visualize by phosphor-imaging and measure the intensities using Imagescan 5.2 software.

4 Notes

1. Similarly, m^7GTP cap analog, although more costly, may be used as an initial starting material for Ant-m^7GTP synthesis. This methylated cap analog may be used in Subheading 3.3, instead of Ant-GTP, for synthesis of capped RNA.
2. To separate reaction products, both Sephadex LH-20 or Sephadex G-15 resin packaged in water may be used (1 × 138 cm column).
3. Sephadex LH-20: the fluorescent analog elutes after the unreacted nucleotide; the fluorescent by-product anthranilic acid elutes after the peak of fluorescent nucleotide. Sephadex G-15: the fluorescent derivative elutes first, followed by the unreacted nucleotide and anthranilic acid. The fluorescent ant-analogs have a brilliant blue fluorescence (as monitored by the UV lamp, 360 nm), while anthranilic acid shows a dark violet fluorescence.
4. Two assay eluted fractions, either of the developing systems may be employed: *System A* includes silica gel plates developed in 1-propanol/NH_4OH/water, 6:3:1, v/v, containing 0.5 g/L ETDA; *System B* involves cellulose TLC plates developed in 2-propanol/water/HCl, 65:18.4:16.6, v/v).
5. These reactions can be scales up or down to suit the template requirements. A 1-mL reaction typically produces 2–5 mg of RNA in 2–4 h.
6. Optimal RNA yields depend on a high-quality DNA template. The DNA template must be free of RNase. If the presence of RNase is suspected, treat the DNA with Proteinase K (100 μg/mL) and SDS (0.5 %) in 50 mM Tris–HCl (pH 7.5), 5 mM $CaCl_2$ for 30 min at 37 °C. Purify the DNA further by extraction with TE-saturated (pH 8.0) phenol:chloroform:*iso*-amyl alcohol (25:24:1, v/v) and ethanol precipitation.
7. For convenience, mix equal volumes of the four individual 100 mM rNTPs to produce a solution that is 25 mM for each nucleotide.
8. DNA can precipitate in the presence of spermidine (a component of the Transcription 5× Buffer) if reactions are setup at colder temperatures.
9. DNA templates are usually linearized prior to in vitro transcription to produce RNA of defined length. Linearize the DNA by digestion with an appropriate restriction endonuclease followed by an appropriate cleanup procedure, such as phenol extraction followed by ethanol precipitation. Start with at least 30 % more DNA than is required for the transcription reaction to allow for DNA loss during purification and

visualization by gel electrophoresis. Avoid the use of restriction enzymes that produce 3′ overhangs (e.g., PvuI, KnpI, PstI). Extraneous transcripts have been reported to appear in addition to the expected transcript when such templates are transcribed. If these enzymes must be used, the linearized template ends can be made blunt using DNA Polymerase I Large (Klenow) Fragment prior to transcription (setup a standard in vitro transcription reaction minus the nucleotides and RNA polymerase; add DNA Polymerase I Large (Klenow) Fragment at 5 U/μg and incubate for 15 min at 22 °C; proceed to the transcription reaction by adding the nucleotide mix and RNA polymerase). The purified linear DNA should be examined by agarose or polyacrylamide gel electrophoresis prior to transcription to verify complete linearization and ensure the presence of a clean (nondegraded) DNA fragment of the expected size.

10. Do not freeze the completed transcription reaction. After the transcription reaction is complete, proceed directly to the DNase step or the removal of unincorporated nucleotides.

11. Chromatographic removal of unincorporated nucleotides: If desired, MicroSpin®G-25 columns (GE Healthcare) (allow purification of 25–50 μL of transcription reaction per column) may be used for purification of RNA from small-scale transcription reactions.

12. RNA used for capping reactions should be purified prior to use and suspended in Nuclease-free water. EDTA should not be present and the solution should be free of salts. While RNase Inhibitor is not required, many users prefer to use it to enhance the stability of their RNA in solution.

13. In a poly(A) tailing reaction, the length of the tail depends on the molar concentration of the RNA 3′ OH ends, reaction time, amount of enzyme, and ATP concentration. Tail length can be modified by changing one or more of these factors. As a general guidance, incubation of 5 U of the enzyme with 1–10 μg RNA in a 20 μL reaction at 37 °C for 30 min (with 1× Reaction Buffer and 1 mM ATP) will result in a tail length of greater than 100 bases.

14. RNA used for tailing reactions should be purified prior to use and suspended in Nuclease-Free water. EDTA should not be present and the solution should be free of salts. RNase inhibitor may be added to enhance the stability of the RNA in solution.

15. Because this method does not discriminate between RNA and DNA, it is advisable to first treat RNA samples with RNase-free DNase to remove contaminating DNA. Other contaminants such as residual proteins and phenol can interfere with absorbance readings; care must be taken during RNA purification to remove them completely.

16. Sample readings are made in quartz cuvettes. Dirty cuvettes and dust particles cause light scatter at 320 nm which can impact absorbance at 260 nm. Since neither proteins nor nucleic acids absorb at 320 nm, perform background correction by making readings from a blank (diluent only) at 320 nm, as well as at 260 and 280 nm.
17. Make sure your RNA dilution is within the linear range of your spectrophotometer. Usually absorbance values should fall between 0.1 and 1.0. Solutions that are outside this range cannot be measured accurately. Generally the greatest error occurs at lower concentrations.
18. The A_{260}/A_{280} ratio is dependent on both pH and ionic strength. As pH increases, the A_{280} decreases while A_{260} is unaffected. This results in an increasing A_{260}/A_{280} ratio [28]. Because water often has an acidic pH, it can lower the A_{260}/A_{280} ratio. We recommend using a buffered solution with a slightly alkaline pH, such as TE (pH 8.0) buffer, as a diluent (and as a blank) to assure accurate and reproducible readings.
19. The At5g13570 gene from *Arabidopsis*, which encodes a homolog of the human DCP2 decapping enzyme [17], expressed as a recombinant protein in am *E. coli* expression system, was used in decapping assays.
20. Similarly, the decapping assay may be monitored on the PEI-TLC plates that were predeveloped in water and then 0.225 M ammonium sulfate solution. The plates are developed in 0.225 M ammonium sulfate, and exposed to film [30].

Acknowledgment

We thank Jason A. Domashevskiy for his critical review of the manuscript.

References

1. Koukhareva II, Lebedev AV (2004) Chemical route to the capped RNAs. Nucleosides Nucleotides Nucleic Acids 23:1667–1680
2. Furuichi Y, LaFiandra A, Shatkin AJ (1977) 5′-Terminal structure and mRNA stability. Nature 266:235–239
3. Lewis JD, Izaurralde E (1997) The role of the cap structure in RNA processing and nuclear export. Eur J Biochem 247:461–469
4. Iizuka N, Najita L, Franzusoff A, Sarnow P (1994) Cap-dependent and cap-independent translation by internal initiation of mRNAs in cell extracts prepared from Saccharomyces cerevisiae. Mol Cell Biol 14:7322–7330
5. Cougot N, van Dijk E, Babajko S, Séraphin B (2004) Cap-tabolism. Trends Biochem Sci 29:436–444
6. Carberry SE, Darzynkiewicz E, Goss DJ (1991) A comparison of the binding of methylated cap analogues to wheat germ protein synthesis initiation factors 4F and (iso)4F. Biochemistry 30:1624–1627
7. Carberry SE, Friedland DE, Rhoads RE, Goss DJ (1992) Binding of protein synthesis

initiation factor 4E to oligoribonucleotides: effects of cap accessibility and secondary structure. Biochemistry 31:1427–1432
8. Domashevskiy AV, Miyoshi H, Goss DJ (2012) Inhibition of pokeweed antiviral protein (PAP) by turnip mosaic virus genome-linked protein (VPg). J Biol Chem 287:29729–29738
9. Domashevskiy AV, Cheng SY (2015) Thermodynamic analysis of binding and enzymatic properties of pokeweed antiviral protein (PAP) toward tobacco etch virus (TEV) RNA. J Nat Sci 1:e82
10. Gunawardana D, Domashevskiy AV, Gayler KR, Goss DJ (2014) Efficient preparation and properties of mRNA containing a fluorescent cap analog: Anthraniloyl-m^{7}GpppG. Translation 3:e988538
11. Ren J, Goss DJ (1996) Synthesis of a fluorescent 7-methylguanosine analog and a fluorescence spectroscopic study of its reaction with wheatgerm cap binding proteins. Nucleic Acids Res 24:3629–3634
12. Shuman S (1990) Catalytic activity of vaccinia mRNA capping enzyme subunits coexpressed in Escherichia coli. J Biol Chem 265:11960–11966
13. Higman MA, Christen LA, Niles EG (1994) The mRNA (guanine-7-) methyltransferase domain of the vaccinia virus mRNA capping enzyme. Expression in Escherichia coli and structural and kinetic comparison to the intact capping enzyme. J Biol Chem 269:14974–14981
14. Myette JR, Nieles EG (1996) Domain structure of the vaccinia virus mRNA capping enzyme. Expression in Escherichia coli of a subdomain possessing the RNA 5′-triphosphatase and guanylyltransferase activities and a kinetic comparison to the full-size enzyme. J Biol Chem 271:11936–11944
15. Kyrieleis OJ, Chang J, de la Peña M, Shuman S, Cusack S (2014) Crystal structure of vaccinia virus mRNA capping enzyme provides insights into the mechanism and evolution of the capping apparatus. Structure 22:452–465
16. Liu SW, Jiao X, Welch S, Kiledjian M (2008) Analysis of mRNA decapping. Methods Enzymol 448:3–21
17. Piccirillo C, Khanna R, Kiledjian M (2003) Functional characterization of the mammalian mRNA decapping enzyme hDcp2. RNA 9:1138–1147
18. Li Y, Song MG, Kiledjian M (2008) Transcript-specific decapping and regulated stability by the human Dcp2 decapping protein. Mol Cell Biol 28:939–948
19. Khan MA, Goss DJ (2004) Phosphorylation states of translational initiation factors affect mRNA cap binding in wheat. Biochemistry 43:9092–9097
20. Khan MA, Goss DJ (2005) Translation initiation factor (eIF) 4B affects the rates of binding of the mRNA m7G cap analogue to wheat germ eIFiso4F and eIFiso4F.PABP. Biochemistry 44:4510–4516
21. Domashevskiy AV (2012) Turnip mosaic virus genome-linked protein (VPg) inhibits pokeweed antiviral protein (PAP)-mediated depurination of RNA. Doctorate, Graduate Center and Hunter College, City University of New York
22. Hiratsuka T (1983) New ribose-modified fluorescent analogs of adenine and guanine nucleotides available as substrates for various enzymes. Biochim Biophys Acta 742:496–508
23. Gurevich VV, Pokrovskaya ID, Obukhova TA, Zozulya SA (1991) Preparative in vitro mRNA synthesis using SP6 and T7 RNA polymerases. Anal Biochem 195:207–213
24. Cunningham PR, Ofengard J (1990) Use of inorganic pyrophosphatase to improve the yield of in vitro transcription reactions catalyzed by T7 RNA polymerase. Biotechniques 9:713–714
25. Dasso MC, Jackson RJ (1989) Efficient initiation of mammalian mRNA translation at a CUG codon. Nucleic Acids Res 17:6485–6497
26. Herrmann BG, Frischauf AM (1987) Isolation of genomic DNA. Methods Enzymol 152:180–183
27. Schenborn ET, Mierendorf RC (1985) A novel transcription property of SP6 and T7 RNA polymerases: dependence on template structure. Nucleic Acids Res 13:6223–6236
28. Wilfinger WW, Mackey K, Chomczynski P (1997) Effect of pH and ionic strength on the spectrophotometric assessment of nucleic acid purity. Biotechniques 22:478–481
29. Baldwin AE, Khan MA, Tumer NE, Goss DJ, Friedland DE (2009) Characterization of pokeweed antiviral protein binding to mRNA cap analogs: competition with nucleotides and enhancement by translation initiation factor iso4G. Biochim Biophys Acta 1789:109–116
30. Harder KW, Owen P, Wong LK, Aebersold R, Clark-Lewis I, Jirik FR (1994) Characterization and kinetic analysis of the intracellular domain of human protein tyrosine phosphatase beta (HPTP beta) using synthetic phosphopeptides. Biochem J 298:395–401
31. Baykov AA, Evtushenko OA, Avaeva SM (1988) A malachite green procedure for orthophosphate determination and its use in alkaline phosphatase-based enzyme immunoassay. Anal Biochem 171:266–270

32. Zhang S, Williams CJ, Wormington M, Stevens A, Peltz SW (1999) Monitoring mRNA decapping activity. Methods 17:46–51
33. Jones BN, Quang-Dang DU, Oku Y, Gross JD (2008) A kinetic assay to monitor RNA decapping under single- turnover conditions. Methods Enzymol 448:23–40
34. Blewett N, Coller J, Goldstrohm A (2011) A quantitative assay for measuring mRNA decapping by splinted ligation reverse transcription polymerase chain reaction: qSL-RT-PCR. RNA 17:535–543
35. Shinshi H, Miwa M, Kato K, Noguchi M, Matsushima T, Sugimura T (1976) A novel phosphodiesterase from cultured tobacco cells. Biochemistry 15:2185–2190
36. Shinshi H, Miwa M, Sugimura T (1976) Enzyme cleaving the 5′-terminal methylated blocked structure of messenger RNA. FEBS Lett 65:254–257
37. Lu G, Zhang J, Li Y, Li Z, Zhang N, Xu X, Wang T, Guan Z, Gao GF, Yan J (2011) hNUDT16: a universal decapping enzyme for small nucleolar RNA and cytoplasmic mRNA. Protein Cell 2:64–73
38. Song MG, Li Y, Kiledjian M (2010) Multiple mRNA decapping enzymes in mammalian cells. Mol Cell 40:423–432
39. Deana A, Celesnik H, Belasco JG (2008) The bacterial enzyme RppH triggers messenger RNA degradation by 5′ pyrophosphate removal. Nature 451:355–358
40. Anderson CW, Straus JW, Dudock BS (1983) Preparation of a cell-free protein-synthesizing system from wheat germ. Methods Enzymol 101:635–644
41. Krieg PA, Melton D (1984) Functional messenger RNAs are produced by SP6 in vitro transcription of cloned cDNAs. Nucleic Acids Res 12:7057–7070
42. Weber LA, Hickey ED, Nuss DL, Baglioni C (1977) 5′-Terminal 7-methylguanosine and mRNA function: influence of potassium concentration on translation in vitro. Proc Natl Acad Sci U S A 74:3254–3258

Chapter 5

Intronless β-Globin Reporter: A Tool for Studying Nuclear RNA Stability Elements

Jessica A. Brown and Joan A. Steitz

Abstract

The intronless β-globin reporter, whose mRNA is intrinsically unstable due to the lack of introns, is a useful tool to study RNA stability elements in a heterologous transcript. Insertion of a stability element leads to the accumulation of intronless β-globin mRNA that can be visualized by conventional Northern blot analyses. In this chapter, we explain how to perform the β-globin reporter assay using the ENE (expression and nuclear retention element), a triple-helix-forming RNA stability element that protects reporter mRNA from 3′– 5′ decay. A list of considerations is included for the use of ENEs as a tool to stabilize other RNAs. In this chapter, we provide a brief description of how to insert an ENE sequence into the 3′-untranslated region of an intronless β-globin reporter plasmid using basic cloning technology. Then, we provide a detailed protocol for quantitative measurements of steady-state levels of β-globin mRNA. This entails the transient transfection of mammalian cells with β-globin reporter plasmids, isolation of total cellular RNA, and detection of reporter mRNA via Northern blot. This methodology can be applied for the study of any nuclear RNA stability element using the intronless β-globin reporter.

Key words β-globin, Reporter assay, RNA stability, RNA decay, Triple helix, ENE

1 Introduction

The β-globin reporter system has been adapted to study various facets of messenger RNA (mRNA) processing, particularly mRNA decay in eukaryotic cells. Historically, reporter plasmids encoding the human β-globin gene were instrumental in understanding nonsense-mediated mRNA decay because the β-globin gene from β-thalassemia patients contains naturally occurring nonsense mutations. These mutations create a premature termination codon (PTC) that renders the resulting β-globin mRNA susceptible, or sometimes resistant for rare mutations, to nonsense-mediated decay (Fig. 1, β-PTC) [1, 2]. The β-globin reporter system has also been used to study RNA elements that stabilize or destabilize mRNA. Destabilizing RNA elements are typically studied in the context of a wild-type (WT), intron-containing β-globin reporter

Robert E. Rhoads (ed.), *Synthetic mRNA: Production, Introduction Into Cells, and Physiological Consequences*, Methods in Molecular Biology, vol. 1428, DOI 10.1007/978-1-4939-3625-0_5,

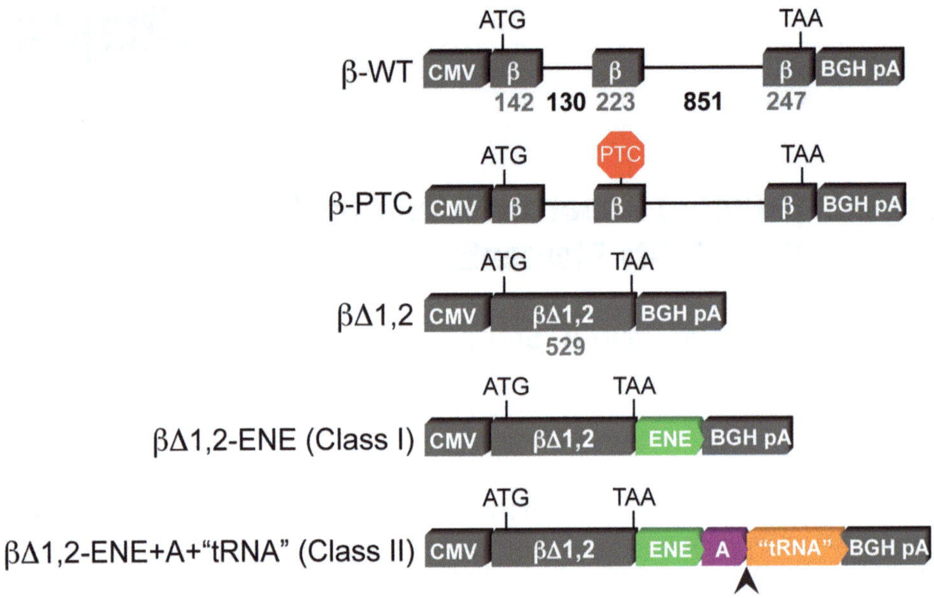

Fig. 1 Schematic of β-globin reporter plasmids. For all constructs, the backbone vector is pcDNA3; it also contains neomycin resistance and ampicillin resistance genes (not shown). Each β-globin sequence is flanked by a human CMV immediate-early promoter at the 5′ end and a bovine growth hormone polyadenylation (BGH pA) signal at the 3′ end. There is a FLAG tag (not shown) at the N-terminus of the β-globin sequence. Introns are represented by *black lines*, exons are represented by boxes, PTC denotes the premature termination codon, and the RNase P cleavage site is denoted by an arrowhead. Lengths (in nts) of introns and exons are given below. All ENEs (*green boxes*) are inserted into the ApaI site in the 3′-UTR. Downstream of class II ENEs are a genomically encoded A-rich tract (*purple box*) and a tRNA-like sequence (*orange box*). All plasmids have been described in detail previously [2, 4, 6, 7]

(Fig. 1, β-WT), which generates a highly stable transcript that localizes predominantly to the cytoplasm [3, 4]. In contrast, RNA stability elements are typically studied in the context of an intronless β-globin reporter (Fig. 1, βΔ1,2), which accumulates to much lower levels than its intron-containing counterpart and exhibits predominantly nuclear localization [4, 5].

Recently, a unique RNA stability element known as an ENE (expression and nuclear retention element) was discovered near the 3′ end of viral and host long noncoding RNAs (lncRNAs), as well as in viral genomic RNAs (Table 1) [4, 6, 7]. To stabilize RNA, the ENE folds into a stem–loop structure containing a U-rich internal loop, which sequesters the downstream 3′-poly(A) tail or a genomically encoded A-rich tract into a triple helix (ENE + A) composed of mostly U•A-U major-groove base triples flanked by A-minor base triples (Fig. 2) [8, 9]. The formation of this RNA triple helix inhibits rapid 3′–5′ RNA decay [9, 10] and can stabilize β-globin, green fluorescence protein (GFP), and mCherry reporter transcripts [7, 11]. Insertion of viral or host ENEs into the 3′-untranslated region (UTR) of intronless β-globin reporter

Table 1

ENE sequences tested in the βΔ1,2 reporter and fold increase in RNA levels

Class I: ENE + poly(A) tail			Ref.
ENE[a]	Relative RNA Levels[b]	Sequence of ENE Insert (5′–3′)[c]	
KSHV PAN	9 ± 2	TGTTTTGGCTGGGTTTTTCCTTGTTCGCACCGGACACCTCCAGTGACCAGACGGCAAGG TTTTTATCCCAGTGTATATT	[6]
RRV PAN	3 ± 1	AGACGTTTGTGTTGGTTTTTATGACCAGCTTGGTACAAAAACCTGCTGGTGATTTTTTA CCCAACAAATAATT	[6]
EHV2 PAN	2.0 ± 0.4	AAAGAATATTTTTAAAGACTTTTTTCCCCAACCTCTGGGTTGGGTTTTTTCTCTTTAAAA TATTCAATC	[6]
C.c. bracovirus	1.7 ± 0.2	TTCATCAAGGAGGTTTTTTCCCAGCCTAGCTGGGTTTTCCTCCTTTGGGG	[6]
PSIV	1.10 ± 0.03	CATTGGCAGTAGAGTTTTTCCCCAGGGAGCTTCACTGTCTGGGTTTTCTCTACTTATCAC	[6]
Class II: ENE + A-rich tract			
Human MALAT1	45 ± 7	TAGGGTCATGAAGGTTTTTCTTTTCCTGAGAAAACAACACGTATTGTTTTCTCAGGTTTT GCTTTTTGGCCTTTTTCTAGCTTAAAAAAAAAAAAAAGCAAAAGATGCTGGTGGTTGGCACTCCTGG TTTCCAGGACGGGGTTCAAATCCCTGCGGCGTCTTTGCTTTGACT	[7]
Human MENβ	6 ± 1	GCCGCCGCAGGTGTTTCTTTTACTGAGTGCAGCCCATGGCCGCACTCAGGTTTTGCTTT TCACCTTCCCATCTGTGAAAGAGTGAGCAGGAAAAAGCAAAA GGCGCTGGTGGTGGCACGTCCAGCACGGCTGGGCCGGGGTTCGAGTCCCCGCAGTGTTGCTGCTTCCTT	[7]

[a]RRV is rhesus rhadinovirus; EHV2 is equine herpesvirus 2; *C.c.* is *Contesia congregata*; PSIV is *Plautia stali* intestine virus

[b]Relative RNA levels are the fold increase in RNA levels relative to WT βΔ1,2. All RNA levels were determined in HEK293T cells. All values are for a single ENE insert

[c]The U-rich internal loops of ENEs are shown in *green*; A-rich tracts are *purple*; and tRNA-like sequences are *orange*

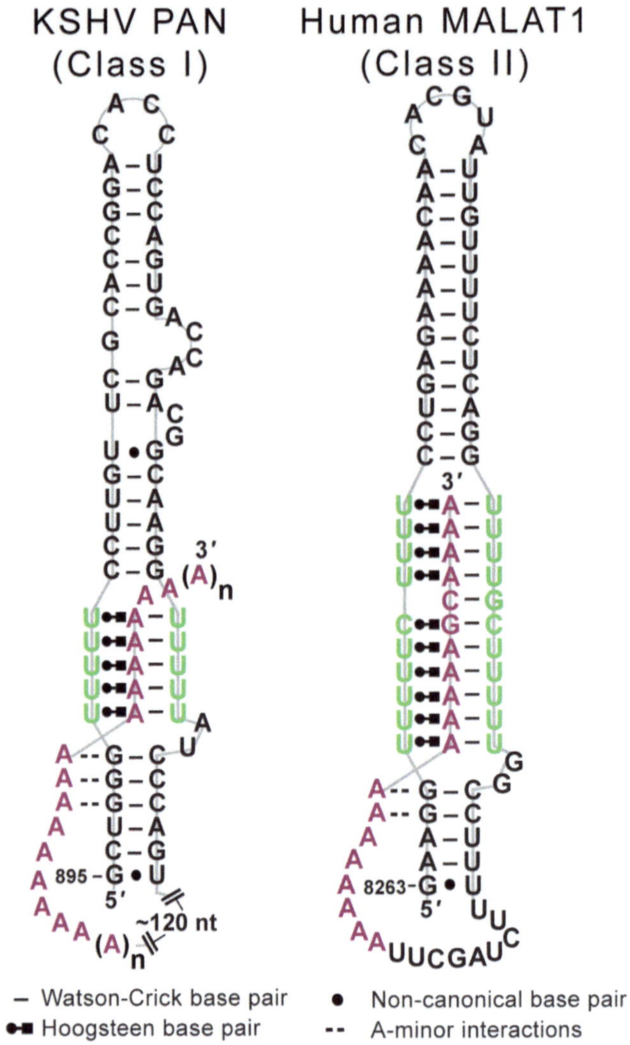

Fig. 2 Schematic of ENE + A structures. Triple-helical ENE + A structures are schematized for representative examples of class I and class II ENEs: KSHV PAN ENE bound to the 3′-poly(A) tail and MALAT1 ENE bound to its A-rich tract. The U-rich internal loops are in *green font*, while the poly(A) tail and A-rich tract are in *purple font*. Base pair and base triple interactions are defined in the legend. Nucleotide numbers at the 5′ end are from the native transcripts

plasmids (Fig. 1, βΔ1,2-ENE and βΔ1,2-ENE + A + "tRNA") increases the accumulation of β-globin mRNA by 1.1- to 45-fold over an intronless reporter without an ENE in HEK293T cells (Table 1) [4, 6, 7].

Currently, two different classes of ENEs have been characterized: class I ENEs, which interact with a posttranscriptionally added 3′-poly(A) tail, and class II ENEs, which interact with a genomically encoded A-rich tract at the 3′ end of transcripts (Fig. 2). Class I ENEs have been discovered in viral lncRNAs as well as in viral single-stranded genomic RNA [4, 6]. The best characterized class I ENE is from polyadenylated nuclear (PAN) RNA produced by the Kaposi's sarcoma-associated herpesvirus (KSHV) [4, 8, 10, 12]. Class II ENEs have been found in cellular metastasis-associated lung adenocarcinoma transcript 1 (MALAT1) and multiple endocrine neoplasia beta (MENβ) lncRNAs whose 3′ ends do not undergo canonical cleavage and polyadenylation [7, 11]. Rather, tRNA-like structures immediately downstream of a genomically encoded A-rich tract are endonucleolytically processed by RNase P to generate the 3′ end of MALAT1 and MENβ [13, 14]. Although both class I and II ENE + A structures inhibit 3′–5′ RNA decay, the decay mechanisms are expected to be different because class I ENE + A structures present a triple helix with a traditional poly(A) overhang whereas class II ENE + A structures present a triple helix with a blunt end [9].

ENE + A structures can be used as a tool to stabilize the 3′ ends of other RNAs in vivo. One successful example demonstrated that five copies of the KSHV PAN ENE increased primary microRNA transcripts for human let-7a-1 and *Caenorhabditis elegans* lin-4 by ~10-fold in HeLa cells [15]. This application of ENE structures has not been tested extensively, but we can offer the following points of advice. (1) ENE + A structures preferentially stabilize intronless RNAs rather than spliced RNAs [4]. (2) Class I ENEs are preferable when the 3′ end of the RNA is generated via cleavage and polyadenylation. These ENEs appear to be most effective within 150 nucleotides (nts) upstream of the poly(A) tail; multiple copies can be inserted for enhanced stabilization [4, 6]. (3) Class II ENEs appear to be effective only when the A-rich tract is at the extreme 3′ end of the transcript; the sequence of the tRNA-like structure must be included so that the 3′ end can be generated via RNase P cleavage (Table 1 and Fig. 1) [7]. These ENEs are useful for studying non-polyadenylated RNAs [16]. (4) Class II ENE + A structures enhance the translation of mRNAs in which they reside [11, 17]. (5) Avoid inserting the ENE adjacent to another highly structured region. (6) Consider testing multiple ENEs in different cell types. We have tested the stabilization activity of only seven ENEs in the β-globin reporter introduced into HEK293T cells via transient transfection. Activity may vary in different sequence and structural contexts as well as cell types. (7) To block 5′–3′ RNA decay, consider using XRN1-resistant RNA (xrRNA) structures derived from flaviviruses [18, 19].

In this chapter, we describe how to insert an ENE sequence into an intronless β-globin reporter plasmid using standard cloning techniques. The β-globin reporter plasmids we describe use pcDNA3 as the vector; therefore, the β-globin gene is flanked upstream by a cytomegalovirus (CMV) immediate-early promoter and downstream by the bovine growth hormone polyadenylation (BGH pA) signal (Fig. 1). Multiple copies of the class I ENE can be inserted, although reporter mRNA levels appear saturated at three or more copies in the case of KSHV PAN ENE [4]. Reporters with ENEs inserted in the reverse orientation serve as negative controls. Lastly, we provide a detailed protocol to measure β-globin mRNA levels under steady-state conditions. Briefly, intronless β-globin reporter plasmids are introduced into HEK293T cells by transient transfection, total cellular RNA is harvested, and mRNA levels are measured by quantitative Northern blot analysis.

2 Materials

2.1 Construction of a βΔ 1,2-ENE Reporter Plasmid

Because standard cloning techniques are used to create these plasmids, only the pertinent reagents are listed. For a complete description of all required materials, please refer to any edition of *Molecular Cloning: A Laboratory Manual* by J. Sambrook and others [20].

1. Reagents and equipment for performing PCR (e.g., Phusion® high-fidelity DNA polymerase), chemically synthesized PCR primers with an ApaI site, and genomic material. When the appropriate genomic material is not available, the ENE insert is created by annealing chemically synthesized oligonucleotides.
2. Intronless β-globin plasmid (βΔ1,2 in Fig. 1, *see* **Note 1**).
3. ApaI.
4. Antarctic phosphatase (New England BioLabs).
5. T4 DNA ligase.

2.2 Transient Transfection of Mammalian Cells with Reporter Plasmid

1. Complete DMEM media: high-glucose Dulbecco's Modified Eagle Medium, 1 % penicillin/streptomycin, 10 % heat-inactivated fetal bovine serum, and 1 % L-glutamine.
2. Low-passage HEK293T cells cultured at 37 °C and 5 % CO_2.
3. 100 × 20 mm tissue culture dishes treated by vacuum gas plasma and 6-well cell culture plates.
4. Opti-MEM media (Gibco).
5. Mirus *Trans*IT®-293 transfection reagent.
6. DNA plasmids: β-globin reporters (*see* Subheadings 2.1 and 3.1), pBluescript, and pmaxGFP.
7. Fluorescence microscope.
8. TRIzol® reagent (Ambion).

2.3 Measurement of β-Globin mRNA Levels by Northern Blot Analysis

1. MOPS (4-morpholinepropanesulfonic acid) buffer (10×): 200 mM MOPS pH 7.0, 5 mM sodium acetate, and 10 mM EDTA. Filter solution (0.22 μm) and store in a light-protected bottle at room temperature (*see* **Note 2**).

2. A 1.2 % agarose-formaldehyde gel: Weigh 3.6 g of agarose into a 500 mL Erlenmeyer flask and add 217.5 mL of distilled water. Dissolve agarose by heating in a microwave for ~3 min. Let solution cool for ~2–3 min and then add 30 mL of 10× MOPS buffer and 52.5 mL of 37 % formaldehyde. Swirl to mix contents and pour agarose solution into a 24 × 20.5 cm gel casting tray (*see* **Note 3**). Use a pipet tip to remove any air bubbles. The gel surface should be bubble free so that the blotting membrane is in direct contact with the gel during the transfer process. Allow the gel to set for 30 min at room temperature. Caution: Formaldehyde is toxic. Pour gel in a ventilated chemical fume hood and keep covered whenever possible.
3. Gel running buffer (1×): 1× MOPS buffer (*see* **Note 4**) and 0.925 % formaldehyde. Prepare 2.5 L immediately before use.
4. Sample-loading dye (2×): 1× MOPS buffer, 5.6 % formaldehyde, 50 % formamide, 0.5 % bromophenol blue, 0.5 % xylene cyanol, and 10 mM EDTA. Prepare appropriate volume based on number of samples immediately before use (*see* **Note 5**).
5. 0.5–10 kb RNA ladder.
6. Horizontal gel electrophoresis apparatus.
7. Peristaltic pump and tubing to circulate gel running buffer.
8. Power supply.
9. SSC buffer (20×): 3 M sodium citrate pH 7.0 and 0.3 M sodium chloride. Prepare appropriate volume and store at room temperature.
10. Zeta-Probe® blotting membrane (Bio-Rad).
11. Whatman™ chromatography paper 3MM (Whatman).
12. Glass dish (23 × 33 cm), glass plate (30 × 45 cm), glass rod, parafilm, paper towels, plastic wrap, and weight (e.g., textbook).
13. UV Stratalinker 2400 (Stratagene).
14. Methylene blue staining solution (0.3 %): 0.3 M sodium acetate pH 5.2 and 0.3 g methylene blue in water. Prepare 100 mL and store at room temperature. This staining solution can be reused.
15. ExpressHyb™ hybridization solution (Clontech).
16. Hybridization tubes and hybridization oven with a temperature range of 37–65 °C.
17. T4 polynucleotide kinase and [γ-^{32}P]ATP for 5′-[^{32}P]-labeling three DNA oligonucleotide probes against neomycin resistance (NeoR) mRNA: 5′-GCATCAGAGCAGCCGATTGTCTGTT G-3′, 5′-GCATCAGCCATGATGGATACTTTCTCGG-3′, and 5′-CGGCCATTTTCCACCATGATATTCGGCAAGC-3′.

18. SP6 RNA polymerase, DTT, ATP, CTP, GTP, [α-^{32}P]UTP, and NcoI-linearized β-globin plasmid for preparing RNA probe against β-globin mRNA.
19. Illustra MicroSpin G-25 columns (GE Healthcare) to remove unincorporated [^{32}P]NTPs from labeling reactions.
20. Wash buffer I: 2× SSC and 0.1 % SDS in water. Prepare 1 L for each blot and store at room temperature.
21. Wash buffer II: 0.5× SSC and 0.1 % SDS in water. Prepare 1 L for each blot and store at room temperature.
22. Wash buffer III: 0.1× SSC and 0.1 % SDS in water. Prepare 1 L for each blot and store at room temperature.
23. Phosphorimager screen (Molecular Dynamics) and scanner (Storm 860).
24. Stripping buffer: 0.1 % SDS in water. Prepare 1 L for each blot and store at room temperature.
25. Heated shaker at 65 °C.
26. Computer with appropriate software programs (e.g., ImageQuant, Microsoft Excel, GraphPad Prism) to quantitate mRNA levels.

3 Methods

3.1 Construction of a βΔ1,2-ENE Reporter Plasmid

Standard cloning techniques are used to create plasmids; therefore, only the pertinent reagents and details are described. For a complete description of each step, please refer to any edition of *Molecular Cloning: A Laboratory Manual* by J. Sambrook and others [20].

PCR is used to create the insert containing the ENE region-of-interest flanked by ApaI sites. Phusion High-Fidelity DNA polymerase and the supplied GC buffer are used when genomic material is available. Otherwise, complementary oligonucleotides containing the ENE region-of-interest and ApaI sites are chemically synthesized and then annealed.

1. Digest the ENE insert and βΔ1,2 vector using ApaI according to the manufacturer's recommended protocol.
2. Because the βΔ1,2 vector is digested with only ApaI, it is subsequently treated with antarctic phosphatase per the manufacturer's recommended protocol. Removal of the 5′-phosphate minimizes the ligation of an empty vector.
3. Use an agarose gel to isolate the digested insert and the linearized, dephosphorylated βΔ1,2 vector.
4. Ligate the insert and βΔ1,2 vector using T4 DNA ligase per the manufacturer's recommended protocol. Be sure to include a

vector-only control. If multiple copies of the ENE need to be inserted, then set up several ligation reactions, gradually increasing the insert:vector ratio.

5. Immediately after the ligation reaction, transform the reaction mixture into chemically competent DH5α cells.
6. If colonies appear on the carbenicillin-treated LB agar plates, then prepare plasmids for DNA sequencing to determine the orientation of insert(s), the number of inserts, and if insert sequence is correct. The BGH reverse primer is used for DNA sequencing. Remember, βΔ1,2 vectors with inserts in both forward and reverse orientations are needed. Plasmids are prepared using any commercially available Maxi prep kit (*see* **Note 1**).

3.2 Transient Transfection of Mammalian Cells with βΔ1,2 Reporter Plasmids

1. Grow low-passage HEK293T cells in complete DMEM until they are ~85–95 % confluent. About 18–24 h before transfection, seed the HEK293T cells into 6-well plates. The total volume of each well is 2 mL with ~6×10^5 cells/well. Distribute the cells uniformly by shaking the plate in a side-to-side and front-to-back (not circular) motion.
2. Before transfection, check the confluency of the HEK293T cells in each well. Confluency should be ~50–70 %. Remove Opti-MEM media from the refrigerator and warm to room temperature.
3. Prepare plasmids for transfection by adding the following components to a sterile 1.5 mL tube immediately before transfection: 0.4 μg β-globin reporter, 1.4 μg pBluescript, and 0.2 μg pmaxGFP. A total of 2 μg of plasmid DNA is transfected (*see* **Note 6**) and the total volume of plasmid should not exceed 10 μL.
4. For each sample, mix together 200 μL Opti-MEM and 6 μL Mirus *Trans*IT-293 and incubate at room temperature for 15 min (*see* **Note 7**). If preparing a master mix, add an extra sample to allow for pipetting error.
5. Add 200 μL Opti-MEM-*Trans*IT-293 to each DNA sample. Gently pipet up-and-down once to mix. Incubate DNA transfection complex at room temperature for 20 min.
6. Remove the 6-well plate from the incubator and add all of DNA transfection mixture dropwise to each well. Mix gently by slowly rocking the plate back-and-forth in the air, while holding rather than placing the plate on a surface, after *each* addition.
7. Incubate cells at 37 °C for 24–48 h. Examine the transfection efficiency by determining the percent of cells expressing GFP

using a fluorescence microscope. The transfection efficiency of HEK293T cells is typically 80–100 %.

8. Harvest cells 24–48 h after transfection. Remove medium and lyse cells by adding 1 mL of TRIzol® to each well. Isolate RNA according to the manufacturer's protocol for TRIzol®. Resuspend RNA in 10–12 μL RNase-free water and measure the A_{260} value to calculate the amount of RNA in μg/μL. Yields are typically 50–70 μg of total RNA per well.

3.3 Detection of β-Globin mRNA Levels by Northern Blot Analysis

1. In a chemical fume hood, setup a horizontal gel electrophoresis apparatus by preparing a 1.2 % agarose-formaldehyde gel and 1× gel running buffer (*see* **Note 8**).
2. Prepare RNA samples. Combine 10 μg of total RNA and sample-loading dye (*see* **Note 5**). For the RNA ladder, add 3 μL of 0.5–10 kb RNA ladder to sample-loading dye. Heat denature samples, including the ladder, at 95 °C for 5 min.
3. Load gel and resolve RNA at 100 V until the samples enter the gel. Meanwhile, prepare the peristaltic pump by installing the tubing and priming the tubing with leftover 1× gel running buffer. Once the samples have entered the gel, turn off the power supply and place tubing into buffer chambers so that the buffer can circulate. Buffer circulation is necessary to prevent formation of a pH gradient, whereby high pH would degrade RNA. Then, adjust the voltage for the appropriate length of run time; end the run when the bromophenol blue dye is about 1 in. from the bottom of the gel (*see* **Note 9**).
4. Remove gel in casting tray and place in a clean dish. For **steps 4–6**, always keep the gel in the casting tray so that the large gel does not break. Add ~800 mL distilled water to the dish and place it on a shaker at room temperature for 5 min to gently wash away formaldehyde in the gel running buffer.
5. Decant water and add ~600 mL of 20× SSC. Return gel to the shaker to allow it to equilibrate for 45 min. The gel will float because of the high-salt buffer. To keep the gel submerged during equilibration, adjust the shaker RPM so that it is strong enough to keep the gel submerged but gentle enough to keep the gel intact. If gel tearing is a problem, tape the ends of the gel casting tray to restrict mobility of the gel.
6. During the equilibration step, begin to set up the capillary transfer system (Fig. 3). Fill an RNase-free 23 × 33 cm glass dish with ~800 mL of 20× SSC. Set an RNase-free 30 × 45 cm glass plate on top of the glass dish so that its position is perpendicular to the glass dish (*see* Fig. 3). Wet two pieces of 9 × 19 in. Whatman paper by passing the sheets under the glass plate and pulling them over to rest atop the glass plate. Both ends of the Whatman paper should be submerged in the

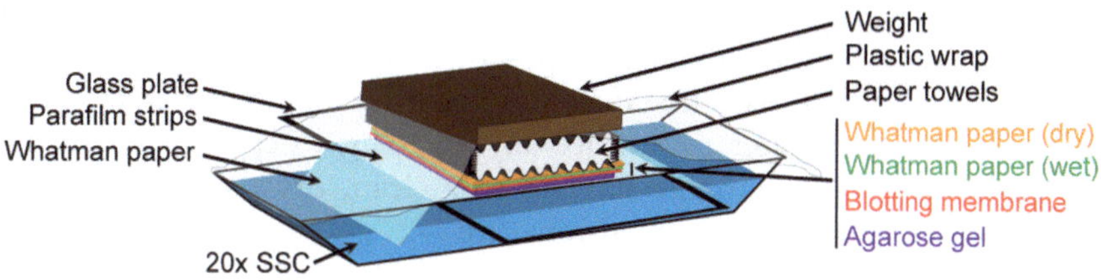

Fig. 3 Schematic of capillary transfer of RNA from agarose gel to blotting membrane. RNA migrates from the agarose gel to the blotting membrane by a moving phase of 20× SSC buffer. Refer to Subheading 3.3, **steps 6–8** for a detailed description of how to assemble this apparatus

buffer. Smooth the Whatman paper surface by rolling a glass rod over any air pockets. Fold a roll of paper towels accordion style and set aside until ready for use. Cut Zeta-Probe® blotting membrane and four pieces of Whatman paper.

7. After the 45-min equilibration period, decant buffer off the gel and use a razor blade to trim away excess gel. The bromophenol blue defines the lower boundary; ~2 in. above the xylene cyanol band is the upper boundary. Cut one corner of the gel to track gel orientation. Carefully transfer the gel to the center of the wetted Whatman paper (*see* **Note 10**). Outline, but do not cover, the gel with strips of parafilm so that the surface of the wetted Whatman paper is completely covered. Importantly, there should be no direct contact between the SSC-wetted Whatman paper and the stack atop the membrane. Otherwise, transfer of RNA from the gel to the membrane will be inefficient due to a short-circuiting effect.

8. Wet the Zeta-Probe® blotting membrane, whose dimensions are cut so that it is ~1 cm wider than the gel, in distilled water. Carefully, center the membrane and place it on top of the gel so that the membrane completely covers the gel. Do not move the membrane once it is in contact with the gel. Cut the membrane in the same corner as the gel to track orientation. Wet two pieces of Whatman paper, whose dimensions are ~1 cm wider than the membrane, in distilled water. Carefully center the wetted papers and place them on top to completely cover the membrane. Roll a glass rod over the wetted papers and membrane to remove any air pockets so that the membrane surface is in direct contact with the gel surface. Carefully center two pieces of dry Whatman paper and place on top of the smooth, wetted Whatman paper. Center the stack of paper towels and place on top of the dry Whatman paper. Drape plastic wrap over the paper towels and glass tray. Place a weight (e.g., textbook) on top of the plastic wrap-covered paper towels. The plastic wrap will keep the weight from absorbing the 20× SSC buffer and minimize evaporation. Allow the transfer to proceed overnight (or at least 8 h).

9. Disassemble the transfer stack by removing each layer until the membrane is exposed. Place the membrane face up in an RNase-free dish and wash the membrane gently with distilled water to remove any salt or gel deposits. Then, transfer the membrane face up to a clean, dry piece of Whatman paper, whose dimensions are ~2 cm wider than the membrane. Allow the membrane to air dry.
10. Once the membrane is dry, UV crosslink RNA to the membrane using a Stratalinker 2400 set to 1200 μJ × 100.
11. Transfer the membrane face up to an RNase-free glass dish and cover the membrane surface with 0.3 % methylene blue stain until the entire surface is stained blue. Then, decant the stain into the original bottle for re-use (*see* **Note 11**). Destain the membrane by rinsing it with distilled water several times. Destain until the ribosomal RNA (rRNA) bands (18 S at 1.9 kb and 28 S at 5 kb) are visible but the blue background is washed away. If rRNA bands are not visible, then the RNA has become degraded and the experiment will need to be repeated, beginning with **step 1** in Subheading 3.2. If rRNA bands are visible, then an image of the blot is captured using a GBox or any digital camera device. Loading can also be examined. Use an ethanol-resistant marker to write the name of the experiment on the blot as well as to mark the 0.5–10 kb ladder bands, rRNA bands, and xylene cyanol and bromophenol blue dyes.
12. Place the blot in a hybridization tube (*see* **Note 12**) and add ~20 mL of ExpressHyb (*see* **Note 13**). Rotate the blot in a hybridization oven at 37 °C for at least 1 h.
13. Prepare radiolabeled probes against NeoR and β-globin mRNAs. 5′-[^{32}P]-label three DNA oligonucleotide probes (1 μM each) against NeoR mRNA using T4 polynucleotide kinase and [γ-^{32}P]ATP according to the manufacturer's protocol. Use SP6 RNA polymerase to synthesize an antisense β-globin ribonucleotide probe in the presence of 20 mM DTT, 3.3 mM ATP, 3.3 mM CTP, 3.3 mM GTP, 5.75 μM [α-^{32}P]UTP, and 100 ng/μL NcoI-linearized β-globin plasmid. Illustra MicroSpin G-25 columns are used to remove unincorporated [^{32}P]NTPs according to the manufacturer's protocol.
14. Heat denature the pooled α-NeoR oligonucleotide probes at 95 °C for 5 min and then add probe to hybridization tube (*see* **Note 14**). Allow probes to hybridize to the membrane overnight at 37 °C.
15. On the following day, wash the membrane with wash buffer I three times for 5 min each at room temperature. After each wash, use a Geiger counter to make sure the radioactive signal

is detectable. A stronger signal should be detected in the 1–1.5 kb range.

16. Next, wash the membrane with wash buffer II three times for 5 min each at room temperature. After each wash, use a Geiger counter to make sure the radioactive signal is detectable.
17. If the background is still high, the membrane can be washed with wash buffer III three times for 5 min each at room temperature. Expose the membrane to a Phosphorimager screen for 24–48 h depending on signal strength and then scan the screen using a Storm 860.
18. To remove the α-NeoR probe, wash the membrane in boiling-hot stripping buffer three times for 15 min each in a heated shaker at 65 °C. After each wash, decant the stripping buffer and check the radioactive signal using a Geiger counter. The radioactive signal should be at or below background after three washes.
19. Repeat **steps 12** and **14–17** to probe for β-globin mRNA. However, since this probe is RNA, the prehybridization and hybridization steps are performed at 65 °C. Also, the washing procedure is modified for wash buffer III (**step 17**). First, wash the blot at room temperature for 15 min in 400 mL of wash buffer III. Then, wash the blot in a heated shaker at 65 °C for 15 min using 600 mL of wash buffer III that has been pre-warmed in a microwave for 3 min.
20. Expose the membrane to a Phosphorimager screen for 12–24 h depending on signal strength. Scan the screen using a Storm 860.

3.4 Quantitation of β-Globin mRNA Levels

1. Use ImageQuant (or another comparable program) to measure the NeoR and β-globin mRNA levels by volumetric densitometry analysis. A box of equal size is placed around the signal in each lane. Then, a box of equal size is dragged to a position below each sample; this box provides the background signal.
2. After subtracting the background signal, the β-globin signal is divided by the signal for NeoR mRNA. NeoR serves as a loading and transfection control as it is transcribed from the same plasmid as the β-globin reporter.
3. All values are normalized relative to βΔ1,2 (or the corresponding WT counterpart if investigating mutants), which is set to an arbitrary value of 1.
4. The assay is repeated at least two more times so that the mean and standard deviation can be calculated for each sample. Neighboring samples can affect relative accumulation levels. For example, a strong signal in one lane may artificially increase the signal in adjacent lanes because of signal diffusion. This problem can be minimized by skipping a lane when loading samples.

4 Notes

1. βΔ1,2 and other plasmids can be obtained from the Steitz lab upon request. When preparing βΔ1,2 plasmids, yields are typically low, even though pcDNA3 (Invitrogen) is a high-copy number plasmid.
2. When preparing MOPS buffer, adjust pH with 10 M NaOH if using the free acid form (MW = 209.3 g/mol) or with concentrated HCl if using the sodium salt form (MW = 231.2 g/mol). MOPS buffer will color with age and light exposure. Straw-colored buffer works OK but anything darker does not.
3. Quantities of gel running reagents are based on a gel system supporting a 24 × 20.5 cm gel, which provides the best resolving power for polyadenylated β-globin mRNA while avoiding overlap with 18S rRNA. Proportionally scale reagents for smaller or larger gel systems.
4. MOPS running buffer as low as 0.1× can be used for gel runs.
5. Load small sample volumes (<12 μL) to maintain high resolution. Try not to exceed 20 μL of total volume. Load 5–10 μg of total RNA (~1–2 μL) in 8 μL sample-loading dye.
6. All plasmid concentrations are checked by agarose gel electrophoresis prior to transfection to ensure that equal amounts of plasmid are transfected.
7. The transfection procedure has been modified slightly from the manufacturer's protocol.
8. Use a level to ensure that the gel is poured and run on an even surface; otherwise, RNA migrating in a thinner region of the gel will be lost.
9. Gel run times will vary depending on the gel box; adjust accordingly. For daytime runs, increase the voltage to 140 V (~7–8 h). For nighttime runs, the voltage can be decreased to 70 V (~18 h).
10. Because the RNA sample sinks to the bottom upon loading a horizontally run gel, the gel is inverted so that the RNA has a shorter distance to travel to the membrane. To invert gel, several gloved fingers are placed on top of both sides of the gel and the casting tray is flipped over onto the wetted Whatman paper. The gel should remain adhered to the casting tray. To peel it off, use one finger to pry away one corner until the gel falls from the casting tray onto the Whatman paper.
11. The 0.3 % methylene blue stain can be reused until the stain becomes too weak to visualize rRNA.

12. To put the membrane in a hybridization tube, roll up the blot so that face-up side faces the interior of the tube and adheres flat against the tube walls.
13. A "homemade" hybridization solution can be used rather than ExpressHyb and results in lower background for the β-globin ribonucleotide probe. Components of the homemade hybridization solution are 50 % formamide, 25 mM sodium phosphate pH 6.5, 6× SSC, 5× Denhardt's solution, 0.5 % SDS, and 0.3 mg/mL carrier RNA.
14. The blot must be probed for α-NeoR mRNA first, then for the reporter mRNA because the antisense β-globin ribonucleotide probe cannot be stripped from the blot.

Acknowledgments

We thank Mei-Di Shu and Kazimierz Tycowski for critical review of the manuscript, Angela Miccinello for editorial work, and all Steitz lab members for thoughtful discussions. This work was supported by NIH grants GM111430 (J.A.B.) and GM026154 (J.A.S.). J.A.S. is an investigator of the Howard Hughes Medical Institute.

References

1. Baserga SJ, Benz EJ Jr (1988) Nonsense mutations in the human beta-globin gene affect mRNA metabolism. Proc Natl Acad Sci U S A 85:2056–2060
2. Lykke-Andersen J, Shu MD, Steitz JA (2000) Human Upf proteins target an mRNA for nonsense-mediated decay when bound downstream of a termination codon. Cell 103:1121–1131
3. Newman TC, Ohme-Takagi M, Taylor CB, Green PJ (1993) DST sequences, highly conserved among plant SAUR genes, target reporter transcripts for rapid decay in tobacco. Plant Cell 5:701–714
4. Conrad NK, Steitz JA (2005) A Kaposi's sarcoma virus RNA element that increases the nuclear abundance of intronless transcripts. EMBO J 24:1831–1841
5. Akef A, Lee ES, Palazzo AF (2015) Splicing promotes the nuclear export of beta-globin mRNA by overcoming nuclear retention elements. RNA 21:1–13
6. Tycowski KT, Shu MD, Borah S, Shi M, Steitz JA (2012) Conservation of a triple-helix-forming RNA stability element in noncoding and genomic RNAs of diverse viruses. Cell Rep 2:26–32
7. Brown JA, Valenstein ML, Yario TA, Tycowski KT, Steitz JA (2012) Formation of triple-helical structures by the 3′-end sequences of MALAT1 and MENbeta noncoding RNAs. Proc Natl Acad Sci U S A 109:19202–19207
8. Mitton-Fry RM, DeGregorio SJ, Wang J, Steitz TA, Steitz JA (2010) Poly(A) tail recognition by a viral RNA element through assembly of a triple helix. Science 330:1244–1247
9. Brown JA, Bulkley D, Wang J, Valenstein ML, Yario TA, Steitz TA, Steitz JA (2014) Structural insights into the stabilization of MALAT1 noncoding RNA by a bipartite triple helix. Nat Struct Mol Biol 21:633–640
10. Conrad NK, Mili S, Marshall EL, Shu MD, Steitz JA (2006) Identification of a rapid mammalian deadenylation-dependent decay pathway and its inhibition by a viral RNA element. Mol Cell 24:943–953
11. Wilusz JE, JnBaptiste CK, Lu LY, Kuhn CD, Joshua-Tor L, Sharp PA (2012) A triple helix stabilizes the 3′ ends of long noncoding RNAs that lack poly(A) tails. Genes Dev 26:2392–2407

12. Conrad NK, Shu MD, Uyhazi KE, Steitz JA (2007) Mutational analysis of a viral RNA element that counteracts rapid RNA decay by interaction with the polyadenylate tail. Proc Natl Acad Sci U S A 104:10412–10417
13. Wilusz JE, Freier SM, Spector DL (2008) 3′ end processing of a long nuclear-retained noncoding RNA yields a tRNA-like cytoplasmic RNA. Cell 135:919–932
14. Sunwoo H, Dinger ME, Wilusz JE, Amaral PP, Mattick JS, Spector DL (2009) MEN epsilon/beta nuclear-retained non-coding RNAs are up-regulated upon muscle differentiation and are essential components of paraspeckles. Genome Res 19:347–359
15. Pawlicki JM, Steitz JA (2008) Primary microRNA transcript retention at sites of transcription leads to enhanced microRNA production. J Cell Biol 182:61–76
16. Muniz L, Davidson L, West S (2015) Poly(A) polymerase and the nuclear poly(A) binding protein, PABPN1, coordinate the splicing and degradation of a subset of human pre-mRNAs. Mol Cell Biol 35:2218–2230
17. Bhandari D, Raisch T, Weichenrieder O, Jonas S, Izaurralde E (2014) Structural basis for the nanos-mediated recruitment of the CCR4-NOT complex and translational repression. Genes Dev 28:888–901
18. Chapman EG, Moon SL, Wilusz J, Kieft JS (2014) RNA structures that resist degradation by Xrn1 produce a pathogenic Dengue virus RNA. Elife 3:e01892
19. Chapman EG, Costantino DA, Rabe JL, Moon SL, Wilusz J, Nix JC, Kieft JS (2014) The structural basis of pathogenic subgenomic flavivirus RNA (sfRNA) production. Science 344:307–310
20. Green MR, Sambrook J, Sambrook J (2012) Molecular cloning: a laboratory manual, 4th edn. Cold Spring Harbor Laboratory Press, Cold Spring Harbor, NY

Chapter 6

Synthetic mRNA with Superior Properties that Mimics the Intracellular Fates of Natural Histone mRNA

Wei Su, Michael K. Slevin, William F. Marzluff, and Robert E. Rhoads

Abstract

Since DNA and histone levels must be closely balanced for cell survival, histone expressions are highly regulated. The regulation of replication-dependent histone expression is mainly achieved at the mRNA level, as the mRNAs are rapidly removed when DNA replication is inhibited during S-phase. Histone mRNA degradation initiates with addition of multiple uridines (oligouridylation) following the 3′ stem-loop (SL) catalyzed by terminal uridyltransferase (TUTase). Previous studies showed that histone mRNA degradation occurs through both 5′ → 3′ and 3′ → 5′ processes, but the relative contributions are difficult to dissect due to lack of established protocols. The translational efficiency and stability of synthetic mRNA in both cultured cells and whole animals can be improved by structural modifications at the both 5′ and 3′ termini. In this chapter, we present methods of utilizing modified cap dinucleotide analogs to block 5′ → 3′ degradation of a reporter mRNA containing canonical histone mRNA 3′ SL and monitoring how oligouridylation and 3′ → 5′ degradation occur. Protocols are presented for synthesis of reporter mRNA containing the histone 3′ SL and modified cap analogs, monitoring mRNA stability and unidirectional degradation either from 5′ or 3′ termini, and detection of oligo(U) tracts from degradation products by either traditional or deep sequencing.

Key words Cap analogs, Oligourdiylation, mRNA stability, Nucleoporation, Deep sequencing

1 Introduction

Histones are the most ancient proteins in eukaryotes. Although initially thought to be involved in chromosomal DNA packaging, histones have been shown to be extensively modified, such epigenetic modifications playing various roles in gene regulation [1, 2]. The amount of histones present in the cell is highly dependent on the cell cycle; excessive accumulation of histones is extremely toxic to the cell and can cause numerous abnormalities to disrupt cellular homeostasis [2]. Regulation of replicative histone synthesis is achieved primarily by changes in mRNA levels, which can increase 35-fold as cells enter S-phase. At the end of S-phase, these mRNAs are rapidly degraded. Thus, histone mRNAs are stabilized during DNA synthesis and destabilized when DNA synthesis ceases.

Robert E. Rhoads (ed.), *Synthetic mRNA: Production, Introduction Into Cells, and Physiological Consequences*,
Methods in Molecular Biology, vol. 1428, DOI 10.1007/978-1-4939-3625-0_6,

The mRNAs encoding replicative histones are unusual in that they are transcribed from genes that do not contain introns. Besides, they contain a stem-loop (SL) structure instead of a poly (A) tract at the 3′ end which is conserved across a wide range of species. The conserved 3′ structure is usually 25- to 26-nt long, consisting of 5 nt before the SL, 6 bp in the stem, a 4-nt loop, and 4–5 nt after the SL [2] (Fig. 1a). The SL is essential both for efficient translation and stability of histone mRNA [3]. Upon transcription, the SL is recognized and bound by a stem-loop binding protein (SLBP) which is also required for translation, stabilization, and degradation of histone mRNA [4–6]. It has been suggested that upon cessation of DNA synthesis, Upf1, a protein involved in both mRNA nuclear export and mRNA surveillance, is recruited to SLBP [7], which leads to the recruitment of a terminal uridyltransferase (TUTase) that catalyzes oligouridylation at the 3′-end of histone mRNA. The oligo(U) tract then serves as a binding site for the Lsm1–7 heptamer, which binds the machinery for decapping and bidirectional degradation of histone mRNA by exoribonucleases [6].

Several questions remain regarding this proposed model. First, oligouridylated fragments of histone mRNA are observed [6], but whether oligouridylation is a cause or effect remains unclear. Second, histone mRNAs can be degraded bidirectionally, but whether there is a sequential order for 5′ → 3′ and 3′ → 5′ degradation is not known. Third, it is not known whether decapping is a requisite step or a default pathway, even in the absence of 3′ → 5′ degradation. In the case of polyadenylated mRNAs, decapping does not normally occur until the poly(A) tract has been almost completely removed but can still occur independently, even if initial deadenylation is compromised. Fourth, the reciprocal relationship between mRNA translation and mRNA degradation is well established for polyadenylated mRNAs [8–10], but the relationship between the two is not known for histone mRNAs. In steady-state systems, any treatments of cells (knockdowns, inhibitors, etc.,) could potentially affect the rate of histone mRNA synthesis or degradation, complicating any interpretation of experimental results.

Previously, we have applied several technologies to study degradation of polyadenylated mRNA that potentially could provide insight into SL-containing mRNA as well. The first is the synthesis of mRNAs in vitro with different structural features and their introduction into mammalian cells by nucleoporation [11–13]. The mRNAs are incorporated into polysomes and efficiently translated within 15 min. Both translation and degradation are sensitive to the nature of the cap and the poly(A) tract. The second technology is the development of cleavage-resistant cap analogs [11, 13–16] (Fig. 1). These cap analogs have substitutions in the triphosphate chain intended to prevent hydrolysis by decapping enzymes. It was found that blocking decapping promoted both stabilization and efficient translation of polyadenylated mRNAs in HeLa cells (Fig. 1b–d). Such modified cap analogs can potentially facilitate dissection of the various steps in the degradation of an mRNA containing the histone SL. Specifically, by blocking 5′ → 3′ degradation using the cleavage-resistant cap, one can observe the events of 3′ → 5′ degradation independent of decapping. Ligation-coupled RT-PCR (LC-RT-PCR) allowed us to trap down oligouridylated histone mRNA degradation fragments in real time (Fig. 2), and RNA sequencing gel, was used to analyze these degradation products (Fig. 3). By applying these two technologies to a reporter mRNA containing the histone SL, we gained new insights into (1) the significance of oligouridylation in SL-containing mRNA turnover, (2) the relationship between decapping and oligouridylation, (3) which TUTases are involved in the degradative pathway, (4) the relationship between translation and degradation of SL-containing mRNA, and (5) the relative contribution of the 5′ → 3′ and 3′ → 5′ pathways. Finally, we developed a high-throughput sequencing (HTS) strategy specifically targeting the 3′ terminus of histone mRNAs that allowed us to trap oligouridylated histone. We found that initial oligouridylation occurs while the histone mRNA in on polyribosomes, and degradation initially proceeds 3′–5′ without decapping, when the mRNA is associated with ribosomes [17] (Fig. 4).

In this chapter, we provide protocols for each stage of the work described above: (1) in vitro synthesis of firefly luciferase mRNA containing the histone SL (Luc-SL) or mutated TL (Luc-TL) in the presence of cleavable ARCA (Anti-Reverse Cap Analog) or uncleavable BTH (Borano Two-Headed) cap; (2) introduction of Luc-SL or Luc-TL mRNAs with modified cap analogs into cultured cells; (3) measurement of translational efficiency and stability of reporter mRNAs; (4) LC-RT-PCR and sequencing gel electrophoresis; and (5) deep sequencing histone mRNA degradation intermediates and detection of 3′ nontemplated additions.

2 Materials

2.1 Cap Analogs

Two types of cap analogs are used in this chapter, the cleavable ARCA cap and uncleavable BTH cap (*see* below):

1. $m_2^{7,2'\text{-}O}$GpppG (cleavable ARCA cap). The synthesis of this ARCA has been described [18], but its translational properties are indistinguishable from an earlier version of ARCA $m_2^{7,3'\text{-}O}$GpppG [18], which is commercially available.
2. m^7Gpp$_{BH3}$pm^7G (uncleavable BTH cap). This analog does not have modifications on either the 2′ or 3′ positions of m^7Guo

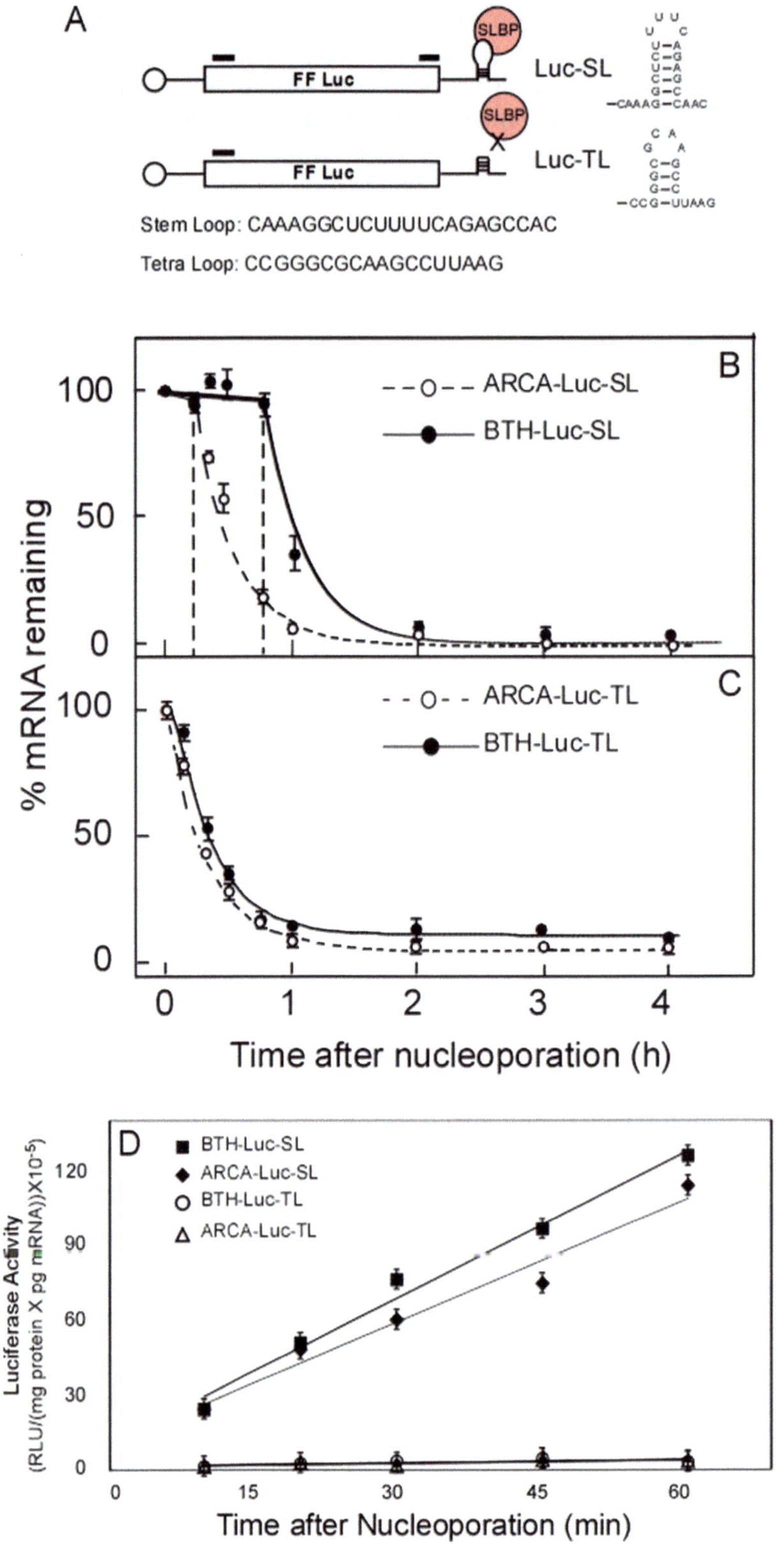

Fig. 1 Stabilization of Luc-SL mRNA is dependent on efficient translation. Luciferase mRNAs containing a wild-type histone mRNA 3′-stem-loop (SL) or a mutated tetra-loop (TL) were synthesized from linearized

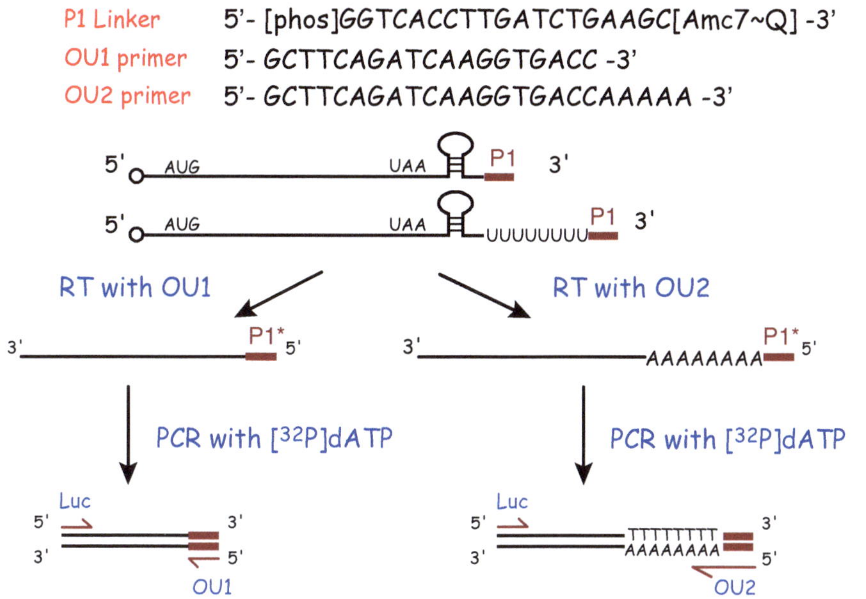

Fig. 2 Illustration of LC-RT-PCR technique used in this study. Briefly, purified total RNAs from cells lysed at indicated times were ligated to a modified oligonucleotide linker P1 using T4 RNA ligase. Two specific primers (OU1 and OU2) were used in separate reverse transcription (RT) reactions: OU1 reaction used a primer that is complimentary to the P1 linker. OU2 reaction used a primer that contains five A residues located 3′ to the sequence complimentary to P1. After RT, two separate PCR reactions were performed using an upstream forward primer near the 3′-end of luciferase coding region and either OU1 or OU2 as the downstream primer. ^{32}P-radio-labeled ATP was included in two PCR reactions, and the amplified PCR products were examined on 5 % denaturing polyacrylamide sequencing gels. (These data were originally published in Su *et al.*, 2013 19:1–16, *RNA*.)

Fig. 1 (continued) pT7-Luc-SL or pT7-Luc-TL in the presence of the indicated cap analogs and delivered into HeLa cells by nucleoporation as described in Subheadings 2 and 3. (**a**) Luc mRNAs with a wild-type SL or mutated TL. Bars indicate nt 90–279 and nt 1468–1616 that are amplified during qRT-PCR with 5′- and 3′-terminal primers, respectively. (**b**) Decay pattern of Luc-SL capped with either ARCA or BTH cap analogs. (**c**) Decay pattern of Luc-TL capped with either ARCA or BTH cap analogs. Cells were lysed at the indicated times and luciferase mRNA was measured by qRT-PCR. Data are plotted as a percentage of luciferase mRNA present immediately after nucleoporation. In (**b**), there is a lag before the initiation of rapid decay. The data for the postlag period were fit to single-exponential function and the $t_{1/2}$ calculated as described in Subheadings 2 and 3. *Vertical dashed lines mark* the boundary between lag phase and first-order decay phase. The data for each transcript represent average from at least two experiments. The error bars represent duplicate luciferase mRNA determinations. (**d**) Luciferase activity for Luc-SL and Luc-TL mRNAs synthesized in the presence of the indicated cap analogs was measured at the indicated times after nucleoporation. Data were normalized for the amount of luciferase mRNA present in the cells. The data were averaged from at least three individual experiments for each transcript. (These data were originally published in Su *et al.*, 2013 19:1–16, *RNA*.)

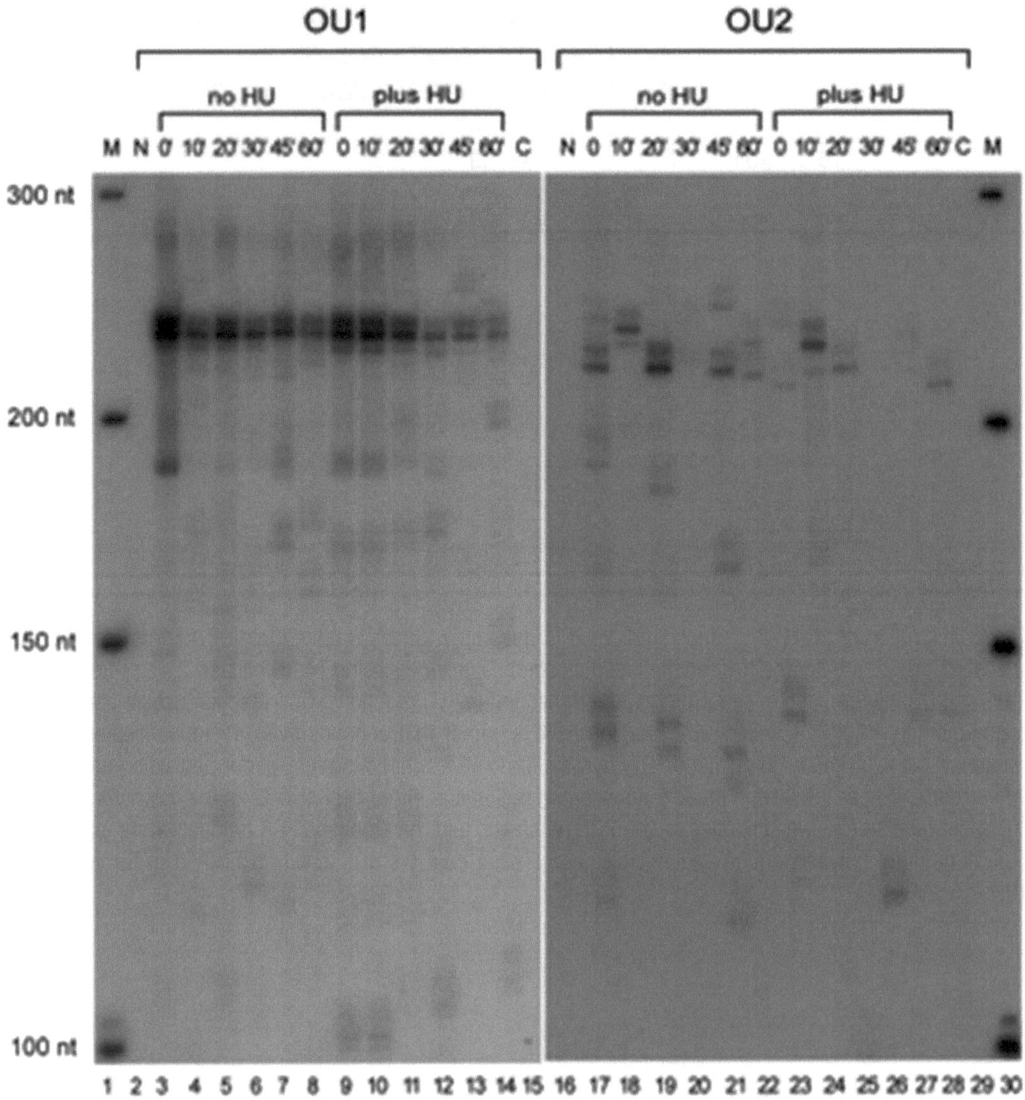

Fig. 3 Analysis of oligouridylated RNAs. BTH-Luc-SL was introduced into synchronous S-phase HeLa cells by nucleoporation and amplified by LC-RT-PCR with either the OU1 (*lanes 3–14*) or OU2 (*lanes 17–28*) primer except that [α-^{32}P]dATP was included during the PCR reaction. Cells were treated without (*lanes 3–8, 17–22*) or with HU (*lanes 9–14, 23–28*). DNA markers are indicated as *M* (*lanes 1, 30*). LC-RT-PCR products from RNA isolated from HeLa cells into which no Luc-SL mRNAs had been introduced are indicated as *N* (*lanes 2, 16*). PCR reactions that did not receive any cDNA (nontemplate control) are indicated as *C* (*lanes 15, 29*). The image represents a single gel, but the *left half* was exposed to film for 16 h, whereas the *right half* was exposed for 32 h. (These data were originally published in Su *et al.*, 2013 19:1–16, *RNA*.)

but nonetheless serves as an ARCA because both nucleoside residues contain m^{7}G. When incorporated into mRNA, this analog and $m_2^{7,2'\text{-}O}Gpp_SpG$, D2, confers the highest translational efficiency of all the cap analogs tested [19].

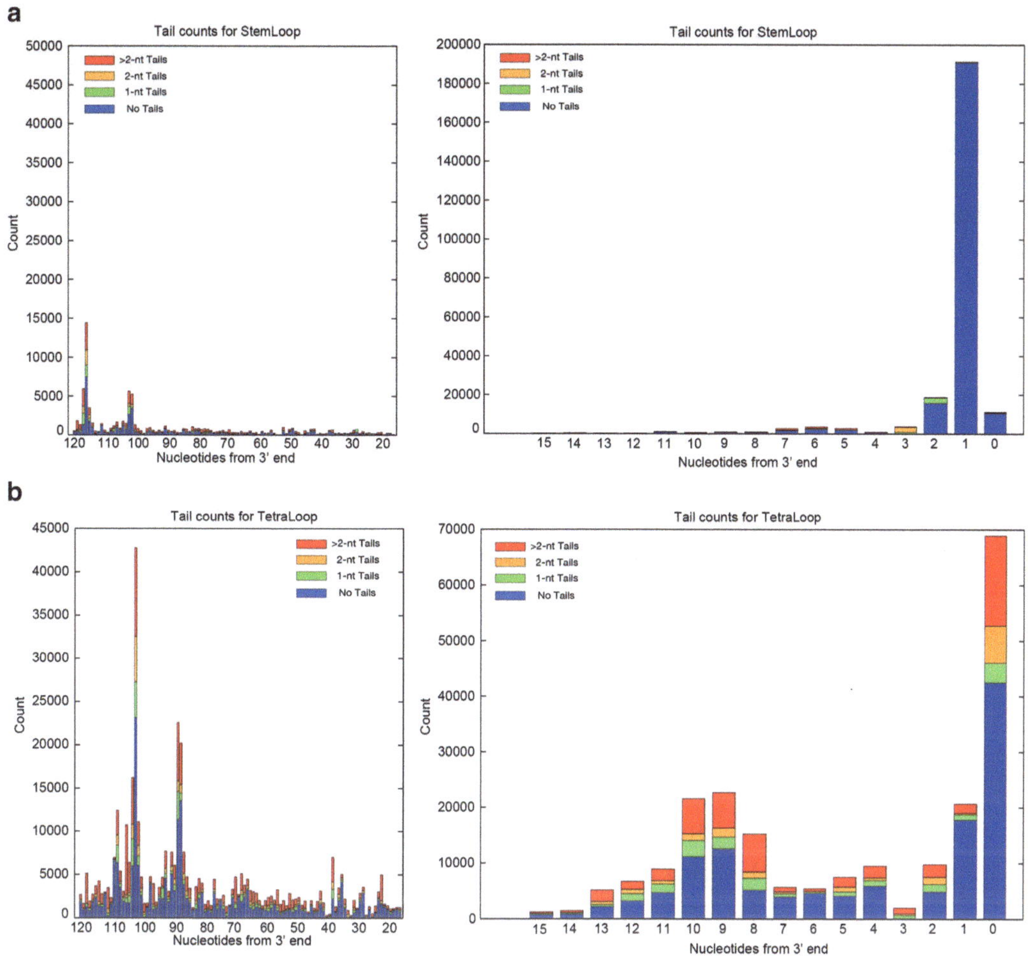

Fig. 4 Sample HTS data on injected total mRNAs. BTH-Luc-SL (**a**) or BTH-Luc-TL RNA (**b**) was introduced into synchronous S-phase HeLa cells by nucleoporation. RNA was harvested and libraries prepared and analyzed on an Illumina MiSeq. The data were processed using AppEnD [25] and plotted with the number of reads that started at the indicated position in the input RNA, where 0 is the 3′ end of the RNA. On the *right* are nts 1–15 which include most of the SL or TL, and on the *left* is nts 15–120, which are molecules which have been partially degraded from the 3′ end. The *blue* bars represent molecules which do not have nontemplated tails added and the *green, orange*, and *red* bars indicate molecules that have 1, 2, or >2 uridines added to the 3′ end. Note that the Luc-SL-RNA shows tailed degradation intermediates at nts 5–7 into the SL, the same intermediates observed in endogenous histone mRNAs [17], while the Luc-TL RNA shows a large number of RNAs nibbled at the 3′ end by an exonuclease, many of which contain nontemplated U-tails. The sequence of part of the SL (**a**) and TL (**b**) is shown. Subsequent degradation into the ORF of both mRNAs produces a large range of intermediates which presumably result from stalling of the exosome, and some of these intermediates have nontemplated U-tails, consistent with reinitiation of degradation requiring uridylation of the RNA

The concentrations of cap analog solutions are determined by UV absorbance at pH 7.0 using the extinction coefficient $\varepsilon_{255nm} = 22.6 \times 10^3/M/cm$.

2.2 In Vitro Synthesis of Firefly Luciferase mRNA with Various 5′- and 3′-Ends

1. DEPC-treated water.
2. 5× Transcription buffer: 200 mM Tris–HCl, pH 7.9, 30 mM $MgCl_2$, 10 mM spermidine (*see* **Note 1**).
3. 100 mM dithiotheritol (DTT).
4. 10 mg/mL bovine serum albumin (BSA).
5. RNase Inhibitor, 20 units/μL.
6. ATP, 10 mM.
7. UTP, 10 mM.
8. CTP, 10 mM.
9. GTP, 10 mM.
10. Cap dinucleotide, 10 mM (*see* Subheading 2.1).
11. 1 μg/μL DNA template (pT7-Luc-SL or pT7-Luc-TL), linearized (*see* **Note 2**).
12. T7 RNA polymerase, 10 units/μL.
13. RNase-free DNase RQ1, 1 unit/μL.
14. E.Z.N.A. Total RNA Kit I, consisting of HiBind RNA spin columns, collection tubes, RNA Wash Buffer I, and RNA Wash Buffer II (Omega Bio-Tek, Cat. No. R6834-01).
15. 100 % ethanol.
16. 70 % ethanol.
17. Horizontal electrophoresis system.
18. Agarose.
19. Formaldehyde.

2.3 Cell Culture, Synchronization, and Cell Cycle Analysis

1. HeLa cell line.
2. Cell culture medium: DMEM high glucose containing 10 % fetal bovine serum and 1× penicillin/streptomycin antibiotics.
3. Thymidine, 2 mM.
4. 0.05 % trypsin with 2 mM EDTA.
5. 70 % cold ethanol (*see* **Note 3**).
6. PBS.
7. Propidium iodide, stock solution 50 μg/mL.
8. Triton X-100.
9. BD LSRII Flow Cytometry Analyzer (BD Biosciences).

2.4 Introduction of Synthetic mRNAs into Cultured Mammalian Cells by Nucleoporation

1. Cell culture medium: DMEM high glucose containing 10 % fetal bovine serum and 1× penicillin/streptomycin antibiotics.
2. Nucleofector™ II nucleoporator (Lonza, Cat. No. AAD-1001).
3. Nucleofector™ Solution R (Lonza, Cat. No. AAF-1001B).
4. Phosphate-buffered saline (PBS): Dissolve 8 g of NaCl, 0.2 g of KCl, 1.44 g of Na_2HPO_4, and 0.24 g of KH_2PO_4 in distilled water. Adjust pH to 7.4 with HCl. Make up to 1000 mL. Sterilize by autoclaving.
5. 1.5-mL Eppendorf tubes and cell culture dishes.

2.5 Analysis of Translational Efficiency in Cultured Cells

1. Cell Culture Lysis Reagent, 5× (Promega, Cat. No. E153A). Dilute to 1× before use.
2. Luciferase Assay System.

2.6 Analysis of mRNA Stability in Cultured Cells

1. E.Z.N.A. Total RNA Kit I.
2. PBS (*see* Subheading 2.4).
3. 70 % ethanol.
4. 10× RQ1 DNase buffer.
5. RNase-free DNase RQ1, 1 unit/μL.
6. 10× Reverse transcription buffer: 100 mM Tris–HCl, pH 8.3, 500 mM KCl.
7. 25 mM $MgCl_2$.
8. 10 mM of dNTPs (2.5 mM for each of the four dNTPs).
9. 50 μM Random Hexamers.
10. RNase Inhibitor, 20 units/μL.
11. MultiScribe™ Reverse Transcriptase, 50 units/μL.
12. RNase-free water (*see* **Note 4**).
13. SYBR Green PCR SuperMix.
14. Mixture of primers to amplify 5′ end of Firefly Luciferase cDNA: forward primer, 5′-GGATGGAACCGCTGGAG AG-3′, reverse primer, 5′-GCATACGACGATTCTGTGATT TG-3′, 10 μM each.
15. Mixture forward primers to amplify 3′ end of Firefly Luciferase cDNA, 5′-ATCGTGGATTACGTCGCCAGTCAA-3′ and 5′-TTTCCGCCCTTCTTGGCCTTTATG-3′, 10 μM each.
16. 10 μM mix of primers to amplify human 18S rRNA cDNA: forward primer, 5′-CGAGCCGCCTGGATACC-3′; reverse primer, 5′-CAGTTCCGAAAACCAACAAAATAGA-3′.
17. RNA samples isolated from HeLa cells at different time points.

2.7 Ligation-Coupled RT-PCR (LC-RT-PCR)

1. DEPC-treated water.
2. Commercially designed modified P1 linker [5′-(Phos) GGTCACCTTGATCTGAAGC(AmC7-Q)-3′].
3. Specific primers: OU1 contains the antisense sequence of P1 oligos (5′-GCTTCAGATCAAGGTGACC-3′); OU2 contains additional 5 As at the 5′ terminal prior to the antisense P1 sequence (5′-GCTTCAGATCAAGGTGACCAAAAA-3′).
4. T4 RNA ligase and 1× ligation buffer.
5. Heat block.
6. 0.1 M Dithiotheitol (DTT).
7. Superscript III reverse transcriptase and 5× first strand buffer.
8. [α-^{32}P]dATP.
9. 50 mM $MgCl_2$.
10. 10 mM dNTPs.
11. Platinum® Taq Polymerase and Platinum® PCR reaction buffer.

2.8 Sequencing Gel Electrophoresis

1. Base Runner Nucleic Acid Sequencer Apparatus (International Biotechnologies).
2. Polyacrylamide.
3. Fixing solution containing 5 % acetic acid and 5 % methanol.
4. Whatman 3 MM filter paper.
5. Gel dryer.
6. Blue X-ray film.
7. Photo Scanner.
8. ImageQuant TL software.

2.9 Deep Sequencing Analysis

1. Preadenylated Linkers
   NEB-Extended 5′-CTGTAGGCACCATCAATCTCACTCCG-NH2-3′ (500 ng/μL).
2. Reverse transcription primers
   NEB-Extended RT 5′-CGGAGTGAGATTGATGGTGCCTA CAG-3′ (25 μM)
   NEB-Extended A(3) RT 5′-CGGAGTGAGATTGATGGTGCC TACAGAAA-3′ (25 μM).
3. First round PCR primers
   Primer 1 5′-GGTTCAGAGTTCTACAGTCCGACGATC-NNNN-CGGAGTGAGATTGATGGTGCCTACAG-3′
   Primer H2a 5′ GTGACTGGAGTTCAGACGTGTGCTC TTCCGATCT-NNNNN-CTGGCGGGCAACGCGGC-3′
   Primer H2b 5′ GTGACTGGAGTTCAGACGTGTGCTCTT CCGATCT-GGTCCACCCCGACACCGGCATCT-3′
   Primer Luc 5′ GTGACTGGAGTTCAGACGTGTGCTC TTCCGATCT NNNNN AACACCCCAACATCTTCGAC.

4. Second round PCR primers
   Primer 3 5′ AATGATACGGCG ACCACCGAGATCTACAC-CGACAGGTTCAGAGTTCTAC AGT
   Primer 4 5′ CAAGCAGAAGACGGCATACGAGAT-CGT GAT-GTGACTGGAGTTCAGACGTGTGCTCTTCCGATCT.
5. Sequencing primer
   Primer S 5′ CGACAGGTTCAGAGTTCTACAGTCCGA CGATC.
6. RiboLock RNAse Inhibitor (Life Technologies, Cat. No. EO0381).
7. T4 RNA Ligase 2, Truncated K227Q (NEB, Cat. No. 0381S).
8. 100 and 75 % ethanol.
9. dH_2O.
10. GlycoBlue™ coprecipitant (15 mg/mL).
11. 3 M Sodium Acetate, pH 5.2.
12. UltraPure™ Phenol:Chloroform:Isoamyl Alcohol (25:24:1, v/v), pH 8.05–8.35.
13. SuperScript® III Reverse Transcriptase.
14. 10 mM dNTPs.
15. Phusion® High-Fidelity DNA Polymerase.
16. Thermocycler.
17. Agencourt AMPure XP PCR Purification (Beckman Coulter, Cat. No. A63880).
18. Magnetic particle concentrator.)
19. Agilent Bioanalyzer 2100 (Agilent Technologies, Cat. No. G2939AA).
20. Agilent DNA 1000 Kit (Agilent Technologies, Cat. No. 5067-1504).
21. Qubit® 3.0 Fluorometer (Life Technologies, Cat. No. Q33216.
22. Qubit® dsDNA BR Assay Kit (Cat. No. Q32853).
23. MiSeq Reagent Kit v2 (300-cycle) (Illumina, Cat. No. MS-102-2002).
24. MiSeq Instrument.

3 Methods

3.1 In Vitro Synthesis of Firefly Luciferase mRNA with Various 5′- and 3′-Ends

1. Prepare a 200-μL transcription reaction mixture in a 1.5-mL centrifuge tube by adding the following reagents at room temperature in the order listed: 40 μL 5× transcription buffer, 20 μL DTT, 2 μL BSA, 5 μL RNase Inhibitor, 10 μL ATP, 10 μL UTP, 10 μL CTP, 2 μL GTP, 20 μL cap analog (ARCA or BTH), 3 μL DNA template, and 10 μL of RNA polymerase.

2. Mix the solution and incubate the reaction 1 h at 37 °C.
3. To increase the yield of RNA, add an additional 10 μL of RNA polymerase and incubate for another 1 h at 37 °C.
4. Treat the reaction mixture with 10 units of RNase-free DNase RQ1 for 30 min at 37 °C to remove the DNA template.
5. Purify RNA using an E.Z.N.A. Total RNA Miniprep Kit I as follows (*see* **Note 5**).
6. Add 500 μL of TRK Lysis Buffer to the reaction and mix thoroughly by vortexing or pipetting.
7. Add an equal volume (700 μL) of 70 % ethanol and mix thoroughly by pipetting up and down a few times.
8. Apply the sample to a RNA column inserted into a 2-mL collection tube.
9. Centrifuge at 10,000 × *g* for 60 s at room temperature and discard the flow-through.
10. Add 500 μL of RNA Wash Buffer I and repeat centrifugation.
11. Add 500 μL of RNA Wash Buffer II and repeat centrifugation (*see* **Note 6**).
12. Repeat this step one more time.
13. Remove the flow-through and centrifuge the RNA column for 2 min at maximum speed to completely dry the matrix.
14. Transfer the column into a new 1.5-mL centrifuge tube and elute the RNA by adding 50 μL of DEPC-treated water (*see* **Note 7**).
15. Centrifuge for 1 min at 10,000 × g.
16. Check RNA concentration by UV absorbance at 260 nm (*see* **Note 8**) and verify RNA integrity by electrophoresis on a 1.2 % agarose gel containing formaldehyde.

3.2 Cell Culture, Synchronization, and Cell Cycle Analysis

1. Maintain HeLa cells in complete DMEM medium.
2. Seed cells in multiples culture plates at a density of 1 × 10^6.
3. Synchronize cells by double-thymidine block described as follows [20].
4. Add thymidine into cell medium to a final concentration of 2 mM.
5. Incubate cells for 18 h in tissue culture incubator.
6. After 18 h, remove thymidine by washing the plates with 1 × PBS and adding fresh medium.
7. Incubate cells for 9 h in the tissue culture incubator.
8. After 9 h, add thymidine to final concentration of 2 mM.
9. Incubate cells for 15 h in tissue culture incubator.

10. Release cells by washing the plates with 1 × PBS and adding fresh medium.
11. Collect one plate of 1×10^6 cells at time 0, 3, 6, and 9 h after releasing cells from DTB, aspirate the medium, wash with PBS 3 times, and fix cells in 70 % cold ethanol for at least 2 h.
12. Spin down cells, wash with PBS, and resuspend them in propidium iodide/Triton X-100 staining solution with RNase A until analysis.
13. If analyzing cell cycle progression after nucleoporation, follow the nuceloporation method in Subheading 3.3, replate cells back to the cell culture dish, and collect cells at time 0, 3, 6, and 9 h after nucleoporation. Follow **steps 10–11** to fix the cells.
14. Analyze DNA content by FACS instrument (BD LSRII Special Order System, BD Biosciences) at core facility of LSUHSC-S. Populations of G1 (2N DNA content), S-phase (2N < S < 4 N DNA content), and G2/M (4N DNA content) can be quantified using FACS DIVA software (version 6.1.3) and are represented as a percentage of the total population.

3.3 Introduction of Synthetic mRNAs into Cultured Mammalian Cells

1. Release cells from double-thymidine block on the day of nucleoporation, and introduce mRNAs into cells 3 h after release (middle of S-phase).
2. Prewarm the supplemented Nucleofector™ Solution R (Lonza) to room temperature.
3. Prewarm an aliquot of culture medium at 37 °C in a 50-mL tube.
4. Harvest the cells when they have reached ~70 % confluency as follows. Remove the medium from the cell culture dish. Wash cells once with PBS. Add 1.5 mL of 0.05 % trypsin and 2 mM EDTA in PBS and incubate cells 5 min at 37 °C in an incubator with 5 % CO_2. Add 5 mL of fresh medium and collect cells by centrifugation 800 × *g* for 5 min. Count the cells, and incubate them in fresh media at 37 °C for 30 min.
5. Pellet cells after incubation by centrifugation and remove the medium. Wash with PBS once and pellet by centrifugation again.
6. Resuspend cells in 100 μL of Nucleofector™ solution R. Use 5×10^5–5×10^6 cells per nucleoporation. Using fewer cells leads to a major increase in cell mortality; using more cells causes inefficient delivery of RNA.
7. Add 1 μg mRNA to the cells in nucleoporation solution and mix well.

8. Transfer the total mix containing cells and RNA into the nucleoporator cuvette and carry out nucleoporation with a Nucleofector™ II, using the program I-13 for HeLa cells (*see* **Note 9**).
9. Add 500 μL prewarmed medium to the cuvette and transfer contents to a 15-mL tube containing 5 mL prewarmed medium. Spin down the cells and remove the medium.
10. Resuspend cells in fresh warm medium and withdraw 0.5×10^6 cells for each time point, e.g., every 15 min. For time points up to 1 h, keep cells in 1.5-mL Eppendorf tubes shaking at 37 °C water bath. For time points greater than 1 h, seed each aliquot of 0.5×10^6 onto a 35-mm dishes and place in a 37 °C incubator with 5 % CO_2. At the designated time points, wash cells with 1× PBS and add lysis buffer (*see* **Note 10**).

3.4 Measurement of Translational Efficiency of Luc-SL and Luc-TL in HeLa Cells

1. Shake aliquots of 0.5×10^6 cells in 1.5-mL Eppendorf tubes at 37 °C for various times after nucleoporation up to 60 min, e.g., 12, 24, 36, 48, and 60 min.
2. Spin down cells at 800 × *g* for 1 min, aspirate off the medium, and wash cells with PBS.
3. Spin down cells to remove PBS, resuspend the pellet in 200 μL 1× Cell Culture Lysis Reagent, vortex for 10 s, and place on ice for 10 min.
4. Spin down the cell debris at 4 °C at 9000 × *g* for 2–3 min.
5. Transfer the supernatant to a new Eppendorf tube and freeze at −80 °C until use.
6. Determine the protein concentration in the lysate using the Bradford Assay.
7. Add 5–10 μL of protein extract into the luciferase assay reagent [13].
8. Transfer the sample to a luminometer and determine luciferase activity (*see* **Note 11**).
9. Normalize luciferase activity by protein concentration of each sample.
10. To compare translational efficiencies between different nucleoporated groups, normalize luciferase activity to the amount of luciferase mRNA present in the cells at 0 min (*see* Subheading 3.5).

3.5 Measurement of Luc-SL and Luc-TL mRNA Stability in Cells

1. Plate cells onto 35-mm dishes after nucleoporation and incubate at 37 °C and 5 % CO_2 until harvest.
2. For hydroxyurea (HU)-treated cells, 10 mM freshly made HU were added to cell resuspended in culture media after

nucleoporation and immediately separated onto dishes, except that the aliquot for time zero did not receive any HU.

3. Extraction of total RNA following the manufacturer's instructions and determine OD under a Nano drop spectrophotometer (Thermo Scientific).
4. Measure luciferase mRNA levels by quantitative RT-PCR (qRT-PCR) as described previously [21].
5. The amount of luciferase mRNA at different time points can be expressed as a percentage of the mRNA at time zero.

3.6 Ligation-Coupled RT-PCR (LC-RT-PCR) and Sequencing Gel Electrophoresis

1. A 10-μL LC-RT-PCR reaction contains 1 μg of purified total mRNAs, 50 μM modified P1 linker [5′-(Phos)GGTCACCTT-GATCTGAAGC(AmC7-Q)-3′, synthesized from MWG-Operon], 1 μL of T4 RNA ligase (NEB), and 1× ligation buffer.
2. Mix mRNAs and P1 linker, heat to 65 °C for 10 min, and cool on ice for 2 min prior to addition of T4 RNA ligase and buffer.
3. Perform the reaction at 37 °C for 30 min and inactivate at 65 °C for 15 min.
4. For reverse transcription, mix a reaction of 2 μL of ligated RNA, 1.8 μL diethylpyrocarbonate (DEPC)-treated H_2O, 1 μL of specific OU1 or OU2 primer (10 mM, *see* Subheading 2.7) and 0.2 μL of dNTPs (10 mM) and heat at 65 °C for 5 min, and then cool on ice for 2 min.
5. Then add 2.3 μL DEPC-treated H_2O, 0.5 μL of DTT (0.1 M), 2 μL 5× first strand buffer (Invitrogen), and Superscript III reverse transcriptase (Invitrogen) to a final volume of 10 μL.
6. Perform reverse transcription reaction according to manufacturer's instructions.
7. PCR products can be labeled with ^{32}P in a 25-μL reaction containing 1 μL of cDNA, 2.5 μL of Platinum® PCR reaction buffer, 0.75 μL of 50 mM $MgCl_2$, 0.5 μL of 10 mM dNTPs, 0.5 μL of 10 mM upstream primer, 0.5 μL of 10 mM OU1 or OU2 primer, 0.5 μL of [α-^{32}P]dATP, and 0.1 μL Platinum® Taq Polymerase (Invitrogen). The PCR reaction is perform by using 35 cycles at 94 °C for 30 s, 55.1 °C for 1 min, and 72 °C for 1 min. A forward upstream luciferase primer, Luc (5′-ATCGTGGATTACGTCGCCAGTCAA-3′) is included in OU1 or OU2 PCR reactions.
8. ^{32}P-labeled PCR products can be analyzed on a sequencing gel as described previously [13] except that the gel contains 5 % polyacrylamide and 7 M urea.
9. Stop the reactions by adding two volumes of Precipitation/ Inactivation buffer (Ambion), and precipitate amplified products at −20 °C for 20 min and collect by centrifugation at 13,000 × *g* at 4 °C for 20 min.

10. Resuspend the precipitated in 5 μL of Sequencing Gel Loading Buffer (Ambion), denature at 95 °C for 5 min and load onto the gel.
11. Run RNA sequencing gels (10 % polyacrylamide) at 45–70 W for ~3 h on a Base Runner Nucleic Acid Sequencer apparatus (International Biotechnologies) [22].
12. Fix the gel in 5 % acetic acid, 5 % methanol for 10–15 min, dry onto Whatman 3MM filter paper (Fisher Scientific), and expose to Blue X-ray film (Kodak).
13. Quantify the radioactivity in individual bands by either of two methods: analyzing scanned film using ImageQuant TL program (GE Health Care, version 7.0), cutting them out and determining Cherenkov radiation.

3.7 Cloning and Sequencing of LC-RT-PCR Products

1. Insert 1 μL of LC-RT-PCR product into the linearized TOPO® vector and ligate according to manufacturer's protocol (Invitrogen).
2. Screen colonies for correct inserts by colony hybridization using radio-labeled probe according to molecular cloning protocol [23].
3. Send plasmids with possibly correct inserts for sequencing (Iowa State DNA facility or Arizona State DNA facility) and analyze the results.

3.8 Application of Deep Sequencing Technique to Analyze Histone mRNA Degradation Products

3.8.1 Preadenylated Linker Ligation

1. In a volume of 8.5 μL mix 2 μg of RNA with 0.5 μg of preadenylated linker (*see* **Note 12**) and 1 μL of 10 × T4 RNA Ligase 2 reaction buffer (NEB).
2. Denature the RNA by incubating at 65 °C for 15 min. Following denaturation immediately place the reaction on ice for 3 min.
3. Add 1 μL of RiboLock and 0.5 μL of T4 RNA Ligase 2 (NEB). Mix contents by pipetting.
4. Ligate overnight at 16 °C (*see* **Note 13**).
5. Bring the volume up to 100 μL with DEPC-treated dH_2O or equivalent.
6. Add an equal volume of phenol:chloroform:isoamyl alcohol and 1/10th volume of 3 M sodium acetate. Vortex and then centrifuge at 12,000 RCF for 10 min at 4 °C.
7. Extract the aqueous phase to a new centrifuge tube and add an equal volume of chloroform. Vortex and centrifuge at 12,000 RCF for 10 min at 4 °C.
8. Extract the aqueous phase to a new centrifuge tube and add 1 μL of glycoblue and 2.5 volumes of 100 % ethanol. Place at −80 °C for 10 min.

9. Extract the aqueous phase to a new centrifuge tube and add 750 μL of 75 % ethanol. Vortex and centrifuge at 12,000 RCF for 15 min at 4 °C.
10. Remove the supernatant taking care not to disturb the pellet (if visible). Add 1 mL of 75 % ethanol, vortex, and centrifuge at 12,000 RCF for 15 min at 4 °C.
11. Remove the supernatant and air dry the pellet (5–10 min). Resuspend in 10 μL of DEPC-treated dH_2O or equivalent. Store at −20 °C.

3.8.2 Linker Mediated Reverse Transcription

1. Mix 1 μL of 10 dNTPs and 2 μL of 25 μM RT primer (*see* **Note 14**) with the 10 μL ligation resuspension. Mix by pipetting.
2. Denature at 65 °C for 15 min.
3. Immediately place reaction on ice for 3 min.
4. Add the following reagents in the order listed: 4 μL 5× reaction buffer, 1 μL 0.1 M DTT, 0.5 μL Ribolock, 0.5 μL DEPC-treated dH_2O or equivalent, and 1 μL of SuperScript III reverse transcriptase. Mix by pipetting.
5. Incubate reaction at 50 °C for 1 h.
6. Heat inactivate reverse transcriptase for 20 min at 70 °C.

3.8.3 PCR Round 1: Molecular Barcoding

1. Prepare a 50 μL PCR reaction for the first round PCR by adding the following reagents in the order listed: 31.5 μL dH_2O, 5 μL 5× Phusion polymerase buffer, 5 μL of cDNA, 1 μL dnTPs, 1 μL primer 1, 1 μL primer 2 (*see* **Note 15**), and 0.5 μL Phusion polymerase (*see* **Notes 16** and **17**).
2. For the first round of PCR use the following cycling parameters (*see* **Notes 18**): 1 cycle at 95 °C/3 min, 10 cycles at 95 °C/15 s–56 °C/15 s–72 °C/15 s, and 1 cycle at 72 °C/2 min. Hold at 16 °C.

3.8.4 AMPure XP PCR Purification of First Round PCR

1. Add 40 μL of AMPure XP beads to 50 μL of PCR product. Mix by pipetting ten times.
2. Bind for 10 min at room temperature then move the tube to a magnetic stand and concentrate the beads for 5 min.
3. Remove the supernatant and wash with 200 μL of 100 % ethanol. Remove the wash and repeat.
4. Air dry the pellet (~5–10 min) until the pellet begins to dry (no longer appears “wet” or “glassy”).
5. Resuspend in 20 μL of TE.
6. Quantitate by taking a spectrophotometer reading at 260 nm.

3.8.5 PCR Round 2: Addition of Illumina Flow Cell Adapters

1. Prepare a 50-μL PCR reaction for the first round PCR by adding the following reagents in the order listed: 75 ng of the first round PCR product, 5 μL 5× Phusion polymerase buffer, 1 μL dNTPs, 1 μL primer 3, and 1 μL primer 4 (*see* **Note 19**). Bring the volume up to 49.5 μL with dH_2O and add 0.5 μL of Phusion polymerase.
   (a) For the first round of PCR use the following cycling parameters: one cycle at 95 °C/3 min, 5 cycles at 95 °C/15 s–56 °C/15 s–72 °C/15 s, and one cycle at 72 °C/2 min. Hold at 16 °C.
2. Repeat steps listed in Subheading 3.8.4

3.8.6 Library Analysis and Quantitation

1. Load 1 μL of the purified library on a DNA 1000 K chip and electrophorese on an Agilent 2100 bioanalyzer.
2. Check for the absence of primer dimers and quantitate the library by selecting the peak corresponding to the size of the expected library (~450 nts) by using the manual integration option.
3. Quantitate the library using a Qubit flourometer and broad range dsDNA kit.
4. Pool libraries that are free of excessive primer dimers and match the expected size distribution (*see* **Note 20**). Into a single tube combine 5 μL of each library and mix by pipetting.
5. Load three lanes of a DNA 1000 K chip with the pooled libraries and electrophorese on an Agilent 2100 bioanalyzer. Quantitate the library by selecting the detected peak using the manual integration option. Take the average of the three lanes for the nM concentration.

3.8.7 Illumina Sequencing: 300 Cycle MiSeq Paired End Sequencing Run

1. Dilute an aliquot of the pooled library to 2 nM with TE.
2. Combine 7.5 μL of the pooled library with 2.5 μL of 2 nM Illumina PhiX (this is to increase the complexity of the library and facilitate correct cluster generation).
3. To the library/PhiX mix add 10 μL of freshly prepared 0.1 N NaOH. Incubate 5 min at room temperature then add 980 μL of Illumina HT1 buffer. The concentration is now 20 pM.
4. Dilute an aliquot to 8.5 pM by mixing 425 μL of the denatured library/PhiX with 575 μL of Illumina HT1 buffer. Place on ice until sequenced.
5. Place a MiSeq cartridge in a container and fill with dH_2O up to the line indicated on the cartridge. Thaw for ~1 h. Mix the contents by inverting several times and check there are no air bubbles at the bottom of the reagent reservoirs.

6. Remove a flow cell and rinse with dH_2O. Remove excess dH_2O then wipe down the flow cell with 70 % ethanol until the glass is clean and streak free. Load the flow cell in the MiSeq.
7. Pipet 600 μL of the sequencing library into the indicated well position.
8. Open the sequencing well and inject the sequencing primer to a final concentration of 0.5 μM. If the volume is 600 μL add 3 μL of a sequencing primer whose concentration is 100 μM.

4 Notes

1. Store this and all buffers at −20 °C.
2. pT7-Luc-SL and pT7-Luc-TL plasmids were previously constructed [3, 24] and linearized by restriction enzyme AflII at the site immediately downstream of the SL or TL.
3. Store 70 % ethanol in −20 °C before use.
4. Use DEPC-treated water to prepare solutions of PBS, 10× reverse transcription buffer, KCl, $MgCl_2$, and dNTPs.
5. The procedure described is given in more detail in the manufacturer's instructions. Alternatively, RNA may be purified with one of the following methods: phenol:chloroform extraction, followed by LiCl precipitation; Trizol Reagent (Invitrogen); or columns for RNA purification from other companies such as Omega Bio-Tek.
6. RNA Wash Buffer II must be diluted with 100 % ethanol before use.
7. Add the water directly onto the center of the column matrix.
8. The concentration of the RNA should be ~500 ng/μL.
9. Electroporation can also be used to deliver synthetic mRNAs into cells, but it requires more reagents and cells since it causes more cell death.
10. Time points to follow RNA decay can be taken every 15 min until 1 h and then every 30 min or 1 h afterwards. For luciferase production, samples can be collected every 12 min up to 60 min.
11. Measure luciferase activity in duplicate or triplicate, include blanks in the same experiment (5 μL lysis buffer added into luciferase assay reagent).
12. The sequence for the preadenylated linker is derived from the New England Biolabs microRNA cloning linker. The 5′ end of the linker was extended (NEB-Extended) to increase the Tm associated with the first round PCR primer (primer 1). This adapter in combination with its upstream counterpart

(primer H2a and H2b) targets the histone mRNAs coding sequence. Preadenylated linkers were made in house according to the Bartel labs protocol.

13. Ligation of the preadenylated reaction can be done at 25 °C for 1–2 h.
14. Reverse transcription is done with the reverse complement of the ligated linker (NEB-Extended RT). Specific targeting of oligouridylated RNAs can be done by appending three adenines to the 3′ end of the reverse transcription primer (NEB-Extended RT A(3)).
15. Primer 1 contains part of Illumina's V1.5 Small RNA sequencing primer followed by a five random nucleotide sequence and the complement to the preadenylated linker sequence. The five random nucleotides increases library complexity for cluster formation. Primer 2 begins with the reverse complement for Illumina's Read 2 sequencing primer followed by a thymidine (Illumina's TruSeq adapters occur as annealed pairs and the P5 arm ends with an overhanging T which facilitates ligation to the A-tailed dsDNA products. Therefore, the thymidine is "added" during ligation and is required for the read 2 sequencing primer to function) and a histone specific consensus sequence (sense orientation).
16. Q5 polymerase (NEB) is equally effective for PCR steps.
17. The first round of PCR is done to attach the initial halves of the 5′ and 3′ adapters required for Illumina sequencing. Splitting library creation into two steps reduces the cost of the primers which in their full form are > 90 nts in length. However, when possible it is preferable to perform a single round of PCR.
18. Performing 10 cycles of PCR during the first round prevents over amplification; the number of cycles can be modified if required.
19. Primer 3 contains both Illumina's P5 flow cell sequence and V1.5 Small RNA sequencing primer. Primer 4 contains the reverse complements for Illumina's P7 flow cell sequence a 6 nt index from their LT index series and read 2 sequencing primer.
20. In the absence of nontemplated additions the targeted consensus sequences in histones H2A and H2B will give an insert of ~220 nts. Addition of Illumina adapter sequences will produce a library of ~400 nts though nontemplated additions and degradation intermediates will give a smear ranging from ~250–420 nts.

Acknowledgments

The authors were supported by NIH grants GM29832 to W.F.M. NIH grant HG06272 and NSF grant ABI/EF08 50237 to J.F.P. and NIH grant GM20818 to R.E.R. M.K.S. was supported by NIH grant F32GM87059. J.D.W. was supported by NSF Graduate Research Fellowship DGE-1144081. W.S. was supported by an Ike Muslow predoctoral fellowship from the LSUHSC-S Research Council.

References

1. Henikoff S, Ahmad K (2005) Assembly of variant histones into chromatin. Annu Rev Cell Dev Biol 21:133–153
2. Marzluff WF, Wagner EJ, Duronio RJ (2008) Metabolism and regulation of canonical histone mRNAs: life without a poly(A) tail. Nat Rev Genet 9(11):843–854
3. Gallie DR, Lewis NJ, Marzluff WF (1996) The histone 3′-terminal stem-loop is necessary for translation in Chinese hamster ovary cells. Nucleic Acids Res 24(10):1954–1962
4. Sanchez R, Marzluff WF (2002) The stem-loop binding protein Is required for efficient translation of histone mRNA in vivo and in vitro. Mol Cell Biol 22(20):7093–7104
5. Cakmakci NG et al (2008) SLIP1, a factor required for activation of histone mRNA translation by the stem-loop binding protein. Mol Cell Biol 28(3):1182–1194
6. Mullen TE, Marzluff WF (2008) Degradation of histone mRNA requires oligouridylation followed by decapping and simultaneous degradation of the mRNA both 5′ to 3′ and 3′ to 5′. Genes Dev 22(1):50–65
7. Kaygun H, Marzluff WF (2005) Regulated degradation of replication-dependent histone mRNAs requires both ATR and Upf1. Nat Struct Mol Biol 12(9):794–800
8. Schwartz DC, Parker R (2000) mRNA decapping in yeast requires dissociation of the cap binding protein, eukaryotic translation initiation factor 4E. Mol Cell Biol 20(21):7933–7942
9. Coller J, Parker R (2005) General translational repression by activators of mRNA decapping. Cell 122(6):875–886
10. Parker R, Sheth U (2007) P bodies and the control of mRNA translation and degradation. Mol Cell 25(5):635–646
11. Grudzien-Nogalska E et al (2007) Phosphorothioate cap analogs stabilize mRNA and increase translational efficiency in mammalian cells. RNA 13:1745–1755
12. Grudzien-Nogalska E et al (2007) Synthesis of anti-reverse cap analogs (ARCAs) and their applications in mRNA translation and stability. Methods Enzymol 431:203–227
13. Su W et al (2011) Translation, stability, and resistance to decapping of mRNAs containing caps substituted in the triphosphate chain with BH3, Se, and NH. RNA 17(5):978–988
14. Kowalska J et al (2005) Synthesis and properties of mRNA cap analogs containing phosphorothioate moiety in 5′,5′-triphosphate chain. Nucleosides Nucleotides Nucleic Acids 24:595–600
15. Kowalska J et al (2009) Phosphorothioate analogs of m^7GTP are enzymatically stable inhibitors of cap-dependent translation. Bioorg Med Chem Lett 19(7):1921–1925
16. Jemielity J et al (2010) Synthetic mRNA cap analogs with a modified triphosphate bridge—synthesis, applications and prospects. New J Chem 34:829–844
17. Slevin MK et al (2014) Deep sequencing shows multiple oligouridylations are required for 3′ to 5′ degradation of histone mRNAs on polyribosomes. Mol Cell 53(6):1020–1030
18. Jemielity J et al (2003) Novel "anti-reverse" cap analogues with superior translational properties. RNA 9:1108–1122
19. Blahna MT et al (2011) Terminal uridyltransferase enzyme Zcchc11 promotes cell proliferation independent of its uridyltransferase activity. J Biol Chem 286(49):42381–42389
20. Jackman J, O'Connor PM (2001) Methods for synchronizing cells at specific stages of the cell cycle. In: Current protocols in cell biology. John Wiley & Sons, Inc.
21. Su W et al (2013) mRNAs containing the histone 3′ stem–loop are degraded primarily by decapping mediated by oligouridylation of the 3′ end. RNA 19(1):1–16
22. Slatko BE, Albright LM (1992) Denaturing gel electrophoresis for sequencing. In: Current

protocols in molecular biology. John Wiley & Sons, Inc., pp 7.6.1–7.6.13.

23. Sambrook J, Fritsch EF, Maniatis T (1989) Molecular cloning: a laboratory manual, 2nd edn. Cold Spring Harbor Laboratory Press, Cold Spring Harbor, NY
24. Ling J et al (2002) The Histone 3′-terminal stem-loop-binding protein enhances translation through a functional and physical interaction with eukaryotic initiation factor 4G (eIF4G) and eIF3. Mol Cell Biol 22 (22):7853–7867
25. Welch JD et al (2015) EnD-Seq and AppEnD: sequencing 3′ ends to identify nontemplated tails and degradation intermediates. RNA 21 (7):1375–1389

Chapter 7

Engineering WT1-Encoding mRNA to Increase Translational Efficiency in Dendritic Cells

Daphné Benteyn, Carlo Heirman, Kris Thielemans, and Aude Bonehill

Abstract

Dendritic cells (DCs) are the orchestrators of the immune system and are frequently used in clinical trials in order to boost the immune system in cancer patients. Among several available techniques for DC modification, mRNA electroporation is an interesting technique due to the favorable characteristics of mRNA. Antigen expression level and duration can be increased by multiple optimizations of an antigen-encoding mRNA template. Here, we describe different molecular modifications to a WT1-encoding mRNA construct in order to increase antigen expression and the subsequent introduction of mRNA into DCs.

Key words mRNA, Dendritic cells, Immunotherapy, Antigen expression, Wilms' Tumor 1, Vaccination, Cancer

1 Introduction

Dendritic cells (DCs) are the orchestrators of the immune system. Because of their role as professional antigen-presenting cells, DCs are frequently used as cellular tools for immunotherapy in clinical trials for cancer patients [1]. The goal of DC-based cancer immunotherapy is to induce tumor-specific T cells that can recognize and eliminate cancer cells in an antigen-specific way.

Today, many DC-based clinical trials are performed with peptide-loaded DCs, but only sporadic clinical results are obtained. There is still some room for improvement. The use of mRNA is a potent alternative for peptides. mRNA is not immunogenic in itself and does not integrate in the host genome. In addition, mRNA holds the advantages that any protein of choice can be generated and that the full-length antigen can be introduced into the cells, which is desirable to target a broad range of epitopes. In addition, patients can be included in the trial without any restriction on the HLA type. Another advantage of the use of mRNA is that modifications can be made to the mRNA expression vector to

Robert E. Rhoads (ed.), *Synthetic mRNA: Production, Introduction Into Cells, and Physiological Consequences*, Methods in Molecular Biology, vol. 1428, DOI 10.1007/978-1-4939-3625-0_7,

increase the translational efficiency and antigen expression. As a consequence, improving the immunobioavailability of RNA-based treatments could lead to a better clinical outcome.

In this chapter we describe several modifications to optimize the expression of WT1. The WT1 antigen was chosen as our working antigen as this was listed number 1 in a prioritization list of tumor antigens made by the National Cancer Institute. This was based on the favorable immunological characteristics of WT1 [2].

To optimize the expression of the WT1 antigen, several molecular modifications were made to the original WT1-encoding pGEM vector. The first step was flanking the WT1 sequence by the signal sequence of LAMP-1 and the MHC class II targeting sequence of DC-LAMP. Proteins are introduced in the cytoplasm of the cell after electroporation and this results in MHC class I presentation alone. However, optimal antitumor immunity requires both $CD4^+$ and $CD8^+$ T cells to generate a durable and robust immune response. Introducing a TAA-encoding mRNA linked to the targeting sequence of an endosomal or lysosomal protein can circumvent this limitation [3–5]. In addition to the $CD4^+$ T cell presentation, we observed that this modification also results in an enhanced presentation of the antigen to $CD8^+$ T cells [4, 6].

The WT1 protein is predominantly located in the nucleus [7]. Transport of a transcription factor from the cytoplasm to the nucleus is usually driven by a nuclear localization sequence (NLS) in the zinc finger domains. We have shown that deletion of an NLS of zinc finger domain 1 resulted in an enhanced delivery of WT1 protein to the antigen-processing compartments. This modification led to a higher cytoplasmic expression and antigen presentation compared to the previous modification. A different modification is the in silico optimization of the mRNA-encoding vector. In general, protein expression may be limited by the short half-life of the mRNA. Furthermore, translation may be hampered by several characteristics of the mRNA. This includes the use of rare codons, a low GC-content or the presence of negatively *cis*-acting motifs [8, 9]. Codon optimization can lead to an increase in protein expression [10, 11]. Nevertheless, for some genes, sequence modification appears not to have any effect on transgene expression, as this was also the case for the WT1-encoding mRNA vector.

Another way to increase the translation efficiency and expression is the use of another mRNA-encoding vector. The previous adaptions were made in a pGEM vector, but the translation efficiency and stability could further be increased by the use of different mRNA-encoding vector. The pST1 vector is an ideal candidate for this. This vector contains two consecutive 3′ UTR of human β-globin and a free-ending poly(A) tail of 120 adenine residues, which are known to enhance translational efficiency and stability of the mRNA [8, 12]. We were able to confirm that the use of pST1 vector-transcribed mRNA results in a higher and longer-lasting

WT1 antigen expression and presentation by DCs in vitro as compared to the pGEM vector [4].

In this chapter we provide the protocols for the (1) construction of plasmid DNA, (2) generation of in vitro transcribed mRNA, and (3) the electroporation of the mRNA into DCs.

2 Materials

2.1 Construction of Plasmid Template for Synthesis of WT1 mRNA

1. pGEM and pST1 vector.
2. Primers.
3. Restriction enzymes *Sap*I and *Spe*I with buffer recommended by the supplier.
4. 1 M Sodium acetate, pH 5.2 (nuclease-free).
5. 100 % and 70 % ethanol.
6. DEPC water.
7. Horizontal electrophoresis system.
8. Agarose.
9. NanoPhotometer.

2.2 In Vitro Transcription of mRNA

1. Linearized plasmids at 1 μg/μL.
2. pGEM/WT1 plasmid. The pGEM/WT1 plasmid containing the full-length WT1 cDNA flanked by its natural 5′- and 3′-UTRs and a poly(A) tail of 64 adenines has been described before [13] and was kindly provided by Prof. Van Tendeloo.
3. pGEM/ΔUT/sig-WT1-DC-LAMP plasmid.
4. pGEM/ΔUT/sig-WT1-sh-DC-LAMP plasmid.
5. pGEM/ΔUT/sig-WT1-sh-DC-LAMP OPT plasmid.
6. pST1/sig-WT1-sh-DC-LAMP OPT plasmid.
7. mMESSAGE mMACHINE® T7 Ultra Kit (Life Technologies).
8. DEPC water.

2.3 Electroporation of Dendritic Cells

1. OptiMEM reduced serum medium without phenol red.
2. 4-mm electroporation cuvettes.
3. Gene Pulser XCell electroporator (Bio-Rad).
4. DC culture medium: RPMI-1640 medium containing 1 % huAB serum and 100 U/mL penicillin, and 20 mM L-glutamine.
5. 15 mL tubes.
6. Phosphate buffered saline (PBS).
7. In vitro synthesized mRNA.

3 Methods

3.1 Construction of Plasmid Template for Synthesis of WT1 mRNA

The following procedures must be performed to obtain capped mRNA suitable for electroporation of DCs: (1) cloning of the gene(s) of interest into a suitable vector and production of plasmid DNA and (2) linearization of the plasmid DNA.

3.1.1 Cloning of the Genes of Interest into pGEM or pST1 Vector and Production of Plasmid DNA (Fig. 1)

1. For the cloning of pGEM-sig-DC-LAMP, extract cDNA from mature human DCs and use it as template for PCR to amplify DC-LAMP. Use the following PCR primers: DC-LAMP sense, 5′-CACAGGATCC$_{BamHI}$CTCGTCTGACTACACAATTGTG-3′; and DC-LAMP antisense, 5′-CACAAGATCT$_{BglII}$TTA GATTCTCTGGTATCCAGATC-3′. This PCR adds a *Bam*HI site at the 5′ end and a stop codon at the 3′ end.

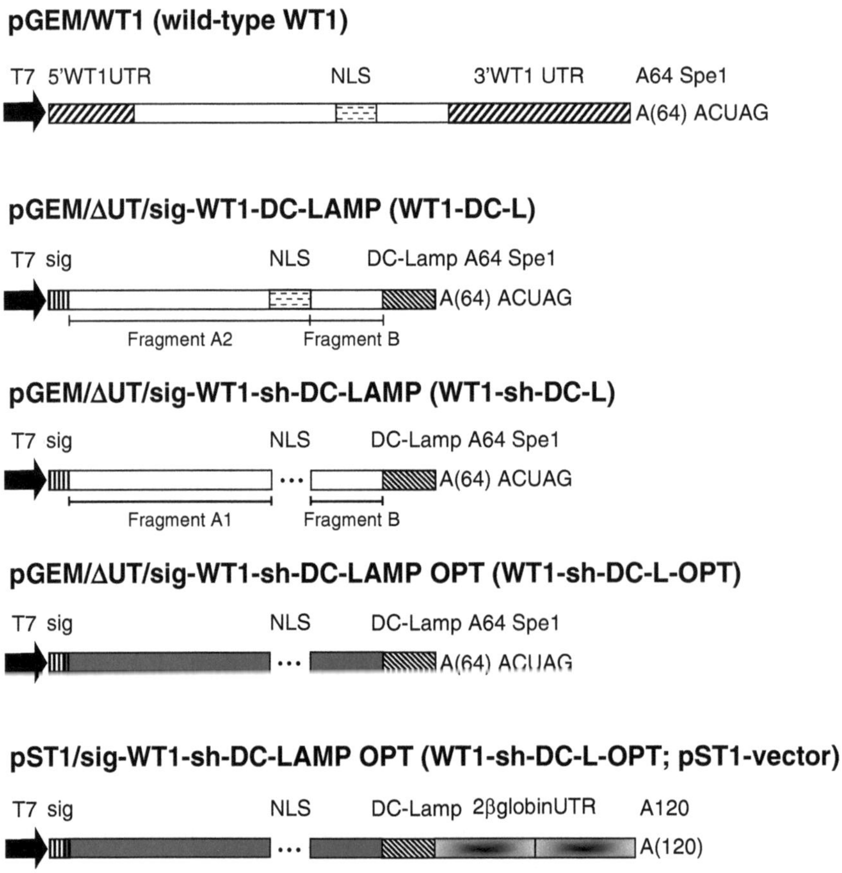

Fig. 1 Schematic representation of the different WT1-encoding vectors. The T7 promoter, 5′ and 3′ untranslated region (UTR), human β-globin UTR, poly(A) tail, the nuclear localization sequence (NLS), signal peptide (sig) of LAMP-1, and the lysosomal targeting sequence (DC-LAMP) are shown. The original WT1 sequence is depicted in *white*, the optimized sequence in *gray*

2. Amplify the signal sequence of LAMP1 (sig) with the following primers: sig sense, 5′-CCCCATGG$_{\mathit{NcoI}}$CGGCCCCCGGC 3′; and sig antisense, 5′-GGGGGATCC$_{\mathit{BamHI}}$TCAAA-GAGTGCTGA-3′, adding an *Nco*I site spanning the start codon at the 5′ end and a *BamH*I site.
3. Generate the pGEM/ΔUT/sig-WT1-DC-LAMP (*see* **Note 1**) plasmid by cloning two PCR generated fragments of WT1 into the pGEM/ΔUT/sig-DC-LAMP plasmid. First clone fragment B as a *BamH*I and *Bgl*II fragment in-frame between the signal peptide and the lysosomal targeting sequence of DC-LAMP. Use these primers: WT1-B-sense, 5′-CCC GGATCC$_{\mathit{BamHI}}$CACCGGT$_{\mathit{AgeI}}$GAGAAACCATACCAGTG-3′; and WT1-B-antisense, 5′-GATCAACATGTTGTTAAGAT CT$_{\mathit{BglII}}$AGCGCCAGCTGGAGTTTG-3′. In addition, clone fragment A2 between sig and WT1-B as a *Bgl*II-*Age*I fragment. Use the following PCR primers: WT1-A2-sense, 5′-CCTCAG CACTCTTTGAAGATCT$_{\mathit{BglII}}$cGGCTCCGACGTGCGGG-3′; and WT1-A2-antisense, 5′-TATGGTTTCTCACCGGT$_{\mathit{AgeI}}$GT GCTTCCTGCTGTG-3′.
4. To obtain the pGEM/ΔUT/sig-WT1-sh-DC-LAMP, clone fragment A1 in-frame between the signal peptide and fragment B. Use these primers: WT1-A1 sense, 5′-CCTCAGCACTC TTTGAAGATCT$_{\mathit{BglII}}$CGGCTCCGACGTGCGGG-3′; and WT1-A1-antisense, 5′-TATGGTTTCTCACCGGT$_{\mathit{AgeI}}$GTG CGTGTGTATTCTGTATTGGG-3′.
5. Optimize the pGEM/ΔUT/sig-WT1-sh-DC-LAMP OPT plasmid by GENEART (*see* **Note 2**).
6. Clone the sig-WT1-sh-DC-LAMP OPT fragment into the pST1 plasmid as a *Spe*I-*Xho*I fragment (*see* **Note 3**).

3.1.2 Linearization of Plasmid DNA

This procedure describes the linearization of 200 μg of plasmid (*see* **Note 4**). When a higher amount of plasmid DNA is linearized, the volumes and the incubation times are increased.

1. Add the following compounds in a 1.5 mL Eppendorf tube at room temperature: 200 μg of plasmid, 10 μL of restriction enzyme (*see* **Note 5**), 50 μL of buffer NEB3.1, making up the volume with DEPC water to a total volume of 500 μL.
2. Incubate overnight at 37 °C.
3. Add 1/10 volume of 1 M sodium acetate, pH 5.2 and 2 volumes ethanol.
4. Mix well.
5. Centrifuge for 15 min at maximum speed.
6. Discard supernatant and add 70 % ethanol.
7. Centrifuge for 5 min at max speed.

8. Resuspend the pellet in 200 μL DEPC water.
9. Check DNA concentration by absorbance at 206 nm and verify linearization by electrophoresis on a 1 % agarose gel.

3.2 In Vitro Transcription of Capped mRNA

In vitro transcription of capped mRNA is performed with T7 RNA polymerase by using the mMESSAGE mMACHINE® T7 Ultra Kit according to the manufacturer's instructions. This kit is designed for the in vitro synthesis of large amounts of efficiently and correctly capped mRNA with a poly(A) tail suitable for electroporation (*see* **Note 6**). After transcription, the remaining plasmid DNA is removed by DNase treatment to reduce the risk of introducing foreign DNA into the cells. Size and integrity of the mRNA are checked by gel electrophoresis and quantity and purity are determined by spectrophotometry. Good quality mRNA is then stored at −20 °C in small aliquots (*see* **Note 7**).

1. Prepare a 200 μL reaction buffer by adding the following components to a 1.5 mL Eppendorf tube: 10 μg linearized plasmid, 40 μL 5× reaction buffer, 70 μL NTP/CAP, 17 μL T7 enzyme mix, making up the volume with RNase-free water.
2. Incubate for at least 2 h at 37 °C.
3. Add 20 μL DNase I (mMESSAGE mMACHINE® T7 Ultra Kit) and incubate for 15 min at 37 °C.
4. Stop the reaction by adding 300 μL DEPC water and 250 μL LiCl precipitation solution (mMESSAGE mMACHINE® T7 Ultra Kit).
5. Purify the RNA by centrifugation for 15 min at max speed.
6. Discard supernatant and add 1 mL 70 % ethanol and centrifuge for 5 min at max speed.
7. Carefully take off the supernatant and resuspend the pellet in 200 μL DEPC water.
8. Check the RNA concentration by absorbance at 260 nm.

3.3 Electroporation of Dendritic Cells (See Note 0)

1. Prepare a 15 mL tube with 5 mL of DC culture medium (at 37 °C).
2. Adjust the physical parameters of the Gene Pulser XCell electroporator as follows: voltage, 300 V; capacitance, 150 μF; and resistance, ∞ Ω.
3. Wash 8 × 10^6 DCs with 10 mL OptiMEM (*see* **Note 9**).
4. While performing the washing step, prepare 20 μg mRNA in a final volume of 200 μL OptiMEM.
5. Resuspend the washed DCs in the mRNA electroporation mix and transfer into a 4-mm electroporation cuvette.

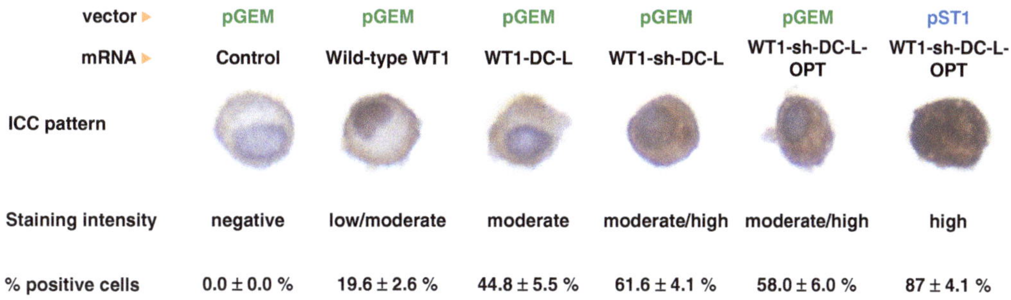

Fig. 2 WT1 expression in dendritic cells. Immunocytochemical staining in WT1-electroporated TriMix DCs. Immunocytochemistry was performed 24 h after electroporation of WT1 or control mRNA. Expression data were scored according to the staining intensity and the percentage of WT1-positive cells (data are expressed as mean ± SEM)

6. Insert the cuvette into the electroporation chamber and trigger the pulse.
7. Immediately after the electroporation, transfer the DCs to the 15 mL Falcon tube with 5 mL of DC culture medium and rinse the electroporation cuvette twice with DC culture medium.
8. Incubate the electroporated DCs for 3.5–4 h in a humidified incubator at 37 °C and 5 % CO_2.
9. Harvest the electroporated DCs.
10. Perform quality control and investigate WT1 expression by immunohistochemistry (Fig. 2) or flow cytometry.

4 Notes

1. Both $CD4^+$ and $CD8^+$ T cells are necessary for the induction of an effective and long lasting antitumor immunity. This can be achieved by linking the antigen to a MHC class II sorting signal. Other sorting signals than DC-LAMP are also commonly used, such as the sorting signal of the invariant chain (Ii) or LAMP1. DC-LAMP is a type I transmembrane protein; in order to translocate type I transmembrane proteins to the rough endoplasmatic reticulum, an additional signal sequence at the *N*-terminus is required. Proteins linked to a MHC class II sorting signal are targeted to the MHC class II compartments and are presented in the context of both MHC class I and class II molecules. In conclusion, MHC class II targeting sequences can be used to improve $CD4^+$ T cell presentation without hampering $CD8^+$ T cell presentation, resulting in a better immune response.
2. Our observations report increased expression of CD40L and caTLR4 after in silico optimization (GENEART).

Nevertheless, for some genes, no major impact on gene expression is observed, including CD70 and WT1 expression.

3. The pST1 vector has a free-ending poly(A) tail of 120 adenines, in contrast to the pGEM vector that has a poly(A) tail of 64 adenines where a part of the consensus sequence of the restriction enzyme remained as an overhang [12].
4. Plasmid DNA must be linearized with a restriction enzyme downstream of the insert to be transcribed. Circular plasmid templates will generate extremely long, heterogeneous RNA transcripts. It is important to examine the linearized template DNA on a gel to confirm that cleavage is complete.
5. Vector linearization is essential to avoid transcription of downstream vector sequences. Type II restriction enzymes, such as *Spe*I, *Not*I, and *Nde*I, cut the vector backbone within their recognition site, resulting in a partial overhang of the restriction enzyme extending the poly(A) tail with other nucleotides than adenosine. In contrast, type IIS restriction enzymes such as *Sap*I and *BfuA*I splice the backbone adjacent to their recognition site resulting in a free-ending poly(A) tail, resulting in a higher translational efficiency [12].
6. A modified cap, Anti-Reverse Cap Analog (ARCA), is used which allows T7 RNA polymerase to synthesize RNAs capped exclusively in the correct orientation. Substitution of traditional cap analog with ARCA allows for synthesis of capped RNAs that are 100 % functional, in contrast to transcription reactions using traditional cap analog where only half of the cap analog is incorporated in the correct orientation. As a result, ARCA cap mRNA molecules are more efficiently translated and much higher protein expression levels can be achieved compared to mRNA produced with the standard cap.
7. Gel electrophoresis of the transcribed mRNA should confer one single, sharp band. If not, mRNA might be degraded or improperly digested. mRNA quantity is measured at 260 and 280 nm. Pure RNA has an A_{260}/A_{280} ratio of 1.9–2.1. If not, RNA might be contaminated with protein or DNA. It is recommended to avoid repeated freeze thawing of the mRNA.
8. Several other strategies have been developed in addition to the electroporation technique to deliver mRNA directly to the cytosol, including nucleofection [14], lipofection [15], and sonoporation [16].
9. Up to 5×10^7 DCs can be electroporated simultaneously, which is of interest for high scale DC generation for clinical trials. For 5×10^7 DCs, 60 μg of mRNA should be used to obtain a high antigen expression. The physical parameters of the Gene Pulser XCell electroporater need to be adjusted to: Voltage, 300 V; capacitance, 450 μF; and resistance, ∞ Ω.

Acknowledgments

This work was supported by grants from the Interuniversity Attraction Poles Program—Belgian State—Belgian Science Policy, the National Cancer Plan of the Federal Ministry of Hearlth, the Stichting tegen Kanker, the Vlaamse Liga tegen Kanker, an Integrated Project and a Network of Excellence sponsored by the EU FP-6, an IWT-TBM program, the Fonds voor Wetenschappelijk Onderzoek Vlaanderen (FWO-Vlaanderen), and the Willy Gepts Wetenschappelijk Fonds of the UZ Brussel.

References

1. Benteyn D, Heirman C, Bonehill A et al (2015) mRNA-pulsed dendritic cell vaccines. Expert Rev Vaccines 14:161–176
2. Cheever MA, Allison JP, Ferris AS et al (2009) The prioritization of cancer antigens: a national cancer institute pilot project for the acceleration of translational research. Clin Cancer Res 15:5323–5337
3. Bonehill A, Heirman C, Tuyaerts S et al (2004) Messenger RNA-electroporated dendritic cells presenting MAGE-A3 simultaneously in HLA class I and class II molecules. J Immunol 172:6649–6657
4. Benteyn D, Anguille S, Van Lint S et al (2013) Design of an optimized Wilms' tumor 1 (WT1) mRNA construct for enhanced WT1 expression and improved immunogenicity in vitro and in vivo. Mol Ther Nucleic Acids 2:e134
5. De Keersmaecker B, Heirman C, Allard S et al (2010) Lumenal part of the DC-LAMP protein is not required for induction of antigen-specific T cell responses by means of antigen-DC-LAMP messenger RNA-electroporated dendritic cells. Hum Gene Ther 21:479–485
6. Kreiter S, Selmi A, Diken M et al (2008) Increased antigen presentation efficiency by coupling antigens to MHC class I trafficking signals. J Immunol 180:309–318
7. Niksic M, Slight J, Sanford JR et al (2004) The Wilms' tumour protein (WT1) shuttles between nucleus and cytoplasm and is present in functional polysomes. Hum Mol Genet 13:463–471
8. Kuhn AN, Diken M, Kreiter S et al (2012) Determinants of intracellular RNA pharmacokinetics: Implications for RNA-based immunotherapeutics. RNA Biol 8:35–43
9. Pascolo S (2008) Vaccination with messenger RNA (mRNA). Handb Exp Pharmacol 183:221–235
10. Fath S, Bauer AP, Liss M et al (2011) Multiparameter RNA and codon optimization: a standardized tool to assess and enhance autologous mammalian gene expression. PLoS One 6: e17596
11. Turksma AW, Bontkes HJ, Ruizendaal JJ et al (2013) Increased cytotoxic capacity of tumor antigen specific human T cells after in vitro stimulation with IL21 producing dendritic cells. Hum Immunol 74:506–513
12. Holtkamp S, Kreiter S, Selmi A et al (2006) Modification of antigen-encoding RNA increases stability, translational efficacy, and T-cell stimulatory capacity of dendritic cells. Blood 108:4009–4017
13. Van Driessche A, Van de Velde ALR, Nijs G et al (2013) Clinical-grade manufacturing of autologous mature mRNA-electroporated dendritic cells and safety testing in acute myeloid leukemia patients in a phase I dose-escalation clinical trial. Cytotherapy 11:653–668
14. Melhem NM, Gleason SM, Liu XD et al (2014) High-level antigen expression and sustained antigen presentation in dendritic cells nucleofected with wild-type viral mRNA but not DNA. Clin Vaccine Immunol 15:1337–1344
15. De Haes W, Rejman J, Pollard C et al (2013) Lipoplexes carrying mRNA encoding Gag protein modulate dendritic cells to stimulate HIV-specific immune responses. Nanomedicine 8:77–87
16. De Temmerman M-L, Dewitte H, Vandenbroucke RE et al (2011) mRNA-Lipoplex loaded microbubble contrast agents for ultrasound-assisted transfection of dendritic cells. Biomaterials 32:9128–9135

Part III

Introduction of Synthetic mRNA into Cells

Chapter 8

Electroporation of Alphavirus RNA Translational Reporters into Fibroblastic and Myeloid Cells as a Tool to Study the Innate Immune System

Christina L. Gardner, Derek W. Trobaugh, Kate D. Ryman, and William B. Klimstra

Abstract

The ability to transfect synthetic mRNAs into cells to measure processes such as translation efficiency or mRNA decay has been an invaluable tool in cell biology. The use of electroporation over other methods of transfection is an easy, inexpensive, highly efficient, and scalable method to introduce synthetic mRNA into a wide range of cell types. More recently, coupling of noncoding RNA sequences or protein coding regions from viral pathogens to fluorescent or bioluminescence proteins in RNA "reporters" has permitted study of host–pathogen interactions. These can range from virus infection of cells to translation of the viral genome, replication and stability of viral RNAs, or the efficacy of host antiviral responses. In this chapter, we describe a method for electroporating viral RNA reporters into both fibroblastic and myeloid cells that encode firefly or *Renilla* luciferase, whose reaction with specific substrates and light emitting activity is a measure of viral RNA translation efficiency. We have used this method to examine host interferon-dependent responses that inhibit viral translation along with identifying secondary structures in the 5′ nontranslated region (NTR) and microRNA binding sites in the 3′ NTR that are responsible for antagonizing the host innate immune responses and restricting viral cell tropism.

Key words Alphavirus, Translation, Electroporation, Neon transfection, Luciferase

1 Introduction

The introduction of synthetic RNA into cells, especially viral RNA, can be a useful technique in molecular virology to generate recombinant viruses and to also study various aspects of host–pathogen interactions. Alphaviruses are mosquito-borne, positive-strand RNA members of the virus family *Togaviridae*. The ability to convert virus RNA into an infectious DNA clone has enabled the construction of full-length replication-competent alphaviruses that encode fluorescent or bioluminescent proteins [1–14]. These "reporter viruses" aid in studying host–pathogen interactions both

Robert E. Rhoads (ed.), *Synthetic mRNA: Production, Introduction Into Cells, and Physiological Consequences*, Methods in Molecular Biology, vol. 1428, DOI 10.1007/978-1-4939-3625-0_8,

in vitro and in live animals by providing a quantifiable readout of either fluorescence or bioluminescence activity.

After infection of a cell, the virus RNA genome, which resembles host cell messenger RNA, is translated by host cell translation complexes producing the virus nonstructural proteins (nsPs) and initiating virus replication [15]. The host translation machinery is able to efficiently translate alphavirus RNA due to the presence of a type 0 7-methylguanosine cap (m^7G) structure at the 5′ end of the virus genome and a 3′ poly(A) tail [16, 17]. Virus translation is enhanced by conserved sequence elements (CSE) encoded in the virus' 5′ nontranslated region (NTR) as well as two stem-loop secondary structure elements within the coding region of nsP1 and a 3′ NTR adjacent to the poly(A) tail [15].

To specifically study translation of the incoming alphavirus RNA, a viral translation reporter is constructed that encodes the 5′ NTR and nsP1 CSE fused in-frame with the firefly luciferase gene followed by the virus 3′ NTR and poly (A) (Fig. 1). RNA synthesized from this template contains the translation control structures in the parental virus genome thus mimicking translation of the incoming viral genome [6, 18]. Such alphavirus translation reporters are being used to study interferon-dependent host responses that inhibit translation of the virus genome and host factors that restrict cellular tropism of viruses [1, 6, 18–20]. Importantly, the construction of alphavirus translation reporters enables the study of aspects of virus life cycle in a biosafety level (BSL) 2 environment instead of the more restrictive BSL3 biocontainment required for some alphaviruses.

The recent work on the NTRs of alphaviruses has led to the discovery that these genetic regions are important for more than

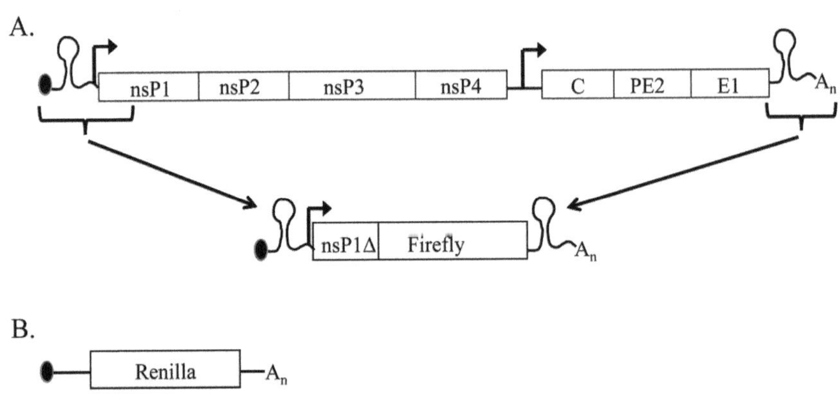

Fig. 1 Diagram of the alphavirus RNA genome and translation reporter and the host mimic. (**a**) The construction of the translation reporter from the alphavirus genome. The alphavirus RNA translation reporters has the m^7G-capped alphavirus 5′ NTR, nsP1 CSE with an in-frame fusion to firefly luciferase, alphavirus 3′ NTR, and polyA tail. (**b**) The host mimic as an m^7G-capped 5′ NTR (random sequence), *Renilla* luciferase, 3′ NTR (random sequence), and polyA tail

just replication of the virus (review [21]) and contribute directly to virus pathogenesis. The secondary structures in the 5′ NTR of many alphaviruses can antagonize host innate immune responses that inhibit translation of nonself mRNAs in the cytoplasm [20]. Additionally, microRNA binding sites in the 3′ NTR of some alphaviruses can limit immune sentinel cell tropism greatly suppressing induction of the innate immune response [19]. There are currently no antiviral therapies or FDA licensed vaccines for alphaviruses and the use of synthetic mRNAs has been suggested as a possible therapeutic. It can now be understood that the NTRs have a tremendous impact on modulating cell tropism, replication efficiency, and induction of innate immune responses. Synthetic RNAs can be designed that target virus sequences known to efficiently suppress virus replication *in vivo* and that RNA drug resistance mutations will specifically reduce the virulence of resultant viruses. Furthermore, knowledge gained regarding virulence loci in coding and noncoding regions can be used for rational design of live-attenuated vaccines (LAV). A crippling problem with current LAV candidates for the alphaviruses that are given as investigational new drugs (INDs) is the high rate of adverse symptoms that resemble infection by the virulent parental virus, presumably due to reversion of the attenuating mutation(s) (review [22]). The current IND LAVs have only one or two attenuating mutations leading to increase likelihood of the attenuated vaccine reverting to virulence (review [22]). Currently, there are several LAV candidates that are being tested that have mutations at multiple loci, including the 5′ NTR and 3′ NTR [22]. These mutations have known mechanisms of action and, thus, can be engineered (e.g., by deletion or multiple site mutation) to resist reversion.

An important factor when using the alphavirus translation reporters is determining the optimal electroporation system and electroporation conditions. We utilize the Dual-Luciferase Assay system (Promega), which requires the electroporation of the test luciferase reporter along with a host mimic reporter that expresses *Renilla* luciferase. The host mimic *Renilla* reporter serves as an internal control for transection efficiency between experiments and is used to normalize the test luciferase reporter results. Two electroporation systems are utilized in this protocol: the Bio-Rad Gene Pulser Xcell™ system and the Invitrogen Neon® transfection system. The Neon® transfection system is reported to be more efficient for primary cells and other hard to transfect cells lines due to a pipette chamber that helps generates a more uniform electric field [23]. We have achieved higher transfection efficiencies of myeloid-lineage cells with the Neon® transfection system compared to the Bio-Rad system [1, 19]. Both systems allow for manipulation of the electroporation conditions for an optimal balance of transfection efficiency and cell viability. However, the Neon® transfection system is limited by the volume of cells per electroporation cuvette.

This chapter describes a detailed step-by-step protocol for electroporating both fibroblastic and myeloid-lineage cells. The Promega Dual-Luciferase assay system is used to quantify changes in the abundance of luciferase protein produced by translation of *in vitro* synthesized alphavirus reporter and control RNAs in the different cell types. Additionally, this chapter describes electroporation conditions for both the Bio-Rad Gene Pulser Xcell system and the Invitrogen Neon® transfection system. For easy-to-transfect fibroblastic cells, we recommend using the Bio-Rad Gene Pulser Xcell system. However, for more difficult cells such as myeloid-lineage cells and primary cells, we recommend using the Neon® transfection system with the understanding that this system is limited in scale but offers more consistent results compared to the Bio-Rad Gene Pulser Xcell system.

2 Materials

2.1 Media/Components for Cell Harvest

1. Dulbeccos' Phosphate-Buffered Saline (DPBS), 1× without calcium and magnesium.
2. Opti-MEM® Reduced Serum media.
3. 0.05 % trypsin with 0.53 mM EDTA, 1× [−] sodium bicarbonate.
4. Growth media for BHK-21 cells: RPMI with 10 % heat-inactivated donor calf serum, 10 % tryptose phosphate broth (TPB; Moltox), 1 % L-glutamine, and 100 U penicillin/streptomycin (per 500 mL of media).
5. Growth media for RAW264.7 cells: DMEM with 10 % heat-inactivated fetal bovine serum, 1 % L-glutamine, and 100 U penicillin/streptomycin (per 500 mL of media).
6. Cell scrapers, 25 cm.
7. Trypan blue.
8. Hemocytometer.
9. 1.5 mL Eppendorf tube.

2.2 Translation Reporter RNA

1. Translation reporter RNA-encoding firefly luciferase: *In vitro* transcribed 5′-capped RNA encoding the firefly luciferase gene fused to the alphavirus translational control sequences as described by Ryman et al. and Tesfay et al. (Fig. 1) [6, 18] (*see* **Note** 1).
2. Translational reporter RNA-encoding *Renilla* luciferase. Host mRNA mimic encoding the *Renilla* luciferase gene fused to short 5′ and 3′ NTRs (Fig. 1b) [8]. The *Renilla* reporter will serve as an internal control to normalize luciferase relative light units (RLUs) to *Renilla* RLUs in each cell.

2.3 Electroporation Components

1. Bio-Rad Gene Pulser II with capacitance extender or Bio-Rad Gene Pulser Xcell.
2. Prewarmed complete media.
3. 0.4 cm gap sterile electroporation cuvette.
4. 1.5 mL Eppendorf tubes or 96-well U-bottom tissue culture treated plate.

2.4 Luciferase Assay Components

1. Orion or similar microplate luminometer with injectors (sensitivity range 300–630 nm).
2. Dual-Luciferase® Reporter Assay System (Promega).
3. HyClone biology grade water.
4. Passive lysis buffer (Promega): Dilute the 5× passive lysis buffer included in the Dual-Luciferase kit to 1× with biology grade water
5. Luciferase assay substrate: Resuspend the lyophilized luciferase assay substrate with 10 mL of the luciferase assay buffer II. Luciferase assay substrate can be stored at −20 °C for up to 1 month or −70 °C for 1 year (*see* **Note** 2).
6. 1× Stop & Glo® substrate (50×): Dilute the Stop & Glo substrate 1:50 in the Stop & Glo buffer. Stop & Glo reagent should be prepared prior to use each time.
7. White polystyrene, flat bottom, nontreated 96-well plates.

2.5 Alternate Neon Electroporation Components

1. Neon transfection system (Invitrogen).
2. Neon 100 μL kit (Invitrogen).
3. Prewarmed complete cell media without antibiotics.

3 Methods

3.1 Harvesting Cells

1. Grow Baby hamster kidney cells (BHK-21; ATCC# CCL-10), or the monocyte/macrophage cell line, RAW264.7 (ATCC #TIB-71), to 80–90 % confluency in a T-175 cm^2 flask or 150 mm dish (*see* **Note** 3).
2. Remove the media from the cells.
3. Wash cells with DPBS to remove remaining traces of media.
4. For BHK cells, add 5 mL of trypsin to the dish. Incubate at 37 °C until cells become nonadherent (~2–5 min). For RAW 264.7 cells, use a cell scraper to gently remove the cells from the flask or dish. For RAW 264.7 cells or other primary cells and myeloid cells, proceed to Subheading 3.3.
5. Add 10 mL of complete media to the flask or dish to transfer the cells to a 50-mL conical (*see* **Note** 4).

6. Count the cells using hemocytometer. Add 10 μL of cells to 90 μL of Trypan blue in a 1.5 mL Eppendorf tube. Transfer 10 μL of cells to hemocytometer to count the cells.
7. Transfer ~1.2×10^7 cells to a 1.5 mL Eppendorf tube.
8. Centrifuge the cells at 200 g for 5 min at 4 °C to gently pellet.
9. Decant the supernatant and disrupt the cell pellet by gently flicking the tube.
10. Add 1 mL of Opti-MEM® and centrifuge at 200 g for 5 min at 4 °C. Repeat 2 times.
11. Decant the supernatant and resuspend the cells in 0.5 mL of Opti-MEM® for each electroporation.

3.2 Electroporation of Cells using a Bio-Rad Electroporator

1. Add 7.5 μg of the viral RNA firefly luciferase translation reporter and 0.75 μg of the host mimic *Renilla* translation reporter to the BHK cells in the Eppendorf tube and mix well.
2. Transfer the cells and RNA to a 0.4 cm gap sterile electroporation cuvette.
3. For BHK cells, electroporate the cells and RNA at 220 V, 1000 μF capacitance. Repeat electroporation for a total of two pulses (*see* **Note 5**) (Table 1).
4. Transfer the cells to 15-mL conical tube containing 13 mL of prewarmed complete media Rinse the cuvette with complete media to recover remaining cells.
5. Invert the conical tube twice to gently mix the cells and then place 1 mL of cells per 1.5 mL Eppendorf tube (*see* **Note 6**) and place at 37 °C.

Table 1
Electroporation conditions for various cell types

Electroporator	Cell type	Voltage[a] (V)	Capacitance (uF)	Pulse width (ms)	Pulses	Alphavirus translation reporter concentration (μg)	Host translation reporter concentration (μg)
Bio-Rad Electroportor	BHK	220	1000	N/A[b]	2	7.5	0.75
	MEF	220	1000	N/A	2	7.5	7.5
	C7/10	360	1000	N/A	1	ND[c]	ND
Neon® Electroporator	BHK	1200	N/A	30	1	ND	ND
	RAW264.7	1750	N/A	25	1	7.5	0.75
	RAJI	1375	N/A	30	1	ND	ND

[a] *See* **Note 3**
[b] Not applicable
[c] Not determined

3.3 Alternate Cell Harvesting Protocol for using Neon® Transfection System

1. The optimal electroporation conditions and amount of RNA will need to be determined prior to the experiment (Table 1).
2. Harvest cells using steps 1–6 from Method 3.1.
3. Transfer 6×10^6 cells to a 1.5 mL Eppendorf tube (*see* **Note** 7).
4. Centrifuge the cells at 200 g for 5 min at 4 °C to pellet.
5. Decant the supernatant and disrupt the cell pellet by gently flicking the tube.
6. Add 1 mL of DPBS and centrifuge at 200 g for 5 min at 4 °C. Repeat 2 times.
7. Resuspend the cells in 110 μL of Buffer R per electroporation (*see* **Note** 8).

3.4 Electroporation of Cells Using Neon® Transfection System

1. In the electroporation cuvette, add 3 mL of Buffer E2 and insert the cuvette into the electroporation station (*see* **Note** 9).
2. Load the 100 μL tip on the pipette by pressing the plunger to the second stop and add the tip to the pipette according to the manufactures guidelines (*see* **Note** 10).
3. In the Eppendorf tube that has either BHK or RAW264.7 cells resuspended in Buffer R, add 7.5 μg of the viral RNA translation reporter and 0.75 μg of the host mRNA translation reporter. Volume of RNA should not exceed 10 % of the total Buffer R volume.
4. Add the cells and RNA into the 100 μL tip, making sure to eliminate all air bubble from the tip (*see* **Note** 11).
5. Enter the appropriate protocol on the Neon® transfection system (Table 1) and then press Start.
6. When the electroporation is complete, a message will appear on the Neon® screen, then eject the cells by depressing the plunger to the second stop into 13 mL of prewarmed antibiotic free complete media (*see* **Note** 6).

3.5 Harvesting Cells for Dual-Luciferase Assay

1. At predetermined times postelectroporation, harvest the cells for the luciferase assay. For translation reporters, time points vary from 30 min to 4 h postelectroporation.
2. Centrifuge the cells at $200 \times g$ for 5 min at room temperature to pellet the cells.
3. Decant the supernatant and add 1 mL of DPBS to wash the cells, making sure to resuspend the pellet. Repeat 2 additional times.
4. After the last wash, decant the supernatant and resuspend the cells in 100 μL of 1× passive lysis buffer (*see* **Note** 12).
5. Store the lysates at −80 °C and allow to freeze for 1 h.
6. Freeze-thaw the samples an additional time to completely lyse the cells (*see* **Note** 13).

3.6 Dual-Luciferase Assay

1. Add 25 μL of the cell lysate into a white polystyrene 96-well plate (*see* **Note** 14).
2. Calculate the volume of the luciferase assay substrate and the Stop & Glo substrate you will need for the luciferase assay. Use 25 μL of each substrate per sample and then add 10 % to total volume to account for priming the injectors with the substrates.
3. To measure firefly luciferase activity, prime luminometer injector 1 with the luciferase assay II substrate. To measure *Renilla* luciferase activity, prime injector 2 with the Stop & Glo substrate. The Stop & Glo substrate will quench the firefly luciferase substrate and subsequently react with the *Renilla* luciferase protein (*see* **Note** 15).
4. Adjust the luminometer to inject 25 μL of the luciferase assay substrate per well with a 10 s delay before measuring luciferase activity (*see* **Note** 16).
5. After measuring firefly luciferase activity, inject 25 μL of the Stop & Glo substrate per well with a 10 s delay before measuring the *Renilla* luciferase activity (*see* **Note** 16).
6. The data can be analyzed with two different methods. First, RLUs of firefly luciferase can be normalized to the *Renilla* luciferase RLUs in each sample to generate a ratio of firefly RLUs to *Renilla* RLUs that can be used to compare samples between independent experiments. A second method is to calculate the change in the RLU ratio between the initial and final time point for each reporter (*see* **Note** 17).

4 Notes

1. The translation reporters encode either a T7 or SP6 promoter driving *in vitro* transcription. The *in vitro* transcription kit (Ambion) must contain a type 0 cap analog to generate 5′-capped RNA for efficient translation of the alphavirus reporters.
2. The luciferase assay substrate can be freeze-thawed up to 6 times before losing luciferase activity. The resuspended luciferase assay substrate can be stored at −20 °C for 1 month or −80 °C for up to 1 year.
3. If more than one T175 cm^2 flask or 150 mm dish is being used for the electroporation at one time just combine the flask or dishes into the same 50-mL conical with up to 3 flask/dishes into the same conical (i.e., electroporating BHK cells with multiple mRNA translation reporters, when harvesting the cells all the cells can be combined together to ensure each electroporation receives the same number of cells).

4. Adding complete media to the BHK cells after typsinizing will inactivate the trypsin. Inactivation of trypsin is important to ensure that the cells maintain optimal viability.
5. Electroporation conditions must be optimized for each Bio-Rad electroporator because the actual electroporation voltage may change during the electroporation resulting in suboptimal conditions.
6. A smaller volume can be used if a 96-well U-bottom plate will be used instead of 1.5 mL Eppendorf tubes.
7. Each of the 100 μL electroporation tips can hold up to 6×10^6 cells, and each tip can be used for two electroporations for a total of 12×10^6 cells.
8. For two electroporations, double the volume of Buffer R to 220 μL. The extra volume helps in removing air bubbles from the tip.
9. Replace the E2 buffer in the cuvette or replace the entire cuvette between electroporations with different translation reporters.
10. If the tip is properly loaded onto the pipette, the gold electrode will move to allow for the cell-RNA mixture to enter the electroporation chamber. Remember that the pipette has two stops when discharging liquid and the second stop ejects the tip.
11. If air bubbles remain in the tip, a spark will occur during electroporation and cells can be ejected from the tip. Additional Buffer R can be added to help in eliminating air bubbles.
12. To ensure better lysis of the cells, pipet up and down several times.
13. During the thaw, place the tubes or plates on a shaker to further facilitate the lysis of the cells.
14. Contamination from neighboring wells can occur if a plate other than the white polystyrene 96-well plates is used to measure luciferase activity.
15. In the injectable luminometer, rinse out the injectors with water before priming the injectors with the substrates.
16. The luminometer will report the luciferase activity as relative light units (RLU).
17. Different cell types will have different levels of translation and/or electroporation efficiency resulting in variable luciferase assay results (e.g., BHK cells will give higher values than RAW264.7 cells) and generally the host mRNA translation reporter will have higher translation compared to the viral RNA translation reporter and thus, less of the host mimic reporter is used compared to the viral reporter.

Acknowledgments

This work was supported by 1R01AI095436 (WBK) and 1R01AI081886 (KDR).

References

1. Gardner CL, Burke CW, Tesfay MZ, Glass PJ, Klimstra WB, Ryman KD (2008) Eastern and Venezuelan equine encephalitis viruses differ in their ability to infect dendritic cells and macrophages: impact of altered cell tropism on pathogenesis. J Virol 82:10634–10646
2. Ryman KD, Gardner CL, Burke CW, Meier KC, Thompson JM, Klimstra WB (2007) Heparan sulfate binding can contribute to the neurovirulence of neuroadapted and nonneuroadapted Sindbis viruses. J Virol 81:3563–3573
3. Ryman KD, Gardner CL, Meier KC, Biron CA, Johnston RE, Klimstra WB (2007) Early restriction of alphavirus replication and dissemination contributes to age-dependent attenuation of systemic hyperinflammatory disease. J Gen Virol 88:518–529
4. Ryman KD, White LJ, Johnston RE, Klimstra WB (2002) Effects of PKR/RNase L-dependent and alternative antiviral pathways on alphavirus replication and pathogenesis. Viral Immunol 15:53–76
5. Sun C, Gardner CL, Watson AM, Ryman KD, Klimstra WB (2014) Stable, high-level expression of reporter proteins from improved alphavirus expression vectors to track replication and dissemination during encephalitic and arthritogenic disease. J Virol 88:2035–2046
6. Tesfay MZ, Yin J, Gardner CL, Khoretonenko MV, Korneeva NL, Rhoads RE, Ryman KD, Klimstra WB (2008) Alpha/beta interferon inhibits cap-dependent translation of viral but not cellular mRNA by a PKR-independent mechanism. J Virol 82:2620–2630
7. Thomas JM, Klimstra WB, Ryman KD, Heidner HW (2003) Sindbis virus vectors designed to express a foreign protein as a cleavable component of the viral structural polyprotein. J Virol 77:5598–5606
8. Bick MJ, Carroll JW, Gao G, Goff SP, Rice CM, MacDonald MR (2003) Expression of the zinc-finger antiviral protein inhibits alphavirus replication. J Virol 77:11555–11562
9. Donnelly ML, Hughes LE, Luke G, Mendoza H, ten Dam E, Gani D, Ryan MD (2001) The 'cleavage' activities of foot-and-mouth disease virus 2A site-directed mutants and naturally occurring '2A-like' sequences. J Gen Virol 82:1027–1041
10. Brault AC, Foy BD, Myles KM, Kelly CL, Higgs S, Weaver SC, Olson KE, Miller BR, Powers AM (2004) Infection patterns of o'nyong nyong virus in the malaria-transmitting mosquito, Anopheles gambiae. Insect Mol Biol 13:625–635
11. Huang HV, Rice CM, Xiong C, Schlesinger S (1989) RNA viruses as gene expression vectors. Virus Genes 3:85–91
12. Pugachev KV, Mason PW, Shope RE, Frey TK (1995) Double-subgenomic Sindbis virus recombinants expressing immunogenic proteins of Japanese encephalitis virus induce significant protection in mice against lethal JEV infection. Virology 212:587–594
13. Tsetsarkin K, Higgs S, McGee CE, De Lamballerie X, Charrel RN, Vanlandingham DL (2006) Infectious clones of Chikungunya virus (La Reunion isolate) for vector competence studies. Vector Borne Zoonotic Dis 6:325–337
14. Xiong C, Levis R, Shen P, Schlesinger S, Rice CM, Huang HV (1989) Sindbis virus: an efficient, broad host range vector for gene expression in animal cells. Science 243:1188–1191
15. Strauss JH, Strauss EG (1994) The alphaviruses: gene expression, replication, and evolution. Microbiol Rev 58:491–562
16. Castello A, Sanz MA, Molina S, Carrasco L (2006) Translation of Sindbis virus 26S mRNA does not require intact eukariotic initiation factor 4G. J Mol Biol 355:942–956
17. Berben-Bloemheuvel G, Kasperaitis MA, van Heugten H, Thomas AA, van Steeg H, Voorma HO (1992) Interaction of initiation factors with the cap structure of chimaeric mRNA containing the 5'-untranslated regions of Semliki Forest virus RNA is related to translational efficiency. Eur J Biochem/FEBS 208:581–587
18. Ryman KD, Meier KC, Nangle EM, Ragsdale SL, Korneeva NL, Rhoads RE, MacDonald MR, Klimstra WB (2005) Sindbis virus translation is inhibited by a PKR/RNase L-independent effector induced by alpha/beta interferon priming of dendritic cells. J Virol 79:1487–1499

19. Trobaugh DW, Gardner CL, Sun C, Haddow AD, Wang E, Chapnik E, Mildner A, Weaver SC, Ryman KD, Klimstra WB (2014) RNA viruses can hijack vertebrate microRNAs to suppress innate immunity. Nature 506:245–248
20. Hyde JL, Gardner CL, Kimura T, White JP, Liu G, Trobaugh DW, Huang C, Tonelli M, Paessler S, Takeda K, Klimstra WB, Amarasinghe GK, Diamond MS (2014) A viral RNA structural element alters host recognition of nonself RNA. Science 343:783–787
21. Hyde JL, Chen R, Trobaugh DW, Diamond MS, Weaver SC, Klimstra WB, Wilusz J (2015) The 5' and 3' ends of alphavirus RNAs - Non-coding is not non-functional. Virus Res 206:99–107
22. Trobaugh DW, Ryman KD, Klimstra WB (2014) Can understanding the virulence mechanisms of RNA viruses lead us to a vaccine against eastern equine encephalitis virus and other alphaviruses? Expert Rev Vaccines 13:1423–1425
23. Kim JA, Cho K, Shin MS, Lee WG, Jung N, Chung C, Chang JK (2008) A novel electroporation method using a capillary and wire-type electrode. Biosens Bioelectron 23:1353–1360

Chapter 9

GMP-Grade mRNA Electroporation of Dendritic Cells for Clinical Use

Judith Derdelinckx, Zwi N. Berneman, and Nathalie Cools

Abstract

mRNA-electroporated dendritic cells (DC) are demonstrating clinical benefit in patients in many therapeutic areas, including cancer and infectious diseases. According to current good manufacturing guidelines, cell-based medicinal products have to be defined for identity, purity, potency, stability, and viability. In order to comply with the directives and guidelines defined by the regulatory authorities, we report here a standardized and reproducible method for the manufacturing of clinical-grade mRNA-transfected DC.

Key words mRNA, Electroporation, Good manufacturing practices (GMP), Dendritic cells, Cell therapy

1 Introduction

Using cells to cure patients is an attractive and innovative approach currently gaining momentum. Cell therapy is based on the transplantation of living cells in order to repair or restore lost or defective functions and has the potential to treat many conditions. This is especially important for diseases for which present conventional treatments are inadequate, including cardiovascular disease, cancer, infectious diseases, neurodegenerative disorders, and autoimmunity. However, a number of specific challenges associated with their manufacture needs to be overcome to permit their clinical use under existing regulatory frameworks to ensure patient safety. The major risks related to a cell-based product are microbiological contamination, loss of cell function, and cell transformation into malignancies. Hence, appropriately tested and qualified starting materials, as well as definition and characterization of the product during and at the end of the process, are of utmost importance.

In general, a cell-based therapeutic product, whether it is for investigational or commercial use, must comply with current good manufacturing practice (cGMP) requirements for medicinal

Robert E. Rhoads (ed.), *Synthetic mRNA: Production, Introduction Into Cells, and Physiological Consequences*, Methods in Molecular Biology, vol. 1428, DOI 10.1007/978-1-4939-3625-0_9,

products, as defined by both the European Medicines Agency (EMA) [1, 2] and the U.S. Food and Drug Administration (FDA) [3]. GMP regulations and guidelines cover both manufacturing and testing of the final product, including the quality control and assurance system, manufacturing facilities, equipment and devices used in the process, raw materials, media and medium supplements, storage, and shipping [4]. In doing so, optimal defined quality and safety in cell transplantation is guaranteed.

DNA- or RNA-based viral and non-viral vectors have been used to engineer changes in cellular expression profiles. Cellular modifications using mRNA are the most recent ones. mRNA allows strong (high amount of protein translated) and transient expression of a unique antigen, thereby representing the most appropriate genetic vehicle for safe immunization. Several cell therapeutic approaches based on mRNA vaccination are under investigation, with a focus on dendritic cells (DC). DC, professional antigen-presenting cells (APC), have the exclusive capacity to initiate and modulate naïve and memory T cell responses, thereby bridging innate and adaptive immunity. We have previously optimized mRNA electroporation in human DC [5]. In accordance to GMP guidelines, it is also necessary to provide safety, suitability, and biocompatibility data for any additional substance that is administered together with, or as part of, the cell-based medicinal product, such as non-cellular components, biomaterials, chemical substances, and biomolecules including mRNA (Fig. 1). The first clinical use of DC was reported in 1996 [6], but since then several phase I/II clinical trials on the investigational use of mRNA-electroporated DC for the immunotherapy of cancer and infectious diseases have been initiated or completed in order to stimulate patients' immune system [7–9]. These studies have revealed that cell therapy using mRNA-electroporated DC was well tolerated and safe in the patient populations investigated. No discernible adverse events or toxicities could be demonstrated. Importantly, clinical and immunological responses were reported, even though this was not the primary objective of these clinical studies.

In this chapter, we describe in detail the procedure for the manufacturing of mRNA-based DC vaccines under GMP-grade conditions. CD14+ monocytes, collected from a leukapheresis product as source material, are differentiated into DC. Next, monocyte-derived DC are loaded with GMP-grade mRNA by means of electroporation. Following cryopreservation and thawing, the mRNA-electroporated final cell product can be administered to the patient. Introduction of mRNA into DC is highly efficient, transient and clinically safe, and results in efficient antigen presentation.

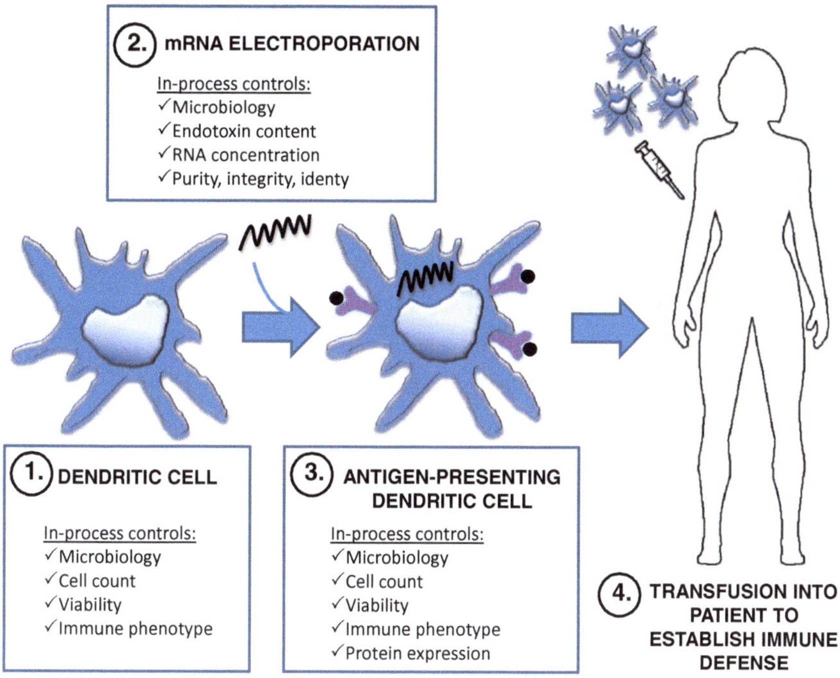

Fig. 1 Simplified schematic overview of the manufacturing process of clinical grade mRNA-transfected dendritic cells according to cGMP guidelines

2 Materials

2.1 General Equipment

1. Centrifuge.
2. 37 °C water bath.
3. (Automated) Hemocytometer.
4. 5 % CO_2 humidified incubator.
5. Transfer pipettes.
6. 5-, 10-, and 25-mL pipettes.
7. Micro pipettes and pipette tips from 5 to 1000 μL.
8. 15- and 50-mL conical tubes.
9. 5-mL tubes.

2.2 In Vitro Generation of Human Monocyte-Derived DC

1. CliniMACS device for fully automated cell-processing and cell-separation procedures from donor apheresis.
2. PBS-EDTA: phosphate-buffered saline (PBS), 1 mM EDTA, pH 7.4.
3. Heat-inactivated human AB serum (30 min at 56 °C).
4. Culture medium: CellGro DC medium, 1 % autologous plasma or human AB serum, 250 U/mL interleukin-4 (IL-4),

800 U/mL granulocyte-macrophage colony-stimulating factor (GM-CSF).

5. Tumor necrosis factor-α (TNF-α).
6. Prostaglandin E_2 (PGE_2).
7. CD14-coupled magnetic beads, for use with the CliniMACS device.
8. Monoclonal antibodies.
9. Viability dye, e.g., propidium iodide (PI).
10. Transfer bags.
11. T175 culture flasks.
12. Flasks for hemocultures.

2.3 mRNA Electroporation of Human Monocyte-Derived DC

1. Electroporation device (Gene Pulser XCell Electroporation Eukaryotic System), including ShockPod cuvette chamber.
2. OptiMEM Reduced Serum Medium without phenol red pH indicator.
3. Culture medium (*see* Subheading 2.2).
4. GMP-grade mRNA of interest, meeting quality controls as specified in **Note 1**.
5. Monoclonal antibodies.
6. Viability dye, e.g., propidium iodide (PI).
7. 4-mm electroporation cuvettes.
8. Filter pipette tips for volumes from 5 to 1000 μL.
9. Ultra-low adherence 6-well plates.
10. Flasks for hemocultures.

2.4 Cryopreservation and Thawing of Electroporated Human Monocyte-Derived DC

1. Freezing medium: human AB serum, 10 % dimethyl sulfoxide (DMSO), 0.4 % glucose (50 % solution).
2. Thawing medium: CellGro DC medium, 1 % human AB serum.
3. Incubation medium: CellGro DC medium, 1 % human AB serum, 250 U/mL IL-4, 800 U/mL GM-CSF, 125 ng/mL TNF-α, 2.5 μg/mL PGE_2.
4. Cryofreezing container (filled with isopropyl alcohol, with a cooling rate of 1 °C/min).
5. 1.5-mL cryovials.

3 Methods

All procedures should be carried out at ambient temperature unless otherwise specified and should be carried out in a clean room following cGMP requirements (*see* **Note 2**).

3.1 In Vitro Generation of Human Monocyte-Derived DC

1. Isolate CD14+ cells after leukapheresis of nonmobilized blood. In our GMP facility, this procedure is being performed automatically through use of the closed CliniMACS system and associated CD14-coupled magnetic beads on day 0.
2. To culture DC from CD14+ monocytes, add 50 mL of Cell-Gro medium to the cell collection bag of the CliniMACS device. Transfer the cell suspension from the collection bag to 50-mL tubes.
3. Take an aliquot of the cell suspension for cell count and viability (*see* **Note 3**).
4. Centrifuge at 480 × *g* for 6 min.
5. Take an aliquot of the supernatant for microbiological testing (aerobic and anaerobic bacterial culture, yeast, and fungi culture) (*see* **Note 4**). Remove the rest of the supernatant.
6. Prepare the culture medium as described in Subheading 2.2, **item 4**.
7. Resuspend the cell pellets in culture medium and transfer the cell suspension to T175 culture flasks (*see* **Notes 5** and **6**).
8. Place the T175 culture flasks horizontally in a 5 % CO_2-humidified incubator at 37 °C.
9. On day 3 or 4, replenish the cells with 250 U/mL IL-4 and 800 U/mL GM-CSF. Replace the culture flasks in a 5 % CO_2-humidified incubator at 37 °C.
10. On day 6, add the maturation stimulus (20 ng/mL TNF-α and 2.5 μg/mL PGE_2). Replace the culture flasks in a 5 % CO_2-humidified incubator at 37 °C.

3.2 mRNA Electroporation of Human Monocyte-Derived DC

3.2.1 Preparation Phase and Harvesting of Human Monocyte-Derived DC

1. Prepare culture medium for collection of cells after electroporation as described in Subheading 2.2, **item 4**. Keep this medium in a 5 % CO_2-humidified incubator at 37 °C.
2. On day 8 of the DC culture, harvest the mature DC by gently shaking the culture flasks and pipetting up and down to detach the cells from the bottom of the flask (*see* **Note 7**).
3. Transfer the cell suspension to 50-mL conical tubes.
4. Rinse the culture flask with 25 mL of OptiMEM medium and add to the 50-mL tubes, using a 25-mL pipette.
5. Centrifuge the 50-mL tubes at 480 × *g* for 6 min.

6. Remove the supernatant and carefully resuspend the cell pellet in OptiMEM medium. Pool the cells from all tubes into two 50-mL tubes.
7. Transfer 500 μL of each tube to one 5-mL tube and mix well by pipetting up and down. Use 100 μL for cell count and viability staining (*see* **Note 8**).
8. In case not all harvested cells are to be electroporated, start the cryopreservation protocol as described further on.

3.2.2 Electroporation of Human Monocyte-Derived DC

1. Put on the power of the electroporation device, adjusting the electroporation settings as follows: protocol = time constant, voltage = 300 V, time = 7 ms, and cuvette = 4 mm.
2. Determine the number of electroporations.
3. Centrifuge the two 50-mL tubes with cell suspension at 480 × *g* for 6 min.
4. Remove the supernatant and gently resuspend the cell pellets in 50 mL OptiMEM per 50-mL tube.
5. Centrifuge cells at 480 × *g* for 6 min.
6. Remove the supernatant and gently resuspend the cell pellets in 25 mL OptiMEM per 50-mL tube. Pool the cells into one 50-mL tube.
7. Centrifuge cells at 480 × *g* for 6 min.
8. Transfer the required volume of warm culture medium to a 50-mL tube (*see* **Note 9**).
9. Remove the supernatant with a pipette and resuspend the cells in the required volume of OptiMEM, taking into account the cell pellet volume (*see* **Notes 10** and **11**).
10. Transfer the cell suspension (minimum 200 μL and maximum 500 μL) in a 4-mm electroporation cuvette (*see* **Note 12**).
11. Thaw the required amount of mRNA (*see* **Notes 1**, **13** and **14**).
12. Add the required amount of mRNA to each electroporation cuvette and mix by gently tapping the cuvette.
13. Put the cuvette in the electroporation device and push the button to start electroporation.
14. Add a small volume of culture medium to the cuvette immediately after electroporation and transfer the cell suspension to a new and empty 50-mL tube (*see* **Note 15**).
15. Write down the obtained values for electroporation settings.
16. Repeat **steps 12–15** for each electroporation.
17. Gently mix the electroporated cells and distribute the cell suspension into ultra-low adherence 6-well plates at 3–4 mL per well.

18. Keep the cells in a 5 % CO_2-humidified incubator at 37 °C for 2 h.
19. After 2 h, harvest the cells by gently pipetting the medium and transfer the cell suspension to a 50-mL tube.
20. Rinse the wells twice with 0.9 % NaCl and add to the 50-mL tube containing the cell suspension. Add 0.9 % NaCl to the 50-mL tube to a final volume of 30 mL.
21. Centrifuge cells at 480 × *g* for 6 min.
22. Remove the supernatant and resuspend the cell pellet in 10 mL 0.9 % NaCl.
23. Take an aliquot of the cells for cell count and viability staining. Add 0.9 % NaCl to the 50-mL tube to a final volume of 30 mL.
24. Centrifuge cells at 480 × *g* for 6 min.
25. Remove the supernatant and resuspend the cell pellet in 0.9 % NaCl into one 50-mL tube.
26. Centrifuge cells at 480 × *g* for 6 min.
27. Take an aliquot of the supernatant for microbiological culture testing (*see* **Note 4**).

3.3 Cryopreservation and Thawing of Electroporated Human Monocyte-Derived DC

1. Centrifuge cells at 480 g for 6 min.
2. Remove the supernatant.
3. Resuspend the cell pellet in cold cryopreservation medium, prepared as stated in Subheading 2.4, **item 1** (*see* **Notes 16** and **17**).
4. Transfer the cell suspension to a cryovial and place in a cryo-freezing container. Immediately transfer to a −80 °C freezer. After 24 h, transfer the cryovials to a liquid nitrogen freezer (*see* **Note 18**).
5. For thawing of cryopreserved DC, prepare 30 mL of thawing medium and 6 mL of incubation medium per aliquot, as stated in Subheading 2.4, **items 2** and **3**.
6. Thaw a vial of electroporated DC directly from the liquid nitrogen freezer in a warm water bath at 37 °C. Remove the cryovial from the warm water bath when only a small clump of frozen suspension is still visible.
7. Add the thawed cell suspension to the thawing medium.
8. Centrifuge cells at 480 × *g* for 6 min.
9. Remove the supernatant and resuspend the cell pellet in incubation medium.
10. Distribute the cell suspension into ultra-low adherence 6-well plates at 3 mL per well.
11. Determine cell count and viability.

12. Place the cells in a 5 % CO_2-humidified incubator at 37 °C for 2 h.
13. Harvest the cells according to **steps 19–27** of Subheading 3.2.2, including cell count, determination of viability, and microbiological testing (*see* **Notes 4** and **19**).
14. Resuspend the cell pellet in the correct volume of 0.9 % NaCl in order for the required number of viable electroporated cells for vaccination to be suspended in 500 μL of 0.9 % NaCl. Take into account the cell pellet volume.
15. Aspirate 500 μL of cell suspension into a 1-mL syringe with a 26G needle. The vaccine is now ready to be administered.

4 Notes

1. For research purposes, mRNA encoding the antigen of interest can be obtained in different ways, including in vitro transcription starting from cDNA using commercially available kits or directly purchased from a specialized company. For clinical purposes, the used mRNA needs to be manufactured according to GMP guidelines taking into account specific quality controls. We refer to Table 1 for the currently applied quality controls for GMP-grade mRNA at our laboratory.
2. For the manufacturing of electroporated DC vaccines for clinical use, all steps of the protocol need to be performed in a clean room facility following GMP guidelines. Operating in compliance with GMP guidelines comprises the basic requirements for medicinal products, including sufficiently qualified personnel manufacturing the cellular vaccine, certified manufacturing premises and equipment, proper documentation of the manufacturing protocol, traceability of all used GMP-grade

Table 1
Currently applied quality controls for GMP-grade mRNA at our laboratory

Quality control	Specification
RNA concentration	1 mg/mL ± 10 %
Endotoxin content	<2 IU/mL
Sterility	– No growth upon aerobic incubation – No growth upon anaerobic incubation – No growth of fungi
Purity, integrity, and identity	– Single band containing >75 % of product on denaturing AGE – No detectable RNA band upon RNAse digestion on denaturing AGE – Residual pDNA <1 % of RNA on DNA-AGE – Total protein <1 % of RNA as assessed by BCA

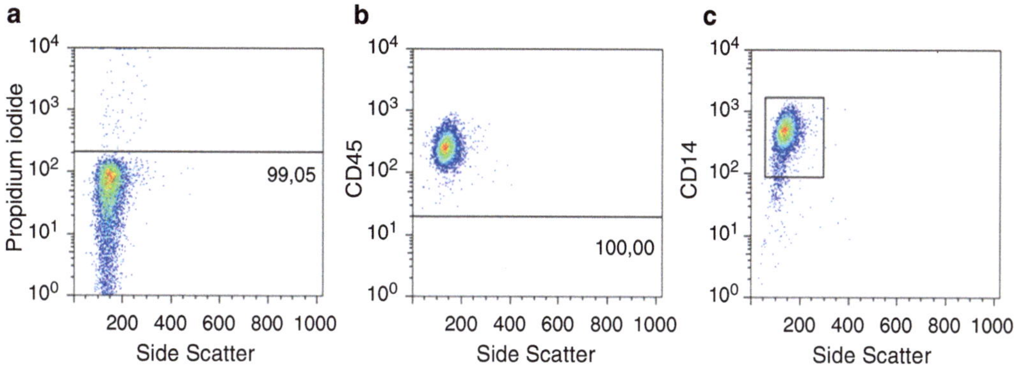

Fig. 2 Immunophenotyping of the CD14-positive cell fraction after isolation using the CliniMACS system ((**a**) viability staining using PI, (**b**) immunophenotyping for CD45, (**c**) immunophenotyping for CD14)

products and adequate quality control of the manufacturing process, including in-process controls, and end product. Additional guidelines are available for active substances as starting material [10]. In this section, the different in-process control measures concerning the manufacturing protocol described are explained.

3. As an in-process control, appropriate CD14+ cell selection can be checked by flow cytometric analysis after staining with CD45 and CD14 monoclonal antibodies. *See* Fig. 2 for an example.

4. Microbiological testing (for the presence of bacteria, fungi, and yeasts) is performed on day 0 of the DC culture, after electroporation of the cultured DC, on day 8 of the DC culture and after thawing of each vaccine aliquot.

5. The optimal concentration for culture of DC from CD14+ monocytes is 70×10^6 cells in 50 mL per T175 culture flask.

6. The use of culture bags could circumvent the use of culture flasks, and a flow-through closed electroporation device with peristaltic pumps would further automate the whole procedure [11].

7. Do not use a detaching agent, e.g., trypsin, to detach adherent cells when harvesting. Fully differentiated DC are in suspension or are only loosely adherent. Therefore, pipetting the medium up and down should be sufficient to harvest all cells of interest.

8. As an in-process control, immunophenotyping using flow cytometry is performed before and 2 h after electroporation. Monoclonal antibodies directed toward the following markers are used: CD83, CD80, HLA-DR, CD14, CD86, and CCR7. *See* Fig. 3 for an example.

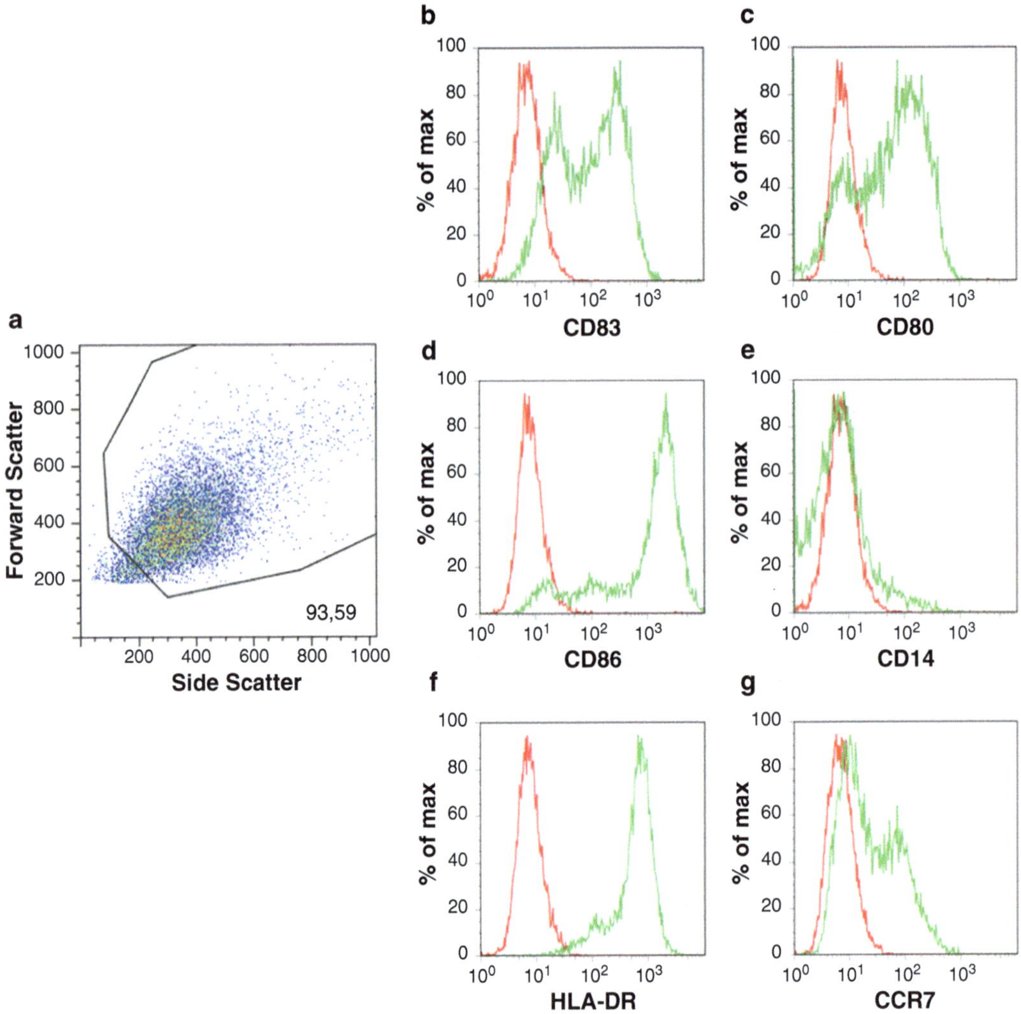

Fig. 3 Immunophenotyping of the harvested mature DC after 7 days of culture, before electroporation [selection of DC based on (**a**) FSC/SSC gating, further immunophenotyping for (**b**) CD83, (**c**) CD80, (**d**) CD86, (**e**) CD14, (**f**) HLA-DR, and (**g**) CCR7 (*in green*)]. Gating for positive cell fraction for each marker was based on staining with the appropriate isotype control (*in red*)

9. The optimal concentration at which cells are to be resuspended after electroporation is 5×10^6 cells per mL, corresponding to 15×10^6 cells per 3-mL per well when using 6-well plates.
10. Use 100 μL OptiMEM per 10×10^6 cells to be electroporated. Use at least 200 μL OptiMEM per electroporation.
11. Ensure a nuclease-free environment when working with mRNA. Wearing gloves and making use of filter tips when working with OptiMEM medium and mRNA is highly recommended, as well as routine treatment of all used material, including pipettors, with RNAse decontaminating reagents.

12. When transferring the cell suspension to the electroporation cuvette, make sure all suspension is placed between the two metal plates to ensure optimal electroporation. Avoid creating air bubbles during pipetting, since these can cause electrical arcing during the electroporation procedure.
13. An optimal working concentration for mRNA is 10 μg/μL. Use 10 μg of mRNA per electroporation when electroporating $\leq 20 \times 10^6$ cells and 20 μg of mRNA per electroporation in case of a higher cell number to be electroporated [12]. It is recommended not to use more than 50×10^6 cells per electroporation.
14. Due to the fast degradation of mRNA at ambient temperature, working quickly during the electroporation procedure is of utmost importance. mRNA can be stored frozen at −80 °C.
15. Adding the collection medium containing serum should be performed immediately after electroporation. A serum-containing environment is necessary for the membrane resealing process after electroporation [13].
16. DMSO is a cryoprotective reagent due to its ability to reduce ice formation and electrolyte fluctuations during the freezing process, but is cytotoxic at room temperature because of its hypertonicity. Therefore, freezing medium containing DMSO should be added slowly and drop-by-drop to the cell suspension. Afterwards, cryovials containing cell suspension should be transferred immediately to a −80 °C freezer in a cryofreezing container for gradual but quick freezing of the cells.
17. Cells are transferred into cryovials, with one cryovial per scheduled vaccination. The final concentration at which cells are frozen is to be based on the number of cells that need to be administered with each vaccination, taking into account a loss of approximately 50 % of cells due to the freezing/thawing procedure.
18. Transfer of the cryovials from the −80 °C freezer to a liquid nitrogen freezer should be performed at the earliest 24 h and at the latest 7 days after placement in −80 °C, since long-term storage at −80 °C results in decreased viability.
19. Additional in-process controls for further characterization of the end product could include flow cytometric immune phenotyping after thawing, immunohistochemical analysis of antigen expression, and quantification of migratory capacity of electroporated DC by migratory assay. All of these controls should be performed at least 2 h after thawing.

Acknowledgments

This work was supported by positive discussion through A FACTT network (Cost Action BM1305: www.afactt.eu). COST is supported by the EU Framework Programme Horizon 2020. Further support was provided by an applied biomedical research project of the Institute for the Promotion of Innovation by Science and Technology in Flanders (IWT-TBM 140191) the Special Research Fund (BOF) from the University of Antwerp, a BOF-GOA grant (ID PS 28313), Medical Legacy Fund (UZA) and the Methusalem Funding Program. Judith Derdelinckx holds a PhD Fellowship of the Research Foundation - Flanders.

References

1. Regulation (EC) No 1394/2007 of the European Parliament and the Council on advanced therapy medicinal products and amending Directive 2001/83/EC and Regulation (EC) No726/2004
2. Directive 2001/83/EC of the European Parliament and of the council of 6 November 2001 on the Community code relating to medicinal products for human use
3. Proposed approach to regulation of cellular and tissue-based products. www.fda.gov/downloads/biologicsBloodVaccines/Guidance/Tissue/UCM062601.pdf
4. Eaker S, Armant M, Brandwein H, Burger S, Campbell A, Carpenito C, Clarke D, Fong T, Karnieli O, Niss K, Van't Hof W, Wagey R (2013) Concise review: guidance in developing commercializable autologous/patient-specific cell therapy manufacturing. Stem Cells Transl Med 2(11):871–883
5. Van Tendeloo VF, Ponsaerts P, Lardon F, Nijs G, Lenjou M, Van Broeckhoven C, Van Bockstaele DR, Berneman ZN (2001) Highly efficient gene delivery by mRNA electroporation in human hematopoietic cells: superiority to lipofection and passive pulsing of mRNA and to electroporation of plasmid cDNA for tumor antigen loading of dendritic cells. Blood 98 (1):49–56
6. Hsu FJ, Benike C, Fagnoni F, Liles TM, Czerwinski D, Taidi B (1996) Vaccination of patients with B-cell lymphoma using autologous antigen-pulsed dendritic cells. Nat Med 2(1):52–58
7. Van Craenenbroeck AH, Smits EL, Anguille S, Van de Velde A, Stein B, Braeckman T, Van Camp K, Nijs G, Ieven M, Goossens H, Berneman ZN, Van Tendeloo VF, Verpooten GA, Van Damme P, Cools N (2015) Induction of cytomegalovirus-specific T cell responses in healthy volunteers and allogeneic stem cell recipients using vaccination with messenger RNA-transfected dendritic cells. Transplantation 99 (1):120–127
8. Van Gulck E, Vlieghe E, Vekemans M, Van Tendeloo VF, Van De Velde A, Smits E, Anguille S, Cools N, Goossens H, Mertens L, De Haes W, Wong J, Florence E, Vanham G, Berneman ZN (2012) mRNA-based dendritic cell vaccination induces potent antiviral T-cell responses in HIV-1-infected patients. AIDS 26(4):F1–F12
9. Van Tendeloo VF, Van de Velde A, Van Driessche A, Cools N, Anguille S, Ladell K, Gostick E, Vermeulen K, Pieters K, Nijs G, Stein B, Smits EL, Schroyens WA, Gadisseur AP, Vrelust I, Jorens PG, Goossens H, de Vries IJ, Price DA, Oji Y, Oka Y, Sugiyama H, Berneman ZN (2010) Induction of complete and molecular remissions in acute myeloid leukemia by Wilms' tumor 1 antigen-targeted dendritic cell vaccination. Proc Natl Acad Sci U S A 107(31):13824–13829
10. European Commission. EudraLex - Volume 4: Good manufacturing practice (GMP) Guidelines. http://ec.europa.eu/health/documents/eudralex/vol-4/index_en.htm
11. Fratantoni JC, Dzekunov S, Singh V, Liu LN (2003) A non-viral gene delivery system designed for clinical use. Cytotherapy 5(3):208–210
12. Cools N, Van Camp K, Van Tendeloo V, Berneman Z (2013) mRNA electroporation as a tool for immunomonitoring. Methods Mol Biol 969:293–303
13. Smits EL, Stein B, Nijs G, Lion E, Van Tendeloo VF, Willemen Y, Anguille S, Berneman ZN (2016) Generation and cryopreservation of clinical grade Wilms' tumor 1 mRNA-loaded dendritic cell vaccines for cancer immunotherapy. Methods Mol Biol 1393:27–35

Chapter 10

Large-Scale mRNA Transfection of Dendritic Cells by Electroporation in Continuous Flow Systems

David Selmeczi, Thomas Steen Hansen, Özcan Met, Inge Marie Svane, and Niels B. Larsen

Abstract

Electroporation is well established for transient mRNA transfection of many mammalian cells, including immune cells such as dendritic cells used in cancer immunotherapy. Therapeutic application requires methods to efficiently electroporate and transfect millions of immune cells in a fast process with high cell survival. Continuous flow of suspended dendritic cells through a channel incorporating spatially separated microporous meshes with a synchronized electrical pulsing sequence can yield dendritic cell transfection rates of >75 % with survival rates of >90 %. This chapter describes the instrumentation and methods needed for the efficient transfection by electroporation of millions of dendritic cells in one continuous flow process.

Key words Dendritic cells, Electroporation, mRNA, Transfection, Microfluidics, Laminar flow

1 Introduction

Electroporation is widely used for cell transfection as an alternative to viral or chemical transfection agents. Clinical applications may particularly benefit from electroporation as a "clean" process that obviates the need for clinically problematic biological or chemical agents, for example in cell-based cancer immunotherapy using dendritic cells [1] or T cells [2]. Conventional electroporation schemes employ centimeter-sized cuvettes with integrated plate electrodes and use potentials in the kilovolt (kV) range to induce the required transient nanopore formation in the cell membranes, typically at electrical field gradients around 1 kV/cm [3]. The high potentials and currents applied during the short electrical pulsing (ms time scale) lead to substantial cell death as well as inhomogeneity in the induced mRNA transfection among the suspended cells. Reducing the chamber dimensions allows for a proportional reduction in the applied potential and possibly more homogeneous transfection based on better control of the electrical field line

Robert E. Rhoads (ed.), *Synthetic mRNA: Production, Introduction Into Cells, and Physiological Consequences*, Methods in Molecular Biology, vol. 1428, DOI 10.1007/978-1-4939-3625-0_10,

distribution [4]. Numerous reports have demonstrated efficient electroporation in microchambers or microfluidic channels using small numbers of cells, either in suspension or adhered to a support surface [5]. Scaling to higher cell numbers using continuous flow systems has also been explored with varying success, e.g., using microshaping of the electrodes used for pulsing [6].

The main technical challenge is to ensure that every suspended cell will experience exactly one electrical pulse since no pulse will clearly not lead to transfection, while two or more pulses will typically result in cell death by the formation of permanent leaky pores in the cell membrane. Normal pressure-driven liquid flow in a closed channel will result in a flow velocity variation with a parabolic profile across the channel diameter. The parabolic velocity profile is caused by the requirement for the liquid to remain stationary (zero velocity) in the immediate vicinity of the channel wall, often termed the no-slip boundary condition in fluidic modeling. Suspended cells transported by the flowing liquid will consequently move at a much lower speed near the channel wall than in the channel center. If electrical pulses are applied at regular intervals along a defined channel length, slow moving cells near the wall will experience numerous (deadly) pulses while fast moving cells in the channel center may not experience a single electroporating pulse before leaving the active part of the channel.

The addition of a microporous mesh spanning the channel cross-section forces the liquid into hundreds or thousands of tiny channel flows, each with a parabolic velocity profile [7]. Upon mixing before and after the mesh, the numerous microflows form a channel-wide flow with an almost constant velocity profile. Electrical pulsing at a constant rate of suspended cells flowing between two such channel-spanning meshes may be synchronized with the liquid flow rate to ensure that the vast majority (>95 %) of the transported cells will experience exactly one pulse during passage [7].

This chapter first describes a procedure to produce the required electroporation device with embedded metal meshes. The procedure is based on assembly of the different channel parts, i.e., flow connections to auxillary tubing, metal meshes for electrical pulsing, and a spacer of defined distance between the meshes, by means of double adhesive tape. The second part of the chapter details the procedure for performing continuous electroporation of mRNA, including important sample preparation steps to reduce the electrical conductivity of the cell medium. The procedure starts by mixing cells and mRNA in a tube segment leading to the electroporation device. Electroporation results from pushing the cell- and mRNA-loaded liquid through the device using an external pump at a constant flow rate while performing electrical pulsing between the metal meshes at a matching frequency, with typical processing times of a few minutes. The specific procedure using human dendritic cells (DC) and mRNA coding for green fluorescent protein (GFP) is reproducibly employed for the transient transfection of 2–20 million DCs from multiple donors at a transfection rate of 75–90 % and a cell viability of 80–95 %.

2 Materials

2.1 Fabrication of Electroporation Device

1. Steel wire mesh (grade AISI 316), 50 μm thick circular wires, 120 μm center-to-center wire spacing in both lateral dimensions (i.e., 70 μm × 70 μm wide openings between wires) (*see* **Note 1**).
2. Scissors capable of cutting in steel wire mesh.
3. Double-sided adhesive tape with protective liner on both sides, 150 μm thickness without liners (ARcare 90106, Adhesive Research, PA) (*see* **Note 2**).
4. Polymer chips parts (diameter 50 mm, thickness 2 mm) with integrated female Luer fitting, aligned through-hole, and a flat back side, produced in cyclic olefin copolymer as described previously [8] (*see* **Note 3**).
5. Hole puncher for holes with a diameter of 2.5–3.0 mm (*see* **Note 4**).
6. Hydraulic press with mounts matching the assembled chip part geometry to enable an even pressure of 3 bar (3 kg/cm^2) across the lateral device dimensions while heated to 40 °C.
7. Rubber gloves.
8. 70 % v/v ethanol/water.

2.2 Electroporation System

1. PC-based pulse generator and oscilloscope (for example HS3, TiePie Engineering, The Netherlands) generating square unipolar pulses with an amplitude up to 6 V, width 1 ms, and slew rate >100 V/ms (*see* **Note 5**).
2. Voltage amplifier that causes a tenfold increase of the input voltage, for a maximum output voltage of 60 V at a slew rate >1000 V/ms and capable of delivering up to 50 mA current during pulsing (*see* **Note 6**).
3. Electrical wire and alligator clips for electrical connection to the steel wire meshes.
4. Syringe pump.
5. Cell-compatible tubing with inner diameter (ID) 0.8 mm (*see* **Note 7**).
6. Fluid 3-way "Y"-connector for tubing with ID 0.8 mm.
7. Female Luer adaptors mating to tubing with ID 0.8 mm.
8. Male Luer adaptor mating to tubing with ID 0.8 mm.
9. Tubing clamps for sealing off the flow.

2.3 Dendritic Cell Preparation

1. Mature human dendritic cells as described previously [9], frozen with 2×10^6 cells per vial.
2. X-Vivo 15 medium (Lonza, Basel, Switzerland).
3. X-Vivo 15 medium without gentamicin or phenol red (Lonza) (*see* **Note 8**).
4. Low conductivity culture medium (Cytoporation Medium Formula T, CytoPulse Sciences, MD) (*see* **Note 9**).
5. Gentamicin.
6. Heating bath at 37 °C.
7. Centrifuge.
8. 50-mL conical tubes.
9. Sterile flow cabinet.

2.4 Electroporation

1. Electroporation system.
2. Mature human dendritic cells in suspension.
3. Messenger RNA coding for Green Fluorescent Protein (GFP), as described previously [9].
4. Low conductivity culture medium.
5. X-Vivo 15 medium.
6. Heat-inactivated human AB-serum (Lonza).
7. GlutaMAX (Life Technologies).
8. 15-mL conical tubes.
9. 1-mL syringes with male Luer connector.
10. Sterile flow cabinet.

2.5 Evaluation of Transfection Efficiency

1. X-Vivo 15 medium.
2. Heat-inactivated human AB-serum.
3. GlutaMAX.
4. Petri dish, diameter 50 mm, tissue culture grade.
5. Humidified CO_2 incubator at 37 °C.
6. Propidium iodide.
7. Fluorescence microscope with filter combinations for visualizing GFP and propidium iodide, as well as bright field or phase contrast microscopy.
8. Sterile flow cabinet.

3 Methods

3.1 Fabrication of Electroporation Device

Wear gloves while performing the following steps.

1. Cut five pieces of double-sided adhesive tape to be approximately 3 mm wider than the perimeter of the polymer chip parts.
2. Punch a hole in the five tape pieces so that the hole approximately is co-located with the through-hole of the chip part.
3. Remove the liner from one side of one piece of tape and align the hole in the tape to the through-hole in one chip part as well as possible by hand (*see* **Note 10**).
4. Repeat **Step 3** using a second piece of tape and a second chip part.
5. Cut two 6 cm × 1 cm rectangular pieces of the steel wire mesh using the scissors.
6. Remove the second liner from the tape on one chip.
7. Apply one piece of steel wire mesh so that the center-line of its long dimension approximately coincides with the center of the hole, one end along the longer dimension is about 5 mm inside the perimeter of the chip part, and the other end is protruding by about 15 mm beyond the chip part perimeter.
8. Remove the liner from one side of two of the remaining tape pieces and attach the exposed sides to each other so that the holes are aligned.
9. Repeat **Step 8** using one more tape piece to form a stack of three attached tape pieces with liners remaining only on the top and the bottom of the stack.
10. Remove one liner from the stack and attach it to the mesh on the first chip part so that the holes are aligned.
11. Repeat **Steps 5** and **6** using the second chip part, the remaining tape piece, and with the wire mesh aligned to protrude by about 15 mm beyond the chip part perimeter to the opposite side of the mesh on the first chip part.
12. Remove the second liner from the tape stack top on the first chip part and attach it to the wire mesh on the second chip part so that the holes are aligned and the wire meshes protruding beyond the chip part perimeters are in opposite directions.
13. Mount the assembled device in the hydraulic press and homogeneously apply 3 bar of pressure on the opposing sides of the assembled chip parts for 10 min while heating the assembly to 40 °C.

14. Remove the device from the hydraulic press and transfer it to a laminar air flow bench.
15. Immerse the device in a beaker with ethanol for 10 min to sterilize it.
16. Place the device on tissue in the bench and leave it overnight in the air flow to dry.

3.2 Electroporation System

1. Connect the output of the pulse generator to the input of the voltage amplifier. Pay attention to matching polarity.
2. Connect the output from the voltage amplifier to the two wire mesh ends protruding on opposite sides of the electroporation device using alligator clips. Either polarity will work.
3. Connect about 10 cm tube to one fork of the Y-piece and attach a terminal female Luer connector at the other end. This piece of tubing is termed the "loading tube" and is used for loading the cell suspension at high concentration.
4. Connect about 20 cm tube to a second fork of the Y-piece and attach a terminal female Luer connector at the other end. This piece is termed the "syringe pump tube" and is used for driving liquid through the electroporation device at constant definable flow rate.
5. Connect about 150 cm tube to the third fork of the Y-piece and attach a terminal male Luer connector at the other end that is inserted on one side of the electroporation device. This piece is termed the "main tube" and holds the concentrated cell suspension prior to initiating electroporation (*see* **Note 11**).
6. Connect about 20 cm tube with a terminal male Luer connector to the other side of the electroporation device. The open tube end is used for collecting the product from the electroporation.

3.3 Dendritic Cell Preparation

Transfection by electroporation should be performed as soon as possible after the dendritic cells have been prepared. Thus, the setup of the electroporation system (Subheading 3.4, **steps 1**–7) should be completed prior to thawing and preparing the dendritic cells.

1. All media in the subsequent steps are preheated to 37 °C in the heating bath.
2. All subsequent procedures are performed under sterile conditions in a flow cabinet and with sterile agents, except for **steps 6** and **8**.
3. Frozen mature human dendritic cells (mDC) result from a previously described procedure [9]. A frozen ampule with 2×10^6 mDC is thawed in the heating bath at 37 °C for as short as possible until it can be pipetted (*see* **Note 12**).

4. Prepare 25 mL X-Vivo 15 medium without gentamicin or phenol red supplemented with 125 μL gentamicin, hereafter termed "Phenol red free X-Vivo 15 medium".
5. Transfer the thawed mDC suspension to a 50-mL tube containing 25 mL X-Vivo 15 medium and mix thoroughly.
6. Centrifuge for 5 min at 400 × *g*.
7. Discard the supernatant, resuspend the cell pellet, add 25 mL Phenol red free X-Vivo 15 medium, and mix thoroughly.
8. Centrifuge for 5 min at 400 × *g*.
9. Discard the supernatant, resuspend the cell pellet in a drop of electroporation medium, add 200 μL electroporation medium, add 40 μg mRNA, and mix (*see* **Note 13**).

3.4 Electroporation

All procedures are performed under sterile conditions in a flow cabinet and with sterile agents.

1. Prepare a 15-mL tube with 5 mL X-Vivo 15 medium supplemented with 5 % v/v heat-inactivated human AB-serum and 1 % v/v GlutaMAX for collecting the electroporated cells.
2. Set the pump rate on the syringe pump to 450 μL/min (*see* **Note 14**).
3. Open the "Arbitrary Waveform Generator" (AWG) in the TiePie software and define a single square pulse of width 1 ms as the pulse form.
4. Set the output amplitude of the AWG to 5.2 V, corresponding to a pulse amplitude of 52 V output to the stainless steel meshes from the tenfold potential amplifier. Set the pulsing frequency of the AWG to 4 Hz and press On followed by Stop.
5. Manually fill the syringe on the syringe pump with 1 mL electroporation medium and mount it in the syringe pump attached to the main tubing.
6. Clamp off the tubing leading to the electroporation device (main tube) right after the Y-piece.
7. Fill the loading tube with electroporation medium using the syringe pump until medium is visible at the Luer connector (*see* **Note 15**).
8. Clamp off the tubing from the syringe pump (syringe pump tube) right before the Y-piece and remove the clamp on the tubing leading to the electroporation device (main tube).
9. Suck the cell suspension into a second 1-mL syringe.
10. Attach this syringe to the Luer connector on the loading tube. Slowly empty the 200 μL cell suspension into the loading tube and further into the main tube (*see* **Note 16**).

11. Clamp off the loading tube right before the Y-piece and remove the clamp on the tube from the syringe pump (syringe pump tube).
12. Press Start on the AWG to initiate electrical pulsing.
13. Turn on the pump at its set rate of 450 μL/min.
14. Collect the dispensed liquid from the electroporation device into a 15-mL tube.
15. Turn off the syringe pump when less than 0.1 mL is left in the mounted syringe.
16. Press Stop and Off on the AWG to stop electrical pulsing.

3.5 Evaluation of Transfection Efficiency

Steps 1 and **2** are performed under sterile conditions in a flow cabinet and with sterile agents.

1. Transfer the suspension of electroporated cells to the petri dish. Add 3 mL X-Vivo 15 medium supplemented with 5 % v/v heat-inactivated human AB-serum and 1 % v/v GlutaMAX. Incubate for 24 h at 37 °C.
2. Add propidium iodide at a concentration of 2 μg/mL and incubate for 30 min at 37 °C.
3. Quantify the transfection efficiency and the cell viability by fluorescence microscopy of the petri dish using a GFP filter set to observe GFP expressing (transfected) cells and a propidium iodide filter set to visualize dead cells. The total number of cells can be quantified from bright field or phase contrast micrographs [7].

4 Notes

1. AISI 316 steel is used for its high resistance to corrosion in saline solution. Other steel types with equivalent or better corrosion resistance may be used instead. The high electrical potential applied and the associated electrical current may lead to release of metal ions from the steel wire surface. AISI 316 steel is an alloy of Cr, Fe, Mn, Mo, and Ni, each of which may be cytotoxic if released at high concentrations. Experiments using 25 electrical pulses per liquid volume at a potential of 65 V, instead of the single pulse per liquid volume at a potential of 52 V used in the described protocol, were performed to enable detection of any released metal. Highly sensitive mass spectrometry with ppb sensitivity only detected increases in Cr and Ni concentrations by <1 ppb [7]. This should be compared to acute toxicity levels in the range of 200–5000 ppb for these metal ions [10, 11].

2. The type of adhesive tape is unlikely to affect the device efficiency as long as the adhesive (1) is nontoxic, (2) can deform to tightly seal around the metal mesh wires, and (3) remains stable in aqueous environments.
3. Any fluid connector with a flat backside of sufficient width to support stable adhesive bonding to the metal mesh should work, including several types of commercial products (cat no 10-1124-0348-02, microfluidic ChipShop, Germany; cat no 81128, ibidi, Germany). The material should remain stable in an aqueous environment.
4. The exact hole diameter is not essential as long as the liquid flow rate is scaled with the actual cross-sectional area of the hole during the electroporation experiments.
5. The pulse generator and the oscilloscope may also be used as two separate units.
6. The pulse generator and voltage amplifier may be combined in one unit that can produce unipolar 1 ms pulses with an amplitude up to 60 V at a slew rate >1000 V/ms and a current of 50 mA.
7. The tubing used is standard plasticized PVC. Due to the short residence time (a few minutes) of the dendritic cells in the pumping system, there is no need for gas permeability or heating of the tubing.
8. X-Vivo 15 medium without phenol red is used for cell preparation to minimize autofluorescence after electroporation. Increased autofluorescence is occasionally observed when traces of phenol red are present in the medium used during electroporation, which compromises the quantification of the GFP produced. However, this may not be an issue when transfecting mRNA coding for nonfluorescent proteins.
9. Equivalent results are obtained using isotonic buffered sucrose solution prepared by dissolving 960 mg sucrose in 9.5 mL ultrapure water and adding 0.5 mL Dulbecco's phosphate buffered saline. The conductivity of the electroporation medium should be reduced sufficiently to avoid visible bubble formation in the device during electrical pulsing. This requirement is fulfilled using the described device and procedure that results in a current of approximately 1 mA during operation.
10. Attachment of the double-sided adhesive tape to the polymer substrates will often result in the unwanted capture of air pockets at the interface. Air pockets can compromise the sealing efficiency of the assembled device so their number and size should be minimized.
11. The Luer adaptors will commonly have a substantial dead volume that may hold a significant fraction of the loaded

volume of cell suspension. Thus, the required length of main tube will depend on the detailed setup.

12. The viability of thawed mDC decreases rapidly with extended residence time in the freezing medium as a liquid.
13. The described method electroporates 2×10^6 mDC suspended in 200 μL electroporation medium. Equally high rates of viability and transfection efficiency are found using 20×10^6 mDC suspended in 2 mL electroporation medium [7].
14. The flow rate, *Q*, should be matched to the actual channel diameter, *w*, through the electroporation device, the wire mesh spacing, *d*, and the pulsing frequency, *f*, so that each liquid volume is pulsed exactly once during passage, i.e., $Q = f \times d \times \pi \times w^2/4$. Reducing the flow rate below the calculated optimum will ensure that all cells experience at least one pulse with the risk of some cells receiving two pulses leading to reduced viability. Increasing the flow rate enhances cell viability at the cost of reduced electroporation efficiency. The flow rate should be experimentally optimized to balance the transfection yield and cell viability.
15. All tubing should be completely filled with electroporation medium prior to the addition of cells as any trapped air will lead to inhomogeneous flow of the cell suspension and may additionally significantly reduce the cell viability.
16. Some cell suspension will be left in the loading tube segment when it is clamped off. If high cell yields are essential, cell-free electroporation medium from an additional syringe may be added after cell loading to push all the suspended cells into the main tubing. However, this comes at a significant risk of introducing unwanted air bubbles in the flow system.

References

1. Met O, Balslev E, Flyger H, Svane IM (2011) High immunogenic potential of p53 mRNA-transfected dendritic cells in patients with primary breast cancer. Breast Cancer Res Treat 125:395–406
2. Choi Y, Yuen C, Maiti SN et al (2010) A high throughput microelectroporation device to introduce a chimeric antigen receptor to redirect the specificity of human T cells. Biomed Microdevices 12:855–863
3. Yarmush ML, Golberg A, Serša G et al (2014) Electroporation-based technologies for medicine: principles, applications, and challenges. Annu Rev Biomed Eng 16:295–320
4. Movahed S, Li D (2010) Microfluidics cell electroporation. Microfluid Nanofluid 10:703–734
5. Geng T, Lu C (2013) Microfluidic electroporation for cellular analysis and delivery. Lab Chip 13:3803–3821
6. Geng T, Zhan Y, Wang J, Lu C (2011) Transfection of cells using flow-through electroporation based on constant voltage. Nat Protoc 6:1192–1208. doi:10.1016/j.jconrel.2010.01.030
7. Selmeczi D, Hansen TS, Met O et al (2011) Efficient large volume electroporation of dendritic cells through micrometer scale manipulation of flow in a disposable polymer chip. Biomed Microdevices 13:383–392. doi:10.1007/s10544-010-9507-1
8. Andresen KOØ, Hansen M, Matschuk M et al (2010) Injection molded chips with integrated

conducting polymer electrodes for electroporation of cells. J Micromech Microeng 20:55010. doi:10.1088/0960-1317/20/5/055010

9. Met O, Eriksen J, Svane IM (2008) Studies on mRNA electroporation of immature and mature dendritic cells: effects on their immunogenic potential. Mol Biotechnol 40:151–160

10. Hallab NJ, Anderson S, Caicedo M et al (2005) Effects of soluble metals on human peri-implant cells. J Biomed Mater Res A 74:124–140

11. Puleo DA, Huh WW (1995) Acute toxicity of metal ions in cultures of osteogenic cells derived from bone marrow stromal cells. J Appl Biomater 6:109–116

Chapter 11

FLT3 Ligand as a Molecular Adjuvant for Naked RNA Vaccines

Sebastian Kreiter, Mustafa Diken, Abderraouf Selmi, Jutta Petschenka, Özlem Türeci, and Ugur Sahin

Abstract

Intranodal immunization with antigen-encoding naked mRNA has proven to be an efficacious and safe approach to induce antitumor immunity. Thanks to its unique characteristics, mRNA can act not only as a source for antigen but also as an adjuvant for activation of the immune system. The search for additional adjuvants that can be combined with mRNA to further improve the potency of the immunization revealed Fms-like tyrosine kinase 3 (FLT3) ligand as a potent candidate. Systemic administration of the dendritic cell-activating FLT3 ligand prior to or along with mRNA immunization-enhanced priming and expansion of antigen-specific $CD8^+$ T cells in lymphoid organs, T-cell homing into melanoma tumors, and therapeutic activity of the intranodally administered mRNA. Both compounds demonstrate a successful combination in terms of boosting the immune response. This chapter describes methods for intranodal immunization with naked mRNA by co-administration of FLT3 ligand, which leads to strong synergistic effects.

Key words FLT3 ligand, IVT-RNA, RNA vaccine, Intranodal injection, Dendritic cell, Cancer vaccination, Immunotherapy, Vaccine adjuvants

1 Introduction

mRNA is a widely recognized and potent tool for generating antigen-specific immunity in the field of vaccine- and immunotherapy. The mRNA technology was first shown in 1990 by Wolff [1], who observed similar protein expression levels after intramuscular injection of either RNA or DNA expression vectors. Vectors for both nucleic acids can be rapidly and accurately synthesized, resulting in expression of the proteins encoded. The major advantage of mRNA is the instant translation after entering the cytoplasm; mRNA, in contrast to DNA, does not need to travel to the nucleus first. In addition, mRNA vaccines can avoid insertional mutagenesis, since no integration into the genome takes place. Moreover, mRNA can act as a potent stimulator of the immune system by activating

Robert E. Rhoads (ed.), *Synthetic mRNA: Production, Introduction Into Cells, and Physiological Consequences*, Methods in Molecular Biology, vol. 1428, DOI 10.1007/978-1-4939-3625-0_11,

intravesicular and cytoplasmic pattern recognition receptors (PRR) [2], rendering itself an adjuvant in addition to serving as an antigen source only. Therefore, mRNA represents a simple, reproducible, and inexpensive tool that can be utilized in high-throughput approaches for drug discovery and drug optimization [3].

Efforts have been made to improve the stability and translation of the mRNA by modifying structural elements of the IVT mRNA vectors such as 5′ cap, 5′- and 3′ UTRs, the poly(A) tail or the coding region [4–6]. Currently, vaccination strategies using mRNA are realized by (1) ex vivo transfection of mRNA into dendritic cells (DCs) and their subsequent transfer to the host and (2) direct injection of mRNA [7, 8]. The efficacy of ex vivo-transfected DCs in cancer immunotherapy and infectious disease vaccines has been shown in preclinical but also in clinical studies (reviewed in [3]).

By 1996, the potency of ex vivo transfection of DCs with tumor-antigen coding mRNA or total RNA of tumor lysate had been shown by the induction of a potent T cell response and controlling tumor growth in a melanoma mouse model [9]. This and other preclinical studies paved the way for clinical use of ex vivo mRNA-transfected DCs in cancer patients (reviewed in [7, 10]). Further refinement of treatment protocols ultimately led to the currently running Phase III clinical study using autologous tumor-derived mRNA-transfected DCs in patients with advanced renal carcinoma (ClinicalTrials.gov identifier: NCT01582672).

Since cell therapies are time- and cost-intensive, many scientists initially focused on other strategies. The most promising method has been the direct injection of mRNA, which has proven feasible in different preclinical settings by the vast amount of tested injection routes like intradermal [11], intranodal [11], subcutaneous [11], intravenous [7], intramuscular [12], and intrathecal [13] injection. Intradermal delivery of naked RNA induced a T cell and an antibody-directed response specific to the antigen [14]. The most effective route to induce a potent T cell response is still the direct injection of naked RNA into the lymph node. The fast induction of a strong antigen-specific T helper cell 1 (T_H1)-response and a potent antitumor immunity has been shown in several preclinical animal models [8, 11, 15]. Moreover, both methods of intradermal and intranodal delivery of naked RNA were transferred to test in clinical studies (ClinicalTrials.gov identifier: NCT01915524, NCT01817738, and NCT01684241). In order to improve the efficiency of vaccination with naked RNA, many adjuvants have already been screened. A combination of naked mRNA with GM-CSF enhanced the humoral response and the polarization to cellular immune responses [16]. In addition, approaches such as pretreatment of the DC administration site with inflammatory cytokines, to provide in vivo stimuli for DC maturation and migration to lymph nodes, have been tested [17]. Interestingly, maturation-inducing adjuvants like poly(inosinic):poly cytidylic acid (PolyI:C), lipoploysaccharide (LPS) and CD40L abrogated the in vivo uptake of mRNA into DCs when injected into the draining area of the lymph node before intranodal immunization [8]. In line with that, in vitro pre-stimulation with those DC-maturing adjuvants prevented human DCs or murine BMDCs to internalize mRNA. It was noted that the reduction in RNA uptake correlated with the maturation state of the DCs as measured by an increase in the expression of the activation marker CD86 [8]. Interestingly, only immature DCs are able to internalize naked mRNA with high efficiency and without reaching saturation by macropinocytosis. An abrogation of this mechanism by DC maturation not only decreased the uptake but also the potency of the intranodally injected naked mRNA vaccine [8]. These studies underlined that an adjuvant should not mature the DCs in order to be combined with naked mRNA vaccination.

In a search for such an adjuvant, Kreiter and colleagues showed that Fms-like tyrosine kinase 3 (FLT3) ligand further enhanced immune responses upon combination with naked mRNA vaccination [18]. FLT3 ligand is a naturally occurring glycoprotein that stimulates early hematopoietic progenitors via the FLT3 receptor [19]. Those hematopoietic progenitors are mobilized from the bone marrow into the peripheral blood and finally accumulate in secondary lymphatic organs. Here, a profound expansion of DCs takes place [20–22]. Kreiter and colleagues revealed that among the DC subsets, plasmacytoid DCs (pDC) play a central role in the adjuvant effects of FLT3 ligand [18], providing unexpected insights into the functional potency of pDCs. Beside the expansion of DCs, FLT3 ligand stimulates natural killer (NK), B and T cells [18]. It has been reported that FLT3-ligand-induced DCs have an immature phenotype [23, 24]. An improved T-cell immunity and antitumor potency could also be observed in vaccination approaches in other mouse models [20–22, 25–27].

In particular we focus to provide protocols for analysis of FLT3 ligand effects as an adjuvant in combination with a naked mRNA vaccine. In detail we focus on the following topics: (1) generation and characterization of FLT3 ligand BMDCs, (2) effect of FLT3 ligand on cellularity and activation of DCs, (3) mRNA transfer into FLT3 ligand BMDCs, (4) effect of FLT3 ligand on luciferase expression and (5) monitoring of elicited immune responses.

2 Materials

2.1 Generation of FLT3 Ligand BMDCs

1. Bone marrow cells from C57BL6/J mice (*see* **Note 1**).
2. DC medium: RPMI1640 medium with 2 mM glutamine, 100 U/mL penicillin, 100 μg/mL streptomycin, 1 mM sodium pyruvate, 1 mM nonessential amino acids, and 10 % heat-inactivated FCS.
3. 6-Well plates.

4. Antibodies for FACS staining: CD11c APC (Miltenyi Biotech, clone N418); CD45RA PE (BD, clone 14.8); PDCA1 FITC (Miltenyi Biotech, clone JF05-1.C2.4.1); 7-AAD (Invitrogen) (*see* **Note 2**).
5. FLT3 ligand (produced in house) (*see* **Note 3**).

2.2 mRNA Transfer into FLT3 Ligand BMDCs

1. Electroporation with Gene Pulser II (Bio-Rad).
2. Electroporation cuvettes (0.4 cm electrode) (Bio-Rad).
3. PBS + 2 mM EDTA.
4. X-Vivo 15 serum-free medium (Bio Whittaker, Belgium).
5. DC medium: RPMI1640 medium with 2 mM glutamine, 100 U/mL penicillin, 100 μg/mL streptomycin, 1 mM sodium pyruvate, 1 mM nonessential amino acids, and 10 % heat-inactivated FCS.
6. RNase-free pipette tips.
7. RNaseZap (Ambion).
8. 0.1 μg/mL 7-AAD (Invitrogen) solution in PBS.
9. IVT mRNA coding for enhanced green fluorescent protein (eGFP-RNA).
10. 4- or 6-color flow cytometry (FACS) system.

2.3 Effect of FLT3 Ligand on Cellularity and Activation of DCs

1. 10 μg FLT3 ligand per mouse.
2. 20 μg mRNA coding for Influenza Hemagglutinin-A (HA-RNA) per mouse.
3. Lymph node and spleen from FLT3 ligand- and mRNA-treated mice or control mice.
4. Digestion buffer for lymph nodes: 1 mg/mL collagenase D in PBS.
5. FACS-Antibodies: CD11c (BD, clone HL3); CD11b (BD, clone M1/70); PDCA1 (Miltenyi Biotech, clone JF05-1C2.4.1); Dec205 (Miltenyi Biotech, clone NLDC-145); CD4 (BD, clone GK1.5), CD8 (BD, clone 53-6.7) CD86 (BD, clone GL1);CD80 (BD, clone 1610-A1) CD40 (BD, clone HM40-3) and MHCII (BD, clone M5-114).

2.4 Effect of FLT3 Ligand on Luciferase Expression

1. FLT3 ligand.
2. IVT mRNA coding for firefly luciferase (luc-RNA).
3. Aqueous solution of D-luciferin (150 mg/kg body weight).
4. In vivo bioluminescence Imaging System.

2.5 Monitoring of Elicited Immune Responses

1. EDTA or heparin coated tube.
2. FACS tubes.

3. APC labeled MHC class I tetramer (iTAg *Tetramer*/APC—H-2 Kb OVA-SIINFEKL, MBL).
4. Lymphocyte-specific monoclonal antibody (FITC conjugated anti-mouse CD8a monoclonal antibody, clone 5H10, Invitrogen).
5. FACS lysis solution (10×) (BD Biosciences).
6. Erythrocyte lysis buffer (ELB): 1× FACS lysis solution made with dH_2O.
7. IVT mRNA coding for SIINFEKL epitope of chicken ovalbumin (SIINFEKL-RNA).
8. 4- or 6-color flow cytometry (FACS) system.

3 Methods

3.1 Generation and Characterization of FLT3 Ligand BMDCs

1. Prepare and count bone marrow cells.
2. Resuspend cells to 1×10^6/mL in BMDC medium and add 200 ng/mL FLT3 ligand.
3. Mix by gentle pipetting.
4. Add 4 mL cell suspension to each well of 6-well plates.
5. On day 2, aspirate 2 mL from each well, centrifuge and discard supernatant.
6. Resuspend again in BMDC medium (2 mL/well) and add double concentration of FLT3 ligand (400 ng/mL).
7. Add 2 mL cell suspension to each well (final concentration of FLT3 ligand should be 200 ng/mL).
8. On day 4, split each well and transfer 2 mL cell suspension into a separate well of a new 6-well plate.
9. Prepare BMDC medium with 200 ng/mL FLT3 ligand and add 2 mL cell suspension to each well (final concentration of FLT3 ligand should be 100 ng/mL).
10. On day 7, harvest, count, and analyze 1.5×10^6 cells by flow cytometry using CD11c, CD45RA, PDCA1, and 7-AAD for viability (Fig. 1, *see* **Note 4**).

3.2 mRNA Transfection into Murine FLT3 Ligand BMDCs

1. Clean all pipettes as well as working space with RNaseZAP to eliminate RNases.
2. Place electroporation cuvettes on ice.
3. Harvest FLT3 ligand BMDCs and centrifuge at $300 \times g$ for 8 min (4 °C).
4. Discard the supernatant and add 40 mL cold PBS with 2 mM EDTA. Resuspend the cells by pipetting up and down (*see* **Note 2**).

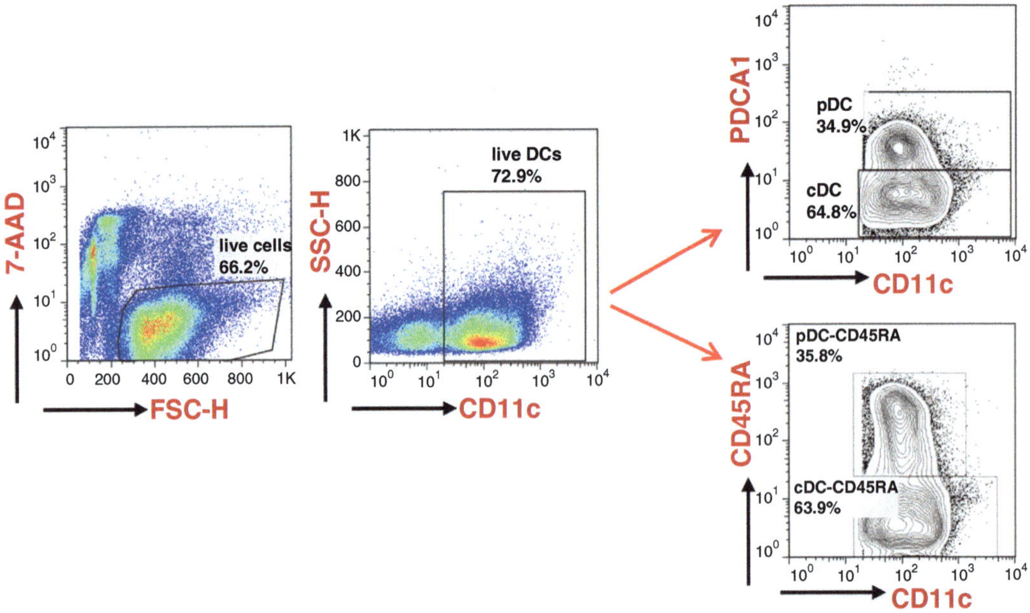

Fig. 1 Characterization of FLT3 ligand BMDCs using FACS. Murine bone marrow cells obtained from bone cavities of C57BL6/J mice were cultured for 7 days in medium supplemented with FLT3 ligand. At day 7 of culturing BMDCs are harvested and characterized for DC-subpopulations (conventional and plasmacytoid DCs) using flow cytometry

5. Centrifuge at 300 × *g* for 8 min (4 °C).
6. Discard the supernatant, add 40 mL cold X-Vivo 15 serum-free medium and resuspend the cells by pipetting up and down.
7. Count cells using the Neubauer chamber while centrifuging the cells.
8. Centrifuge at 300 × *g* for 8 min (4 °C).
9. Discard the supernatant and resuspend cells (1–2 × 10^6/ 250 μL) in cold X-Vivo 15.
10. Transfer 250 μL of cell suspension to electroporation cuvettes.
11. Add synthetic mRNA encoding the antigen of interest (10–20 μg) and mix around 10 s by pipetting up and down with a 100-μL pipette. Avoid air bubbles (*see* **Note 4**).
12. Electroporate the cells by using Gene Pulser II at 276 V and 150 μF (*see* **Note 5**).
13. Transfer the cells to 6-well plates with complete DC medium and incubate at 37 °C and 5 % CO_2.
14. Analyze transfection efficiency and mortality via FACS after 18–24 h (Fig. 2) after electroporation using CD11c and CD45RA antibodies (*see* **Notes 2** and **6**).

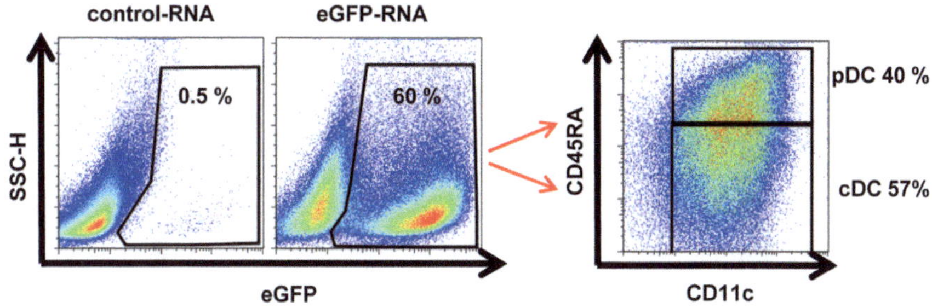

Fig. 2 Electroporation of FLT3 ligand BMDCs with synthetic eGFP-encoding mRNA. BMDCs were electroporated with synthetic eGFP mRNA (10 μg). Transfection efficiency and DC subpopulations (conventional and plasmacytoid DCs) were analyzed via FACS 24 h after electroporation

3.3 Effect of FLT3 Ligand on Cellularity and Activation of DCs

3.3.1 Injection of FLT3 Ligand

1. Formulate the synthetic FLT3 ligand (10 μg of FLT3 ligand for each mouse) in sterile PBS.
2. Store FLT3 ligand solution on ice.
3. Inject FLT3 ligand solution in 200 μL PBS intraperitoneally at day 0 and day 3.

3.3.2 Measuring In Vivo Effect of FLT3 Ligand on Cellularity and Expansion of DCs

1. Excise lymph node and spleen 4 days after the second injection of FLT3 ligand.
2. Homogenize tissues, prepare cell suspension and count cells (*see* **Note 7**).
3. Stain 1×10^6 cells with FACS-Antibodies: CD11c; CD11b; PDCA1; Dec205; CD4 and CD8.
4. Vortex gently (2–3 s) and incubate for 30 min in the dark at 4 °C.
5. Add 3 mL PBS and centrifuge at $450 \times g$ for 6 min.
6. Discard the supernatant.
7. Resuspend the cells in 200–300 μL of PBS and store at 4 °C until data acquisition via FACS and analyze with FlowJo (example provided in Fig. 3).

3.4 In Vivo Effect of RNA on Activation of DCs in Lymph Node After FLT3 Ligand Treatment

1. Inject mice with FLT3 ligand at day 0 and day 3 intraperitoneally.
2. Inject mice with 20 μg of IVT mRNA intranodally at day 4 (*see* **Note 8**).
3. After 24 h, dissect lymph node, homogenize, prepare a cell suspension and count cells.
4. Stain 1×10^6 cells with FACS-antibodies: CD11c; CD11b; PDCA1;CD86;CD80 CD40; MHCII.

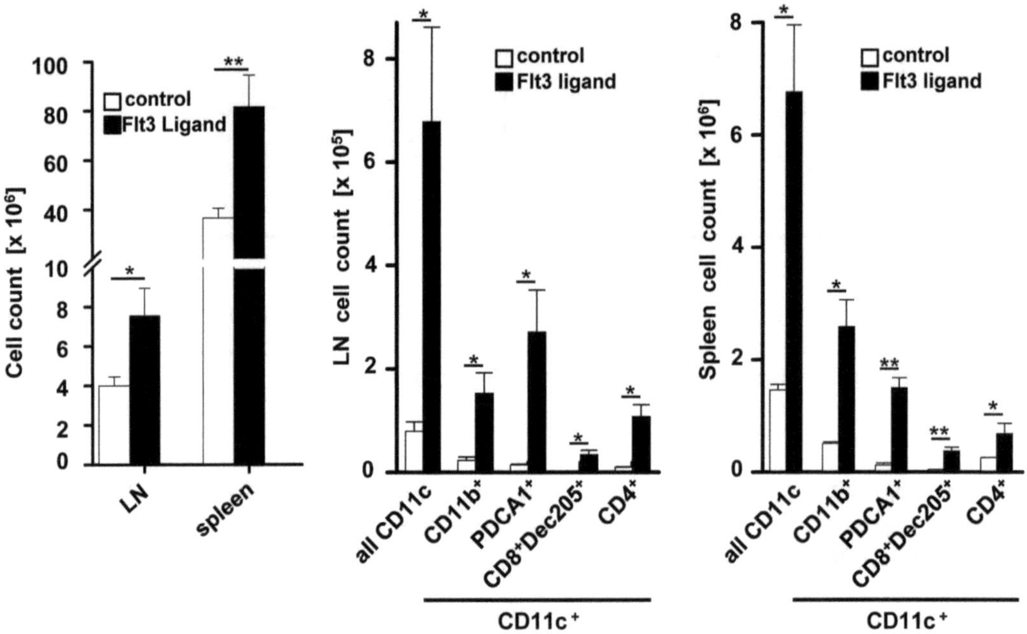

Fig. 3 Effect of FLT3 ligand on cellularity DCs subpopulations in lymph node and spleen. C57BL/6 J mice ($n = 3$–5) received two intraperitoneal injections of FLT3 ligand (10 μg; day 0 and day 3) or PBS (control) and were subjected to analysis on day 10. Cellularity was assessed in lymph node and spleen. DC subpopulations were quantified both in lymph node (*left*) and in spleen (*right*) (Reproduced from Kreiter, 2011 [18])

5. Vortex gently (2–3 s) and incubate for 30 min in the dark at 4 °C.
6. Add 3 mL PBS and centrifuge at 450 × *g* for 6 min.
7. Discard the supernatant.
8. Resuspend the cells in 200–300 μL of PBS and store at 4 °C until acquisition via FACS (example provided in Fig. 4).

3.5 Measuring In Vivo Luciferase Expression After FLT3 Ligand Treatment

1. Inject FLT3 ligand solution in 200 μL of PBS intraperitoneally at day 0 and day 3.
2. Inject 20 μg of IVT mRNA coding for luciferase (Luc-mRNA) intranodally (*see* **Note 8**) at day 10.
3. Intraperitoneal injection of an aqueous solution of D-luciferin (150 mg/kg body weight) 24 h after administration of Luc-mRNA.
4. After 5 min mice were subjected to in vivo imaging (integration time of 1 min).
5. In vivo bioluminescence in regions of interest (ROI) is quantified as total flux (photons per seconds) using IVIS Living Image 3.0 Software (example provided in Fig. 5).

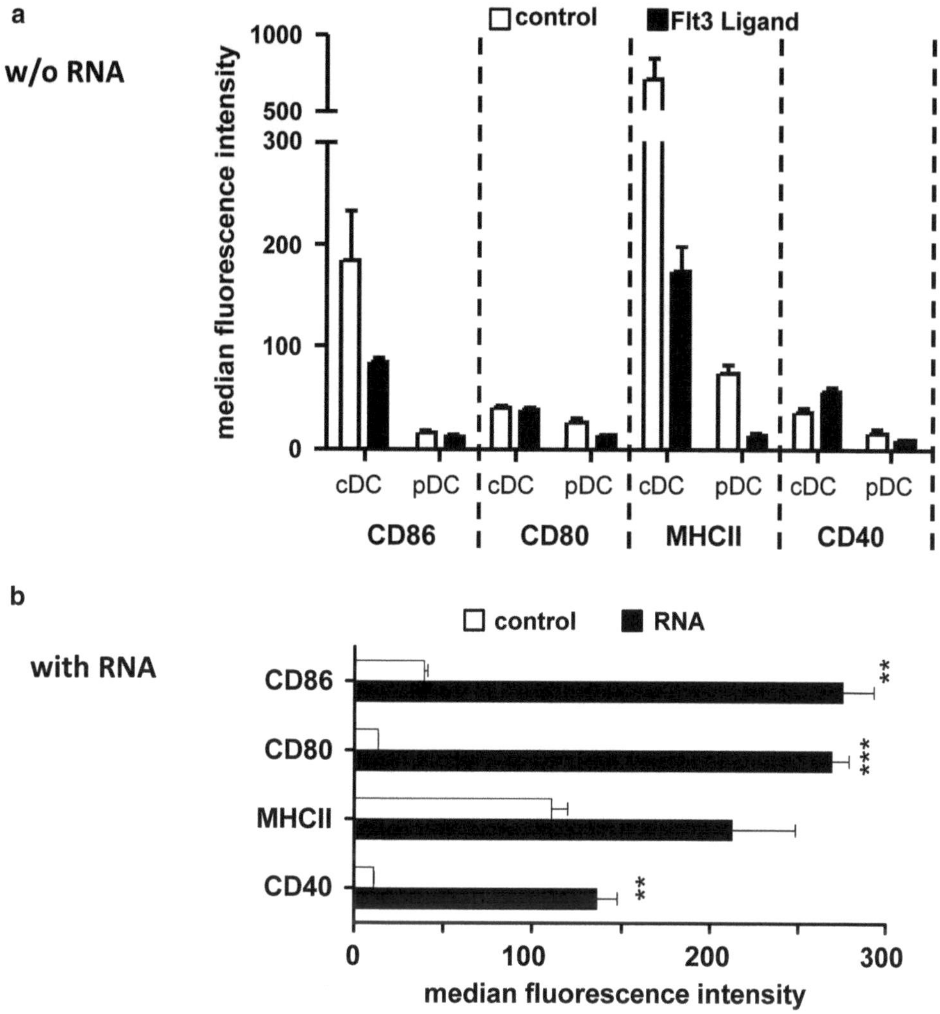

Fig. 4 Effect of FLT3 ligand on activation of DC subpopulations in LN. (**a**) C57BL/6 J mice ($n = 3$–5) received two intraperitoneal injections of FLT3 ligand (10 μg; day 0 and day 3) or PBS (control) and were subjected to analysis of the activation status of cDCs (CD11c+PDCA−) and pDCs (CD11c+PDCA+) by flow cytometry on day 10. (**b**) C57BL/6 J mice ($n = 3$–12) were treated with two cycles of FLT3 ligand (10 μg i.p., day 0 and day 3) followed by intranodal vaccination with influenza hemagglutinin encoding mRNA (20 μg; day 7) or control (in vitro transcription reaction mixture lacking RNA polymerase). Analysis was performed 24 h after mRNA injection (Reproduced from Kreiter, 2011 [18])

3.6 Monitoring of Elicited Cellular Immune Responses After RNA Vaccination

3.6.1 Blood

1. Collect 100–150 μL of blood into an EDTA or heparin coated tube.
2. Transfer 100 μL of blood to a clean FACS tube.
3. Add SIINFEKL-specific MHC tetramer and the anti-mouse CD8 monoclonal antibody (*see* **Note 2**).
4. Vortex gently (2–3 s) and incubate for 30 min in the dark at 4 °C.

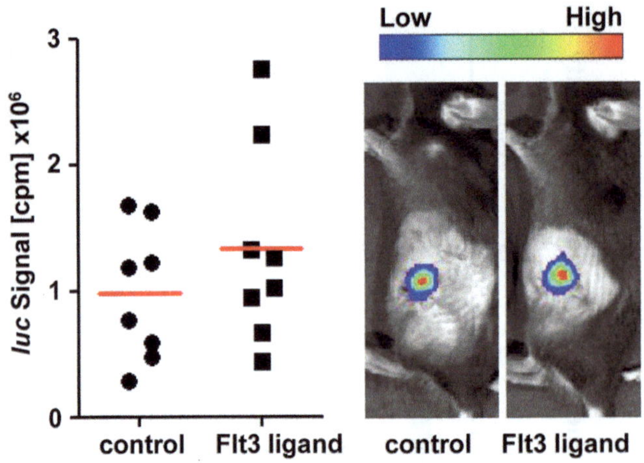

Fig. 5 Effect of FLT3 ligand on luciferase expression. Firefly luciferase encoding RNA (Luc-RNA) (20 μg) was injected on day 10 into lymph nodes of C57BL/6 J mice ($n = 8$) treated with FLT3 ligand (10 μg; day 0 and day 3) or PBS (control) and bioluminescence imaging (BLI) was performed after 24 h (Reproduced from Kreiter, 2011 [18])

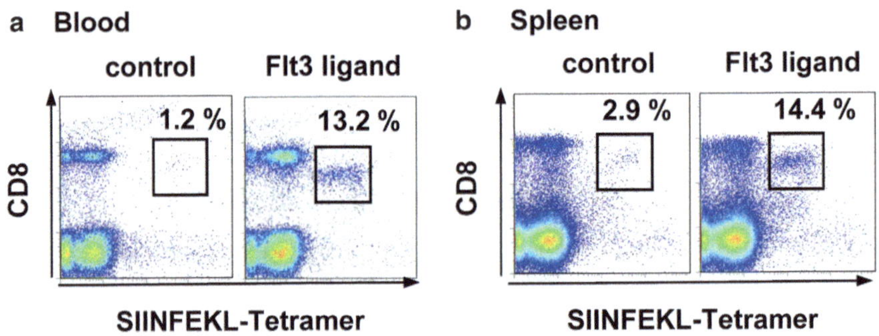

Fig. 6 Monitoring of immune responses elicited by intranodal mRNA immunization combined with FLT3 ligand using tetramer technology. C57BL/6 J mice ($n = 5$) received two intraperitoneal injections of FLT3 ligand (10 μg; day 0 and day 3) or PBS (control) followed by intranodal vaccination (day 7, day 10) with naked synthetic mRNA coding for the immunodominant epitope of chicken ovalbumin (SIINFEKL) (20 μg). SIINFEKL-specific $CD8^+$T cells in peripheral blood (**a**) and spleen (**b**) were quantified by tetramer staining on day 15 (Reproduced from Kreiter, 2011 [18])

5. Add 300 μL of FLB and vortex gently (2–3 s).
6. Incubate in the dark at room temperature for 5 min.
7. Add 3 mL of PBS and centrifuge at 450 × *g* for 6 min.
8. Discard the supernatant.
9. Repeat **steps 7** and **8**.
10. Resuspend the cells in 200–300 μL of PBS and store at 4 °C until data acquisition via FACS (example provided in Fig. 6a).

3.6.2 Spleen

1. Prepare spleen and make cell suspension (*see* **Note 7**).
2. Take 1 × 10^6 cells for FACS staining.
3. Add the SIINFEKL-specific MHC tetramer and the anti-mouse CD8 monoclonal antibody.
4. Vortex gently (2–3 s) and incubate for 30 min in the dark at 4 °C.
5. Add 3 mL of PBS and centrifuge at 450 × *g* for 6 min.
6. Discard the supernatant.
7. Repeat **steps 5** and **6**.
8. Resuspend the cells in 200–300 μL of PBS and store at 4 °C until data acquisition via FACS (example provided in Fig. 6b).

4 Notes

1. Bone marrow cells are obtained by removal of bone marrow from femurs and tibias and preparation of single suspension by cell strainer after lysis of erythrocytes.
2. Fluorophores and amounts of antibodies/tetramers may vary and require optimization.
3. During this study, an in-house produced fusion of FLT3 ligand was used [18]. Recombinant FLT3 ligand can also be used instead.
4. Avoid generation of air bubbles as they reduce transfection efficacy. It is important that **steps 1–10** should be done fast without delays. Otherwise BMDCs will have a reduced vitality compromising the efficacy of the whole procedure.
5. Electroporation parameters can vary based on the instrument used. It is recommended to determine optimal electroporation parameters in advance, which ensure high transfection efficiencies (50–80 %) and low mortality rate (10–15 %).
6. Synthetic mRNA coding for enhanced green fluorescent protein (eGFP) can be used to assess transfection efficiency. Mortality rate is determined via addition of propidium iodide or 7-AAD.
7. For preparation of cell suspension from lymph node, the lymph node is placed in 500 μL digestion buffer containing 1 mg/mL collagenase D in PBS, minced into small fragments with forceps and incubated for 10 min at 37 °C. Digested pieces are mashed against the surface of a 70-μm cell strainer using the plunger of a 3 mL syringe while rinsing with PBS. Filtered cells are washed once with 5 mL PBS and are resuspended in 1 mL of PBS or DC medium. For preparation of cell suspension from spleen, the spleen is mashed within a 70-μm cell strainer using the

plunger of a 3 mL syringe while rinsing with PBS. Filtered cells are washed once with 20 mL PBS, resuspended in 5 mL of lysing buffer and are incubated at room temperature for 5 min. After washing and centrifugation at 450 × *g* for 6 min cells are resuspended in 10 mL PBS.

8. The experimental procedure of intranodal Immunization with naked mRNA is published [28]. Briefly, mice are anesthetized and the inguinal lymph node is surgically exposed. 20 μg of IVT mRNA formulated in 10 μL RNase-free water (Ambion) is injected slowly and the wound is closed.

Acknowledgments

The authors would like to thank Marc Holzmann for helpful technical assistance.

References

1. Wolff JA et al (1990) Direct gene transfer into mouse muscle in vivo. Science 247:1465
2. Chattopadhyay S, Sen GC (2014) dsRNA-activation of TLR3 and RLR signaling: gene induction-dependent and independent effects. J Interferon Cytokine Res 34:427–36
3. Sahin U, Kariko K, Tureci O (2014) mRNA-based therapeutics–developing a new class of drugs. Nat Rev Drug Discov 13:759
4. Holtkamp S et al (2006) Modification of antigen-encoding RNA increases stability, translational efficacy, and T-cell stimulatory capacity of dendritic cells. Blood 108:4009
5. Kallen KJ, Thess A (2014) A development that may evolve into a revolution in medicine: mRNA as the basis for novel, nucleotide-based vaccines and drugs. Ther Adv Vaccines 2:10
6. Kariko K, Kuo A, Barnathan E (1999) Overexpression of urokinase receptor in mammalian cells following administration of the in vitro transcribed encoding mRNA. Gene Ther 6:1092
7. Kreiter S, Diken M, Selmi A, Tureci O, Sahin U (2011) Tumor vaccination using messenger RNA: prospects of a future therapy. Curr Opin Immunol 23:399
8. Diken M et al (2011) Selective uptake of naked vaccine RNA by dendritic cells is driven by macropinocytosis and abrogated upon DC maturation. Gene Ther 18:702
9. Boczkowski D, Nair SK, Snyder D, Gilboa E (1996) Dendritic cells pulsed with RNA are potent antigen-presenting cells in vitro and in vivo. J Exp Med 184:465
10. Van Lint S, Heirman C, Thielemans K, Breckpot K (2013) mRNA: From a chemical blueprint for protein production to an off-the-shelf therapeutic. Hum Vaccin Immunother 9:265
11. Kreiter S et al (2010) Intranodal vaccination with naked antigen-encoding RNA elicits potent prophylactic and therapeutic antitumoral immunity. Cancer Res 70:9031
12. Conry RM et al (1995) Characterization of a messenger RNA polynucleotide vaccine vector. Cancer Res 55:1397
13. Uchida S et al (2013) In vivo messenger RNA introduction into the central nervous system using polyplex nanomicelle. PLoS One 8, e56220
14. Hoerr I, Obst R, Rammensee HG, Jung G (2000) In vivo application of RNA leads to induction of specific cytotoxic T lymphocytes and antibodies. Eur J Immunol 30:1
15. Diken M et al (2013) mTOR inhibition improves antitumor effects of vaccination with antigen-encoding RNA. Cancer Immunol Res 1:386
16. Carralot JP et al (2004) Polarization of immunity induced by direct injection of naked sequence-stabilized mRNA vaccines. Cell Mol Life Sci 61:2418
17. MartIn-Fontecha A et al (2003) Regulation of dendritic cell migration to the draining lymph node: impact on T lymphocyte traffic and priming. J Exp Med 198:615
18. Kreiter S et al (2011) FLT3 ligand enhances the cancer therapeutic potency of naked RNA vaccines. Cancer Res 71:6132

19. Matthews W, Jordan CT, Wiegand GW, Pardoll D, Lemischka IR (1991) A receptor tyrosine kinase specific to hematopoietic stem and progenitor cell-enriched populations. Cell 65:1143
20. Gregory SH, Sagnimeni AJ, Zurowski NB, Thomson AW (2001) Flt3 ligand pretreatment promotes protective immunity to Listeria monocytogenes. Cytokine 13:202
21. Parajuli P et al (2001) Immunization with wild-type p53 gene sequences coadministered with Flt3 ligand induces an antigen-specific type 1 T-cell response. Cancer Res 61:8227
22. Merad M, Sugie T, Engleman EG, Fong L (2002) In vivo manipulation of dendritic cells to induce therapeutic immunity. Blood 99:1676
23. Fong L et al (2001) Altered peptide ligand vaccination with Flt3 ligand expanded dendritic cells for tumor immunotherapy. Proc Natl Acad Sci U S A 98:8809
24. Jefford M et al (2003) Functional comparison of DCs generated in vivo with Flt3 ligand or in vitro from blood monocytes: differential regulation of function by specific classes of physiologic stimuli. Blood 102:1753
25. Fong CL, Hui KM (2002) Generation of potent and specific cellular immune responses via in vivo stimulation of dendritic cells by pNGVL3-hFLex plasmid DNA and immunogenic peptides. Gene Ther 9:1127
26. Pulendran B et al (1998) Prevention of peripheral tolerance by a dendritic cell growth factor: flt3 ligand as an adjuvant. J Exp Med 188:2075
27. Sumida SM et al (2004) Recruitment and expansion of dendritic cells in vivo potentiate the immunogenicity of plasmid DNA vaccines. J Clin Invest 114:1334
28. Diken M, Kreiter S, Selmi A, Tureci O, Sahin U (2013) Antitumor vaccination with synthetic mRNA: strategies for in vitro and in vivo preclinical studies. Methods Mol Biol 969:235

Chapter 12

Transfecting Human Monocytes with RNA

Jens Dannull and Smita K. Nair

Abstract

Targeting monocytes as a delivery system for drugs or nucleic acids, and thereby harnessing their natural tissue-infiltrating capacity, has become an area of intense investigation in both basic and clinical research. Herein we describe an efficient method to deliver mRNA (messenger RNA) or siRNA (small interfering RNA) into human monocytes by electroporation. This method can be applied in the laboratory to monocytes isolated via magnetic bead-based techniques, or in a clinical setting using monocytes that were collected via counterflow centrifugation elutriation using the Elutra® Cell Separation System. We further demonstrate that electroporation of monocytes with RNA represents a robust and highly relevant approach to modify monocytes for cell-based therapies. Last, the procedure described can readily be adapted to monocytes from different species, hence facilitating research in animal models.

Key words Monocytes, Electroporation, RNA transfection, siRNA transfection, Elutriation

1 Introduction

Monocytes belong to the myeloid lineage and represent about 5 % of circulating leukocytes [1]. Generated in the bone marrow from granulocyte/macrophage progenitor cells [2], monocytes circulate in the blood before infiltrating into tissues, where they differentiate to macrophages or dendritic cells (DCs) depending on cues of the microenvironment [3–5]. In situ, monocytes are essential mediators of innate immunity and play a pivotal role in a range of diseases including infection, inflammation, cancer, atherosclerosis, and cerebral ischemia [6–8]. In addition, monocytes have been shown to penetrate the highly selective blood brain barrier [9]. As such, monocytes may represent prime targets for drug or nucleic acid delivery in cell-based therapeutic approaches.

We have recently shown that DCs that were derived from monocytes which had been electroporated with siRNAs targeting immune-proteasome subunits were superior stimulators of antitumor immunity in vitro [10] and in a Phase I clinical trial for patients with metastatic melanoma [11]. Electroporation of monocytes

Robert E. Rhoads (ed.), *Synthetic mRNA: Production, Introduction Into Cells, and Physiological Consequences*, Methods in Molecular Biology, vol. 1428, DOI 10.1007/978-1-4939-3625-0_12,

with siRNA or mRNA may have multiple applications. These include antigen-delivery or strategies to modulate or to program the immune-stimulatory or immune-inhibitory capacity of monocytes [12]. In contrast to alternative methods of gene delivery, such as DNA transfection, viral transduction or transfection with liposomes, electroporation with RNA is safe, highly efficient, and last but not least, comparably easy to achieve. Moreover, this method is superior to other techniques when delivering genetic information into non-dividing cells, since RNA does not have to enter the nucleus to be functional. Lastly, electroporation with RNA represents a safe and efficient means of gene delivery into human monocytes.

In this chapter we describe isolation of human monocytes from blood using magnetic-bead based approaches or counterflow centrifugation elutriation. Next we describe methods to introduce RNA into human monocytes and to monitor protein expression in monocytes by flow cytometry.

2 Materials

2.1 Solutions and Media

1. Flow cytometry (FACS) buffer: Dulbecco's phosphate-buffered saline (PBS) without calcium and magnesium, 2 % FBS (fetal bovine serum), 0.1 % sodium azide.
2. CellGro®/CellGenix™ GMP serum-free dendritic cell medium (Catalog number 20801-0500) (CellGenix USA, Portsmouth, NH).

2.2 *In Vitro* Transcribed RNA Preparation Components

1. pGEM4Z-GFP/A64 [13, 14].
2. Restriction endonuclease: SpeI.
3. DNA purification: QIAquick PCR purification kit (Catalog number 28104) (Qiagen, Valencia, CA).
4. RNA purification: RNeasy Mini kit (Catalog number 74104) (Qiagen).
5. 2-Mercaptoethanol.
6. In vitro transcription reagents: mMESSAGEmMACHINE® T7 kit (Catalog number AM1334) (Ambion, Austin, TX).

2.3 Denaturing Formaldehyde Agarose Gel Electrophoresis Components

1. 10× Formaldehyde gel buffer: 200 mM 3-[*N*-morpholino] propanesulfonic acid (MOPS) (free acid), 50 mM sodium acetate, 10 mM EDTA with pH adjusted to 7.0 using 1 M NaOH.
2. Agarose ultrapure.
3. Ethidium bromide solution 500 μg/mL.
4. Formaldehyde solution 37 %.

2.4 Magnetic Bead-Based Human Monocyte Isolation Supplies

1. Monocyte Isolation Kit II, human (Catalog number 130-091-153) (Miltenyi Biotec Inc., San Diego, CA).
2. MACS buffer: PBS, 0.5 % FBS, 2 mM EDTA (1 week life span at 4 °C).
3. LS columns (Catalog number 130-042-401) (Miltenyi Biotec Inc).
4. MidiMACS or VarioMACS magnet (Miltenyi Biotec Inc).

2.5 Counterflow Centrifugation Elutriation

1. Elutra® Cell Separation System (Terumo BCT, Lakewood, CO).
2. Elutriation Medium: HBSS (Hank's balanced salt solution) without calcium, magnesium, phenol red 4 L—bag.
3. Elutra® system disposable tubing set (Terumo BCT).
4. Transfer pack container with coupler—300 mL (Catalog number 4R2014) (Fenwal, Inc., Lake Zurich, IL).
5. Transfer pack container, 1000 mL (Catalog number 4R2032) (Fenwal, Inc).

2.6 Supplies for Electroporation

1. Electroporation cuvettes 4 mm gap (Model number 640) (BTX, Holliston, MA).
2. ECM 830 square wave electroporator (BTX).
3. MaxCyte electroporation buffer (Catalog number EBR100) (MaxCyte Inc., Gaithersburg, MD).
4. BUMINATE 25 %, albumin (human), USP, 25 % Solution (Baxter Healthcare Corporation, Westlake Village, CA).
5. siGlo (Catalog number, Di-001630-02-20) (Dharmacon Inc., Chicago, IL).

2.7 Antibodies and Reagents for Flow Cytometry (FACS) Analyses

1. Mouse anti-human CD14-PE (Catalog number 347497) (BD Biosciences, San Diego, CA).
2. Mouse anti-human CD15-PE (Catalog number 555402) (BD Biosciences).
3. Mouse IgM, κ Isotype Control (Catalog number PE-CF594) (BD Biosciences).
4. Mouse IgG_{2B} phycoerythrin (PE) isotype control (Catalog number IC0041P) (R&D Systems, Minneapolis, MN).
5. 7-Amino-actinomycin D (7-AAD) in 95 % deionized water, 5 % methanol.
6. Becton Dickinson FACSCalibur flow cytometer.

2.8 Cryopreservation of Monocytes

1. Nalgene Cryo 1 °C freezing container (Catalog number 5100-0001) (Thermo Scientific Nalgene, Lima, OH).
2. Isopropyl alcohol.

3. CryoStor CS10 (Catalog number, 210102) (BioLife Solutions®, Bothell, WA).
4. Cryogenic vials 2 mL.

3 Methods

3.1 Generation of In Vitro Transcribed mRNA

1. Digest 10 μg of the expression plasmid vector with the restriction enzyme SpeI.
2. Purify the linearized transcription vector using a QIAquick PCR purification kit following the manufacturer's instructions and using buffers that are included in the kit.
3. Add 5 volumes of Buffer PB (included in QIAquick PCR purification kit) to 1 volume of the restriction digest and mix.
4. Place a QIAquick spin column in a 2 mL collection tube.
5. To bind DNA, apply the sample to the QIAquick column and centrifuge at 10,000 × *g* for 30–60 s.
6. Discard flow-through. Place the QIAquick column back into the same tube.
7. To wash add 0.75 mL Buffer PE (included in QIAquick PCR purification kit) to the QIAquick column and centrifuge at 10,000 × *g* for 30–60 s.
8. Discard flow-through and place the QIAquick column back in the same tube. Centrifuge the column for 1 min at maximum speed.
9. Place QIAquick column in a clean 1.5 mL microcentrifuge tube.
10. To elute DNA, add 30 μL nuclease-free H_2O to the center of the QIAquick membrane and centrifuge at 10,000 × *g* for 1 min. Determine the quantity and purity of DNAs by reading optical density (OD) at wavelengths of 260 nm and 280 nm using a spectrophotometer (*see* **Note 1**).

3.2 In Vitro Transcription Reaction mMESSAGEm MACHINE® T7 kit

1. To 20 μL with nuclease-free water, 10 μL 2× NTP/CAP, 2 μL 10× reaction buffer, 1 μg linear template DNA, and 2 μL enzyme mix.
2. Incubate reaction mixture for 2 h at 37 °C.
3. Add 2 μL TURBO DNase, mix well and incubate at 37 °C for 15 min to digest template DNA.
4. Isolate mRNA exactly as outlined in the RNeasy Mini Handbook provided by Qiagen, pages 54–55, RNA clean up.
5. Determine the quantity and purity of mRNAs by reading optical density (OD) at wavelengths of 260 nm and 280 nm using a spectrophotometer (*see* **Note 1**).

6. Analyze integrity of RNA using denaturing formaldehyde agarose gel electrophoresis exactly as described in the RNeasy® Mini Handbook, pages 65–66. Formamide loading buffer for electrophoresis is provided in the mMESSAGEmMACHINE® T7 kit.

3.3 Magnetic Bead-Based Isolation of Human Monocytes

1. PBMCs (peripheral blood mononuclear cells) can be isolated from buffy coats or apheresis product (leukapheresis) [15] by Ficoll-Hypaque density gradient (density 1.077 g/mL) centrifugation (*see* **Note 2**).
2. Use Monocyte Isolation Kit II for isolation of 'untouched' monocytes (*see* **Note 3**).
3. Resuspend 10^8 PBMCs in 300 μL of MACS buffer (*see* **Note 4**).
4. Add 100 μL FcR blocking reagent.
5. Add 100 μL of biotin-antibody cocktail.
6. Mix well and incubate for 10 min at 4 °C.
7. Add 300 μL of MACS buffer.
8. Add 200 μL of anti-biotin MicroBeads.
9. Mix well and incubate for an additional 15 min at 4 °C.
10. Wash cells with buffer by adding 1–2 mL of buffer and centrifuge at 300 × *g* for 10 min.
11. Resuspend cells in 500 μL of MACS buffer.
12. Place LS column in the magnetic field of the MACS separator.
13. Prepare column by rinsing with 3 mL MACS buffer.
14. Apply cell suspension to LS column.
15. Allow the cells to pass through and collect effluent as fraction with unlabeled cells, representing the enriched monocyte fraction.
16. Wash column with 3 mL of MACS buffer and collect the entire effluent in the same tube as effluent of step 15.

3.4 Electroporation of Monocytes with mRNA or siRNA

1. Harvest monocytes by centrifugation (300 × *g*) for 10 min and resuspend in electroporation buffer supplemented with 0.1 % human serum albumin at 5×10^7 monocytes/mL (*see* **Note 5**).
2. Transfer up to 500 μL cell suspension into a 4 mm electroporation cuvette and keep cells on ice. Add 5 μg of mRNA/10^6 cells or add siRNA at a final concentration of 500 nM (*see* **Note 6**).
3. Gently tap the cuvette one to two times to mix the RNA and cells.
4. Electroporate monocytes at 500 V and 500 μsec (1,250 V/cm). Using the transfer pipette provided along with the electroporation cuvette, transfer electroporated cells into a 50 mL

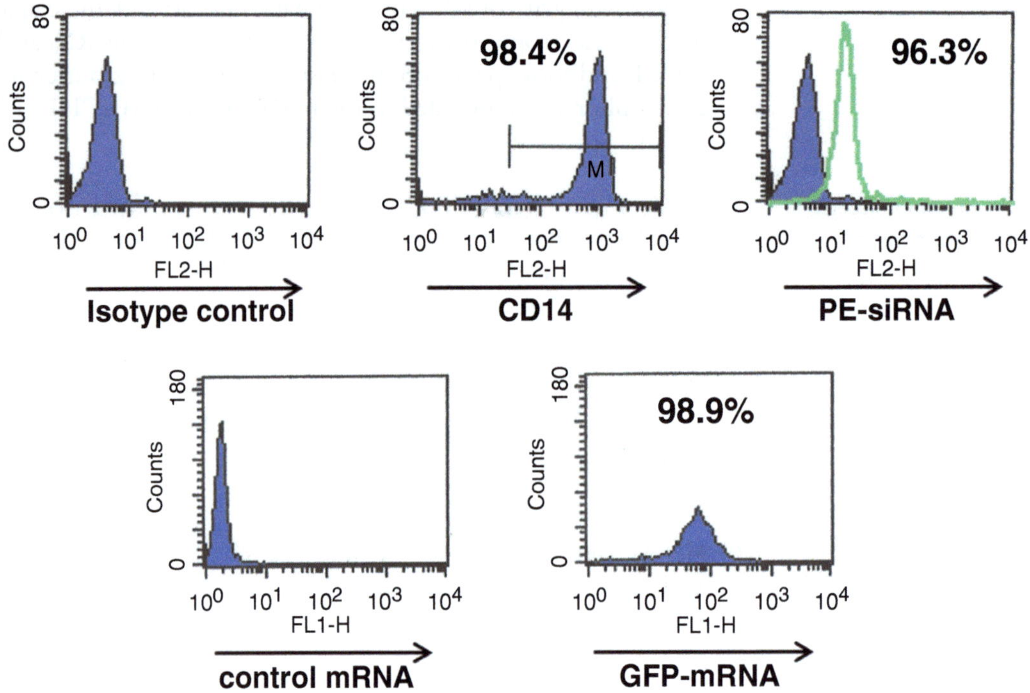

Fig. 1 (*Top panel*) Human monocytes were isolated via magnetic bead-based negative selection and their purity was determined by flow cytometry (FACS) after staining with PE-labeled anti-CD14 monoclonal antibody (*center*). The histogram on the left shows staining with an isotype-matched control antibody. The purity of monocytes was 98.4 % according to FACS analysis. Next, monocytes were either co-incubated with PE-siRNA (*top right, purple histogram*) or electroporated with PE-siRNA (*top right, green histogram*) under optimized electroporation conditions. Cells were electroporated in MaxCyte electroporation buffer supplemented with 0.1 % human serum albumin at 5×10^7 cells/mL using a square wave electroporator (ECM 830, BTX) at 1250 V/cm for 500 μs. The concentration of siRNA was 500 nM. (*Bottom panel*) Monocytes were electroporated with control RNA (*left histogram*) or GFP mRNA (right histogram) as described. 5 μg of RNA was used per 10^6 monocytes. Cells were analyzed 4 h after electroporation by flow cytometry

polypropylene tube containing 20 mL CellGenix medium at room temperature.

5. Centrifuge at 300 × *g* for 10 min and resuspend monocytes at 10^6 cells/mL in CellGenix medium and incubate at 37 °C, 5 % CO_2 (Cells for FACS analyses can be kept in 50 mL tubes and may be incubated at 37 °C, 5 % CO_2 for 4 h). Typical transfection efficacies and the purity of isolated monocytes are shown in Fig. 1.

3.5 Phenotype of Monocytes Using Flow Cytometry

1. Resuspend monocytes at 10^6 cells/mL in ice-cold FACS buffer.
2. Add 100 μL of the cell suspension to round bottom 12 × 75 mm polystyrene tubes.

3. Add the appropriate amount of fluorochrome-labeled monoclonal antibody (refer to antibody data sheet) directly into the cell suspension.
4. Incubate for 20–30 min at 4 °C (or on ice) in the dark.
5. Wash once with 3 mL of FACS buffer, centrifuge at 300 × *g* for 5 min. Resuspend cells in 200 μL FACS buffer. Maintain the cells at 4 °C (or on ice) prior to acquisition and analysis on the flow cytometer.
6. If cells will be maintained for over 2 h and to allow time flexibility for analysis of samples on the flow cytometer the cells should be fixed with 1 % formaldehyde in FACS buffer. Fixed cells can be kept at 4 °C in the dark for up to 2 weeks without significant loss of intensity of fluorescence.

3.6 7-AAD Staining to Assess Cell Viability

1. Dissolve 1 mg 7-AAD in 50 μL absolute methanol and add 950 μL PBS (stock solution, keep at 4 °C in the dark).
2. Prepare staining solution by adding 5 μL stock solution to 1 mL PBS (5 μg/mL 7-AAD).
3. Add 1 mL staining solution to cells and keep samples at 4 °C protected from light for 20 min or until analysis on a flow cytometer.
4. If cells will be maintained for over 1 h, cells should be fixed with 1 % formaldehyde in PBS containing 5 μg/mL 7-AAD. Fixed cells can be stored for up to 2 days without loss in the ability to discriminate dead from live cells.

3.7 Cryopreservation of Monocytes

1. Pellet monocytes by centrifugation at 300 × *g* for 10 min.
2. Resuspend cells in ice-cold Cryostor CS10 solution at 10^7 monocytes/mL. Keep cells on ice and pipette into ice-cold cryopreservation vials (1–2 mL).
3. Transfer vials into precooled Nalgene Cryo 1 °C freezing container. Make sure that the container is filled with isopropyl alcohol according to the manual provided by the manufacturer.
4. Place container into −80 °C freezer overnight.
5. Transfer vials to liquid nitrogen tank for long-term storage.

3.8 Isolation of Human Monocytes by Elutriation

1. Collect mononuclear cells during a 3 h apheresis run on a COBE® Spectra Apheresis System (*see* **Note 2**). Request a 0.5–1 % hematocrit collection. The yield of white blood cells is expected to be around 10^{10}. The capacity of the Elutra® tubing set is 5×10^9 to 2.0×10^{10} mononuclear cells (*see* **Note 7**). The volume of the apheresis product should be between 400 and 500 mL.
2. Configure the Elutra® Cell Separation System profile to 4000 mL/60 min.

3. Set media flow ramp to 0.2 mL/s.
4. Centrifuge ramp 528 rpm/min.
5. Cell to media ratio 1:0.5.
6. Load apheresis product.
7. Chamber rotation speed is maintained at 2400 rpm for fractions 1 through 4.
8. Media flow rate was maintained at 37 mL/min for fraction 1, 97.5 mL/min for fraction 2, 103.4 mL/min for fraction 3 and 103.9 mL/min for fraction 4. Collect 900–1000 mL for each fraction (*see* **Note 8**).
9. Fraction 5 contains remaining monocytes in the chamber and they are collected with the rotor turned off at a flow rate of 103.9 mL/min in a total volume of 250 mL (*see* **Note 9**).
10. Electroporation, phenotypic analysis and cryopreservation of elutriated monocytes are performed exactly as described in Subheadings 3.3–3.6. A representative white blood cell differential, phenotype of elutriated monocytes, and transfection efficacy (as determined by FACS) is presented in Fig. 2.

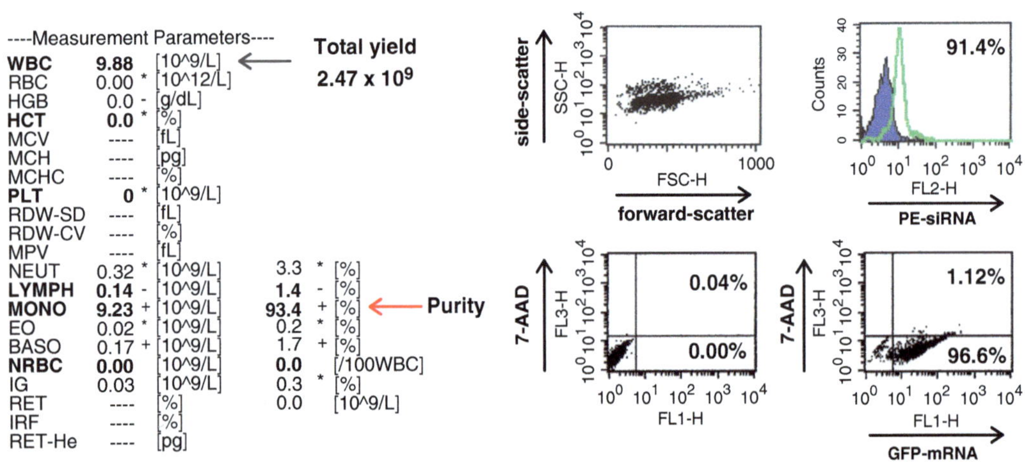

Fig. 2 Monocytes were elutriated from a 3-h leukapheresis obtained that was obtained from a healthy donor. (*Left panel*) Purity of isolated monocytes was assessed by a clinical lab. WBC differential revealed a total yield of 2.47×10^9 cells with 93.4 % of cells representing monocytes. Presence of low numbers of granulocytes 5.5 % (neutrophils, basophils, and immature granulocytes (IG)) and lymphocytes (1.4 %) was detected. (*Right panel*) A forward- side-scatter profile of elutriated monocytes is shown (*top left*). Monocytes were electroporated with PE-siRNA as described (*top right*). The purple histogram shows monocytes co-incubated with PE-siRNA, while the green histogram represents cells electroporated with PE-siRNA. (*Right panel, bottom*). Monocytes were transfected with GFP-mRNA and stained with 7-AAD 8 h after electroporation (*right dot plot*). Non-electroporated monocytes are shown for comparison (*left dot plot*)

4 Notes

1. For single-stranded RNA, 1 A_{260} unit corresponds to 40 μg/mL. The RNA yield can be calculated as follows: $A_{260} \times$ dilution factor $\times 40 =$ μg/mL RNA. For double-stranded DNA, 1 A_{260} unit corresponds to 50 μg/mL. Pure preparations of DNA and RNA have OD_{260}/OD_{280} values of 1.8 or higher. A single in vitro transcription reaction produces 20–30 μg of RNA from 1 μg DNA template.
2. Appropriate precautions must be taken when handling human cells, particularly if from unknown, untested donors. All work must strictly adhere to the Institutional guidelines and all protocols required for such work must be approved before any work with human cells is initiated. Biosafety practices must be followed. All procedures use sterile tissue culture techniques with sterile solutions and equipment in biosafety containment hoods.
3. It is pivotal to isolate 'untouched' monocytes. Electroporation of positively selected monocytes will result in significantly reduced cell viability and transfection efficacy most likely due to cell-bound magnetic beads or co-purified iron compounds.
4. The binding capacity of an LS column is 10^8 cells. The isolation procedure may be scaled up ad libitum.
5. MaxCyte electroporation buffer can be substituted with CellGenix medium or Opti-MEM medium (Invitrogen). However, Opti-MEM medium **cannot** be used when generating monocytes for clinical applications.
6. The amount of mRNA can be increased up to 10 μg/10^6 monocytes, resulting in slightly higher levels of gene expression. If more than one mRNA is used for electroporation adjust the cell density accordingly. Moreover, add RNAs in isotonic solution or adjust with 10× PBS, if necessary.
7. The run kit for the Elutra® is manufactured under good manufacturing practice (GMP) conditions of certified materials and is sterile. A master file is held by Caridian BCT and is available for cross-reference for qualified investigational new drug (IND) applications after Caridian BCT's training of the operating personnel and IQ and OQ (installation and operational qualification) of the device.
8. Lymphocytes can be harvested from Fraction 2 and can be cryopreserved if needed for subsequent immune assays.
9. If the purity of elutriated monocytes is not satisfactory due to the presence of granulocytes, monocytes can be further processed using Ficoll-Hypaque density gradient (density 1.077 g/mL) centrifugation to separate them from granulocytes [16].

References

1. Nichols BA, Bainton DF, Farquhar MG (1971) Differentiation of monocytes. Origin, nature, and fate of their azurophil granules. J Cell Biol 50(2):498–515
2. Akashi K, Traver D, Miyamoto T, Weissman IL (2000) A clonogenic common myeloid progenitor that gives rise to all myeloid lineages. Nature 404(6774):193–197. doi:10.1038/35004599
3. Benoit M, Desnues B, Mege JL (2008) Macrophage polarization in bacterial infections. J Immunol 181(6):3733–3739
4. Mosser DM, Edwards JP (2008) Exploring the full spectrum of macrophage activation. Nat Rev Immunol 8(12):958–969. doi:10.1038/nri2448
5. Varol C, Yona S, Jung S (2009) Origins and tissue-context-dependent fates of blood monocytes. Immunol Cell Biol 87(1):30–38. doi:10.1038/icb.2008.90
6. Ziegler-Heitbrock L (2007) The CD14+ CD16+ blood monocytes: their role in infection and inflammation. J Leukoc Biol 81 (3):584–592. doi:10.1189/jlb.0806510
7. Ghattas A, Griffiths HR, Devitt A, Lip GY, Shantsila E (2013) Monocytes in coronary artery disease and atherosclerosis: where are we now? J Am Coll Cardiol 62(17):1541–1551. doi:10.1016/j.jacc.2013.07.043
8. Eue I (2001) Growth inhibition of human mammary carcinoma by liposomal hexadecylphosphocholine: Participation of activated macrophages in the antitumor mechanism. Int J Cancer 92(3):426–433
9. Tanaka S, Kitagawa K, Sugiura S, Matsuoka-Omura E, Sasaki T, Yagita Y, Hori M (2004) Infiltrating macrophages as in vivo targets for intravenous gene delivery in cerebral infarction. Stroke 35(8):1968–1973. doi:10.1161/01.STR.0000133685.59556.a7
10. Dannull J, Lesher DT, Holzknecht R, Qi W, Hanna G, Seigler H, Tyler DS, Pruitt SK (2007) Immunoproteasome down-modulation enhances the ability of dendritic cells to stimulate antitumor immunity. Blood 110(13):4341–4350
11. Dannull J, Haley NR, Archer G, Nair S, Boczkowski D, Harper M, De Rosa N, Pickett N, Mosca PJ, Burchette J, Selim MA, Mitchell DA, Sampson J, Tyler DS, Pruitt SK (2013) Melanoma immunotherapy using mature DCs expressing the constitutive proteasome. J Clin Invest 123(7):3135–3145. doi:10.1172/JCI67544
12. Kelly C, Jefferies C, Cryan SA (2011) Targeted liposomal drug delivery to monocytes and macrophages. J Drug Deliv 2011:727241. doi:10.1155/2011/727241
13. Boczkowski D, Nair SK, Nam JH, Lyerly HK, Gilboa E (2000) Induction of tumor immunity and cytotoxic T lymphocyte responses using dendritic cells transfected with messenger RNA amplified from tumor cells. Cancer Res 60(4):1028–1034
14. Lee J, Boczkowski D, Nair S (2013) Engineering B cells with mRNA. Methods Mol Biol 969:101–110. doi:10.1007/978-1-62703-260-5_7
15. Nair S, Archer GE, Tedder TF (2012) Isolation and generation of human dendritic cells. Curr Protoc Immunol Chapter 7:Unit7 32. doi:10.1002/0471142735.im0732s99
16. Stroncek DF, Fellowes V, Pham C, Khuu H, Fowler DH, Wood LV, Sabatino M (2014) Counter-flow elutriation of clinical peripheral blood mononuclear cell concentrates for the production of dendritic and T cell therapies. J Transl Med 12:241. doi:10.1186/s12967-014-0241-y

Chapter 13

In Vitro Synthesis, Delivery, and Bioavailability of Exogenous mRNA in Gene Transfer Mediated by *PiggyBac* Transposition

Solenne Bire, Nicole Ishac, and Florence Rouleux-Bonnin

Abstract

Nowadays, nonviral gene transfer is currently of great importance for introducing exogenous genes into genomes and for ensuring that transgene expression is suitable for therapeutic and bioproduction purposes. The *piggyBac* transposon-based system is particularly interesting since it is easy to engineer and has a large cargo capacity, up to 100 kb. In its setup, the system requires only the *piggyBac* transposase protein and the transgene delineated by the two *piggyBac*-specific inverted terminal repeats. Usually the source of transposase is carried by a DNA plasmid. However, the principal drawback of this method is the lasting presence of the transposase, due to episomal persistence or possible integration of the transposase gene vector into the cell's genome. This can lead to genotoxic effects such as multiple genomic integration events and remobilization of the transposon vector once it has been integrated. One alternative to improve the safety of the system is to deliver the transposase as in vitro-synthesized messenger RNA in order to define a very narrow expression window during which a one-shot transposition process would occur. Issues that can be encountered when working on mRNA cell transfer are related to the quality of the synthetic mRNA, the system used to introduce mRNA into the cells and the bioavailability of the mRNA molecules. Here we describe a method to produce mRNA, verify its quality, determine which transfecting reagents can be used and how this mRNA is available to promote the transposition process in HeLa cells. Additionally, we illustrate this method in stromal mesenchymal cell lines in order to support hematopoiesis.

Key words Messenger RNA, In vitro transcription, ARCA, poly(A) tailing, mRNA transfection, Half-life, Bioavailability, Transposition, mRNA trafficking

1 Introduction

Secure strategies to integrate a therapeutic gene into genomes are a major challenge for treatment of many diseases. Nonviral transposon-based systems are particularly well developed for this purpose. They are considered to be less immunogenic and to have a much larger cargo capacity than viral vectors while maintaining highly efficient transgene integration. To date, few transposons have demonstrated their functionality in mammalian cells and

Robert E. Rhoads (ed.), *Synthetic mRNA: Production, Introduction Into Cells, and Physiological Consequences*, Methods in Molecular Biology, vol. 1428, DOI 10.1007/978-1-4939-3625-0_13,

their usefulness for reshaping the genome [1]. One of them is the *piggyBac* transposon [2]. This transposable element is mobile in many different species, including human cells [3]. The basic transposon setup machinery requires two components: a plasmid DNA (pDNA) carrying the gene of interest, and a source of transposase. The architecture of pDNA has been designed to ensure site-directed integration of the transgene and long-term expression through the use of an insulator [4, 5]. But less has been characterized about the source of the transposase that is usually brought as pDNA carrying the transposase cDNA. As we showed previously, the principal drawback encountered using this strategy is the lasting presence of the transposase due to persistence of the episomal pDNA and/or the possible nonspecific integration of the transposase gene into the genome. This could lead to genotoxicity since the transgene could be remobilized once it has been integrated. In the context of gene transfer, integrating stably one copy of a therapeutic transgene per nuclear genome appears to be necessary.

In an attempt to improve the biosafety of gene integration, we chose to deliver the source of transposase as a messenger RNA produced in vitro as a single-strand polymer [6]. We made this choice because mRNAs are labile molecules and are unable to penetrate the nuclear envelope to be integrated in the genome. They are easy to produce and to improve. Consequently, exogenous mRNA is of particular relevance for engineering secure transposon systems with limited transposase expression.

Despite the increasing use of exogenous mRNA, few transfection reagents have been developed for large RNA. Cationic lipids are the most often used [7] since other reagents, including cationic peptides [8] or cationic polymers [9], have poor efficiency in delivering mRNA. We were interested in assaying different commercial chemical reagents for mRNA delivery. At the same time, we were interested in investigating differences in the uptake and intracellular trafficking of exogenous mRNA. This would determine the bioavailability of molecules. Whereas numerous studies have investigated uptake pathways and subsequent intracellular trafficking of pDNA [10–14], few have investigated the intracellular fate of exogenous mRNA. However, exact knowledge of spatial and temporal control of mRNA bioavailability is crucial to achieve improved biosafety of transposon vectors, which is a prerequisite for secure ex vivo gene and cell therapies, but it will also provide the basis for improving transfection efficiency and for designing novel devices for the cytosolic delivery of mRNA, and be beneficial for the development of many other emerging mRNA-based therapies.

Our studies have focused on the pharmacokinetics of the transposition process mediated by the *piggyBac* transposase mRNA transfected in HeLa cells. mRNAs bioavailability was checked both spatially and temporally to understand exogenous mRNAs uptake through clathrin or/and caveolae pathways. Their subsequent intracellular trafficking to endosomes or cytoplasm has been studied and especially their potential storage in Processing Bodies (P bodies) or Stress Granules (SG). P bodies and SG are cytoplasmic foci involved in natural, endogenous mRNA regulation. These aggregates can interact and constitute dynamic sorting sites for storage and/or turnover of endogenous mRNA [15], where intracellular mRNA is routed to sites of translation reinitiation, degradation or storage [16]. We were interested in knowing whether exogenous mRNA would be free in the cytoplasm and not stored in these granules since translationally repressed mRNA can also exit SG and P bodies to reengage in translation [17]. Therefore, the transposition process would occur in bimodal manner, which is not desirable.

Our purpose was to design exogenous mRNA leading to transposase protein production during a very narrow window of expression to promote a one-shot transposition process. This mRNA contained a number of features that would promote high translational efficiency and minimal involvement with P bodies. (1) We "humanized" the codon usage for high expression. (2) We modified the codon usage to limit formation of mRNA duplex structures that could be processed into small RNAs. Indeed, such short RNAs could induce translation repression but also degradation [18–20]. (3) We removed ARE sequences in order to eliminate P bodies addressing ability [21] and produce mRNA "resistant" to P bodies [15]. (4) We had the 5′ and the 3′ UTRs of the β-globin of *Xenopus laevis* whose mRNA naturally shows a half-life of 10 h. Other UTRs could be used to modulate the mRNA half life [22]. (5) We decided to cap and to polyadenylate our exogenous mRNA to improve its expression and stability [23, 24].

Additionally, we illustrated this method in two stromal mesenchymal cell lines in order to support hematopoiesis [25, 26]. Mesenchymal stromal cells are believed to be an essential component of the hematopoietic microenvironment [26]. They regulate the fate of hematopoietic cells via the expression of adherence molecules, cytokines, growth factors, and other factors [25]. In a previous study, we have shown that HS-27a and HS-5 cells represent an attracting model of mesenchymal stromal cells that can be used to study the hematopoietic niche complexity (Ishac et al., in preparation). Here, we used the *piggyBac* transposon system on HS-27a and HS-5 cells to confirm the overall process on cells of therapeutic relevance.

In this chapter, we review the different procedures for mRNA in vitro synthesis including capping and poly(A) tailing. We also describe the transfection of mRNA using different chemical reagents as well as the co-transfection of DNA and mRNA in various mammalian cells. We use in vitro translation and qRT-PCR to assess the functionality and half-life of our in vitro synthesized mRNA. We follow the uptake and subsequent trafficking of mRNA in cells by labeling endocytosis and storage compartments. Finally, we describe gene transfer mediated by *piggyBac* transposition using an mRNA source of transposase in epithelial and mesenchymal stromal cells.

2 Materials

All equipment and reagents must be RNase-free. Surfaces and materials can be treated with RNaseZap® (Cat. Number AM9780, Ambion/Life technologies, Waltham, MA, USA) and solutions can be treated with DEPC to remove RNases contaminations. Gloves should be worn at all times during in vitro transcription of mRNA. Solutions and reagents are kept at −20 °C unless otherwise indicated. Carry out all procedures at room temperature unless otherwise specified.

2.1 Plasmid Vectors

Plasmid vectors containing the genes of interest were constructed as previously described [6] (*see* **Note 1**).

1. Helper plasmid V5PB pDNA: pCS2+ U5V5PBU3 (6.4 kb) plasmid encodes the *piggyBac* transposase fused with the V5 epitope tag (GKPIPNPLLGLDST) surrounded with the 5′ and 3′ UTR (named U5 and U3) of the *Xenopus laevis* β-globin gene.
2. Control plasmid GFP pDNA: pCS2+ U5GFPU3 (5.1 kb) vector encodes the green fluorescent protein. This construct served as a negative control (no transposase).
3. Donor plasmid Neo pDNA: pBSK ITR5′-Neo-ITR3′ (5.2 kbp) was generated by introducing the *piggyBac* 5′ and 3′ Inverted Terminal Repeats (ITRs, 717 bp) into the pBluescript SK− plasmid. The neomycin phosphotransferase gene (1515 bp) under control of the SV40 promoter was then inserted between the ITRs.

2.2 Checking of DNA and RNA Quality

1. Agarose gel electrophoresis system with 7 × 10 cm tray and 8-well 0.75 mm thick combs or any equivalent system.
2. Agarose for molecular biological applications. Store at room temperature (RT).
3. Tris-Acetate-EDTA (1×). Store at RT.
4. Molecular weight marker (i.e., lambda DNA Hind III or single strand RNA ladder).
5. Ethidium bromide solution (10 mg/mL). Store at RT.
6. RNA gel loading dye (2×).
7. DNA gel loading dye (6×).
8. Gel imaging system with ultraviolet transilluminator.

2.3 Preparation of Plasmid DNA Templates

1. Helper plasmid (containing the transposase cassette) or control plasmid (*see* Subheading 2.1).
2. NotI restriction enzyme (*see* **Note 2**).
3. Proteinase K solution (10 mg/mL).
4. Sodium Dodecyl Sulfate (20 %). Store at RT.
5. Phenol–chloroform–isoamyl alcohol (25:24:1), pH 8.0. Store at 4 °C.
6. Chloroform. Store at RT.
7. Ethanol 100 % and 70 %. Store at RT.
8. Sodium acetate (3 M). Store at RT.
9. RNase-free water.

2.4 In Vitro Synthesis of ARCA Capped and Poly(A) Tailed mRNA

1. MEGA Script SP6 kit or any equivalent transcription system (*see* **Note 3**).
2. ARCA (Anti-Reverse Cap Analog) kit (*see* **Note 4**).
3. ChromaTide® Alexa Fluor® 546-14-UTP (red fluorescence, λEx: 550 nm, λEm: 570 nm, Life Technologies) for synthesis of fluorescent mRNA.
4. Poly(A) tailing kit.
5. DNase- and RNase-free ultrapure water. Store at RT.
6. Lithium chloride precipitation solution: 7.5 M LiCl, 50 mM EDTA.
7. NanoDrop 2000 spectrophotometer or similar UV spectrophotometer for determination of RNA and DNA concentrations.

2.5 In Vitro Translation of In Vitro-Synthesized mRNA

1. Rabbit reticulocyte lysate, nuclease treated.
2. Ribonuclease inhibitor (40 μ/μL).
3. RNase-free water.

2.6 Cell Culture

1. Hela cells (ATCC CCL-2) and HS-27a and HS-5 Human stromal cell lines are maintained in stock culture using standard cell culture techniques (*see* **Note 5**).
2. Medium for HeLa cells: supplemented Dulbecco's Modified Eagle's Medium (DMEM, 10 % fetal bovine serum, 10 mL/L penicillin–streptomycin mix). Store at 4 °C.
3. Medium for HS-27a and HS-5 cells: supplemented RPMI 1640 culture medium (RPMI, 10 % fetal bovine serum, 10 mL/L penicillin-streptomycin mix). Store at 4 °C.
4. Trypsin–EDTA (1×). Store at 4 °C.
5. Phosphate buffer saline (PBS 1×).
6. Cell culture-treated flasks 75 cm^2 and 24-well plates.

2.7 Nucleic Acid Transfection

1. Transfection reagents for DNA: jetPEI™ (PolyPlus Transfection); Lipofectamine® 2000 (Invitrogen).
2. Transfection reagents for mRNA: TransMessenger™ (Qiagen); *Trans*IT®-mRNA reagent (Mirus Bio); jetPEI™ (PolyPlus Transfection); Lipofectamine® 2000 (Invitrogen).
3. Sterile NaCl 150 mM solution.
4. Opti-MEM® Reduced serum medium (*see* **Note 6**).
5. RNase-free water.

2.8 Protein Extraction and Western Blot

1. Lysis buffer: 62.5 mM Tris–HCl, pH 6.8, 2 % SDS, 10 % glycerol, 50 mM dithiothreitol; bromophenol blue (*see* **Note 7**).
2. Sonicator.
3. 4–10 % gradient SDS-PAGE gel.
4. Nitrocellulose membrane.
5. 5 % nonfat dry milk.
6. PBS 1×–Tween 20 (0.05 %).
7. Mouse anti-V5 HRP conjugated antibody.
8. Chicken polyclonal ß-Actin antibody and Goat Anti-Chicken IgY H&L (HRP) secondary antibody.
9. ECL™ Western-Blot Analysis System.
10. LAS-4000 apparatus (Fujifilm, Tokyo, Japan) and MultiGauge 4.0 software for protein quantification.

2.9 Gel Retardation Assay of jetPEI™/ mRNA Complexes

1. jetPEI™ (PolyPlus Transfection).
2. RNA gel loading dye (2×).
3. Agarose gel and electrophoresis system as described in Subheading 2.1.

2.10 Half-Life Calculation of mRNA Transcripts

2.10.1 RNA Extraction

1. RNA spin column extraction kit.
2. Ethanol 100 % and 70 %. Store at RT.
3. Reducing agent such as ß-mercaptoethanol or dithiothreitol.

2.10.2 RT-qPCR

1. First Strand cDNA Synthesis kit.
2. qPCR Master Mix.
3. Real-Time PCR System instrument and software.
4. Primers amplifying V5PB mRNA (V5PB Fwd 5′ -cag caa gta cgg cat caa ga- 3′, V5PB Rev 5′ -gtc agc ttg tag ggc tcc tg- 3′) or 18S RNA (18S Fwd 5′ -gtg atg acc tgc agc aga aa- 3′, 18S Rev 5′ -cag gtt ccg cat gaa ctt tt- 3′).

2.11 Uptake and Intracellular Trafficking of mRNA Polyplexes

2.11.1 Cell Transfection, Fixation, and Permeabilization for Immunofluorescence

1. Glass coverslips.
2. Red fluorescent-labeled mRNA (*see* Subheading 2.4).
3. DRAQ5® 5 mM.
4. PBS (1×).
5. Digitonin 5 μg/mL diluted in PBS (1×).
6. 4 % *para*-formaldehyde diluted in PBS (1×).
7. 0.5 % Triton X-100 diluted in PBS (1×).
8. 0.5 % and 1 % bovine serum albumin (BSA) diluted in PBS (1×).
9. Mounting medium.

2.11.2 mRNA/jetPEI™ Co-Localization

1. jetPEI™-FluoF (PolyPlus Transfection).
2. Goat anti-Lamin A/C (N-18) primary antibody (Santa Cruz Biotechnology).
3. Donkey anti-goat IgG-PerCP-Cy5.5 secondary antibody (Santa Cruz Biotechnology).

2.11.3 mRNA Clathrin-Dependent Endocytosis

1. Mouse anti-Clathrin HC primary antibody (Santa Cruz Biotechnology, Santa Cruz CA, USA).
2. Alexa Fluor® 488 goat anti-mouse IgG (H+L) secondary antibody.

2.11.4 mRNA and SG or P Bodies Co-localizations

1. Mouse anti-p70 S6 kinase monoclonal primary antibody (GE-1/Hedls protein marker) (Santa Cruz Biotechnology) for P bodies detection.
2. Goat anti-eIF3η primary antibody (Santa Cruz Biotechnology) for SG detection.
3. Alexa Fluor® 488 goat anti-mouse IgG (H+L) secondary antibody.
4. Donkey anti-goat IgG-FITC secondary antibody.

2.11.5 Confocal Microscopy

1. LSM 510 META scanning device coupled to an Axiovert 200 microscope (Carl Zeiss, Oberkochen, Germany).
2. Excitation and emission wavelengths: Alexa Fluor® 546: excitation 546 nm, emission band pass 560–615 nm; fluorescein or Alexa Fluor® 488: excitation 488 nm, emission band pass 505–530 nm; GFP: excitation 475 nm, emission band pass 509 nm; Cy5: excitation 633 nm, emission band pass 680 nm.
3. Routine software of the LSM 510 and ImageJ.

2.12 Transposition Assay

1. Helper or control plasmid (*see* Subheading 2.1).
2. Donor plasmid (*see* Subheading 2.1).
3. Supplemented DMEM plus G418 sulfate (800 μg/mL). Store at 4 °C.

4. PBS (1×). Store at 4 °C.
5. 70 % ethanol–0.5 % methylene blue.
6. Deionized water.

3 Methods

Some basic precautions should be taken when working with RNA due to presence of RNases. These includes: wearing gloves throughout experiments and frequently change them to prevent contamination from RNases found on human skin, having a dedicated set of pipettors, using guaranteed RNase-free tips and tubes, using RNase-free chemicals and reagents, working in an RNase-free area in the lab away from air vents or open windows and isolated from reagents used for plasmid isolation (a major source of RNase contamination is plasmid DNA purification kits that use RNase A to remove bacterial RNA).

3.1 Checking DNA and RNA Quality

1. Heat 0.8 g of agarose in 100 mL of Tris-Acetate-EDTA (1×) in a microwave until boiling (*see* **Note 8**). This will make a 0.8 % gel. Higher concentrations of agarose can be used depending on the length of the nucleic acid to be check. Usually 1–2 % gel is used when checking RNA molecules.
2. Cool to approximately 60 °C, add 10 μL of ethidium bromide stock solution, pour into a gel tray, and add an appropriate well comb.
3. Cool to room temperature, remove combs, and place gel tray into an electrophoresis apparatus containing Tris-Acetate-EDTA (1×).
4. Apply 0.5 μg of DNA or RNA in loading buffer per well (*see* **Note 9**).
5. Use a standard DNA or RNA ladder.
6. Perform electrophoresis at 90 V during 1 h.
7. Check nucleic acid migration intermittently by UV illumination.

3.2 Preparation of Plasmid DNA Templates

Keep all reagents on ice while assembling the reactions. The nucleotides and enzymes are especially labile. Careful preparation of the plasmid template is essential for high yield in vitro transcription of high quality mRNA.

3.2.1 Plasmid Linearization

1. Digest 10–15 μg of plasmid DNA template (helper plasmid containing the transposase cassette) with 20 U of NotI restriction enzyme in a final volume of 100 μL for 1–2 h or overnight at 37 °C (*see* **Note 10**).
2. Before storing at −20 °C, remove 2 μL of the reaction mixture and check effective DNA linearization with ethidium bromide staining and ultraviolet illuminator.

3.2.2 Proteinase K Treatment

1. Treat 98 μL of the linearized DNA with 2 μL of proteinase K (10 mg/mL), and 0.5 % SDS in a final volume of 200 μL for 30 min at 50 °C (*see* **Note 11**).
2. Cool the reaction on ice.

3.2.3 Phenol–Chloroform Extraction and Ethanol Precipitation

1. Add an equal volume of phenol–chloroform–isoamyl alcohol (25:24:1), pH 8.0 (*see* **Note 12**).
2. Vortex for 1 min, then centrifuge for 5 min at 16,000 × *g*.
3. Carefully extract the aqueous phase (upper phase), which contains the linearized plasmid, and transfer it to a new 1.5 mL tube. Make sure not touching the organic phase to avoid carrying impurity.
4. Add 2 volumes of chloroform.
5. Vortex for 30 s, then centrifuge for 5 min at 16,000 × *g*.
6. Remove the aqueous phase carefully and transfer it to a new 1.5 mL tube (*see* **Note 13**).
7. Add 2.5 volumes of absolute ethanol and 1/10th volume of sodium acetate (3 M).
8. Mix well and chill at −80 °C for at least 1 h.
9. Centrifuge for 20 min at 16,000 × *g* and 4 °C.
10. Remove the supernatant taking care to not disturbing the pellet.
11. Wash the pellet with 150 μL of ethanol (70 %).
12. Centrifuge for 5 min at 16,000 × *g* and 4 °C.
13. Remove the supernatant, re-spin the tube for 30 s and remove the residual fluid with a micropipette not by inverting the tube, since the rehydrated pellet would be lost.
14. Let the remaining ethanol evaporate by leaving the tube open on the bench for about 5 min (*see* **Note 14**).
15. Resuspend in DNase and RNase-free ultrapure water at a concentration of 0.5–1 μg/μL (15–20 μL).
16. Measure the concentration of the linearized DNA using a spectrophotometer.

3.2.4 Quality Control Electrophoresis

1. Prepare an aliquot of 1 μL of treated DNA in loading buffer.
2. Load the sample on a 0.8 % electrophoresis gel and analyze the DNA plasmid quality.

3.3 In Vitro Synthesis of ARCA Capped and Poly(A) Tailed mRNA

In vitro transcription of capped mRNA is performed onto the linearized templates combined with the ARCA allowing capping with a m7(3′-*O*- methyl)G(5′)ppp(5′)G. Capping can also be achieved post-transcriptionally using a capping enzyme derived from vaccinia virus. The addition of a second enzyme that methylates the 2′-OH of the penultimate nucleotide at the 5′-end of the RNA produces a fully functional cap. Here, we add the cap during transcription to limit the number of steps for mRNA preparation.

To increase the mRNA stability and translatability, a poly(A) tail can be added post-transcriptionally or directly encoded in the DNA used to transcribe RNA. However, a large repetition of one nucleotide in the plasmid sequence can result in plasmid instability and intra-recombination leading to poly(A) shortening (23 and personal data). Here, we add the poly(A) tail after transcription to ensure a poly(A) tail of 100–150 bp long.

For mRNA trafficking experiments, a fluorescent UTP is added during the transcription reaction to produce fluorescent mRNA. Green or red labeled-UTPs are available from Life technologies supplier.

3.3.1 In Vitro Transcription Reaction

1. Thaw the frozen reagents and keep all the products on ice except the 10× reaction buffer which should be kept at room temperature while assembling the reaction.
2. Vortex the 10× reaction buffer and the four ribonucleotide solutions (ATP, CTP, GTP and fluorescent UTP).
3. Dilute the GTP solution to 1/5 (*see* **Note 15**).
4. Assemble transcription reaction at room temperature by mixing equal volumes of the four ribonucleotide solutions (2 μL) together with 2 μL of ARCA (final concentration: 4 μM) and 1 μg of the linearized plasmid (helper plasmid) (*see* **Note 16**). To produce fluorescent mRNA, replace non-labeled UTP by 2 μL of ChromaTide® Alexa Fluor® 546-14-UTP.
5. Add 2 μL of the 10× reaction buffer before adding the enzyme mix (2 μL) (*see* **Note 17**).
6. Gently flick the tube, and then microfuge the tube briefly to collect the reaction mixture at the bottom of the tube.
7. Incubate at 37 °C for 4 h (*see* **Note 18**).
8. Add 1 μL TURBO DNase to eliminate the DNA template, and mix well.
9. Incubate at 37 °C for 15 min (*see* **Note 19**)

3.3.2 Polyadenylation of in Vitro-Synthesized RNA

1. In this order, add 36 μL of DNase and RNase-free water to the DNase-treated transcription reaction (21 μL), 20 μL of E-PAP Buffer 5×, 10 μL of $MnCl_2$ 25 mM and 10 μL of ATP 10 mM.
2. Remove 0.5 μL of the mixture and put it in a new microcentrifuge tube (minus-enzyme control of non-polyadenylated RNA).
3. Add 4 μL of E-PAP enzyme.
4. Mix gently, and incubate for 1–2 h at 37 °C.

3.3.3 mRNA Recovery by Lithium Chloride Precipitation

1. Precipitate the RNA by adding 60 μL of cold (4 °C) LiCl 7.5 M EDTA 50 mM precipitation solution.
2. Mix well and incubate at −20 °C for 30 min or overnight.
3. Centrifuge the precipitated RNA at 16,000 × *g* for 15 min at 4 °C.
4. Remove the supernatant, wash the pellet once with 1 mL Ethanol 70 %, and centrifuge for 15 min at 4 °C to maximize removal of unincorporated nucleotides.
5. Carefully remove the supernatant and let the remaining ethanol evaporate by leaving the tube open on the bench for about 5 min. Do not let the RNA pellet completely dry.
6. Resuspend the RNA in 20 μL DNase and RNase-free water.
7. Measure the RNA concentration with a spectrophotometer (*see* **Note 20**).
8. Make aliquots of 1 μg/μL and stock them at −80 °C.
9. Analyze the quality of the RNA by agarose gel electrophoresis and ethidium bromide staining and ultraviolet illumination (Fig. 1).

3.4 Exogenous mRNA Transfection in Culture Cells

Efficiency of three different transfection reagents for mRNA delivery in HeLa cells is assayed. TransIT®-mRNA (Mirus Bio) and TransMessenger™ (Qiagen) are dedicated to mRNA transfection whereas jetPEI™ is commercialized for DNA transfer. Here are described the protocols for the three chemicals. As shown in Fig. 2, jetPEI™ gives the best transfection efficiency in HeLa cells.

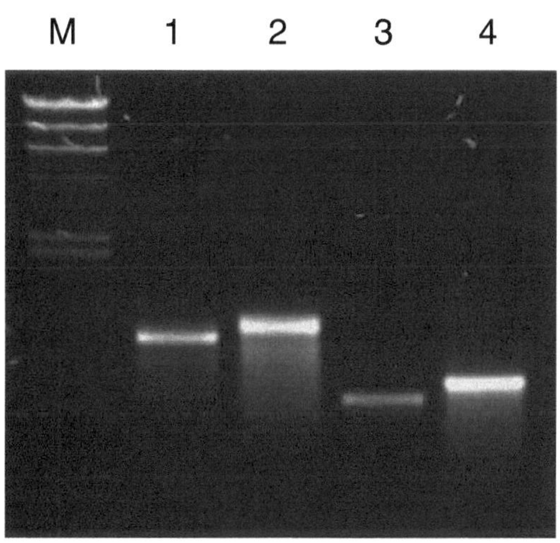

Fig. 1 Analysis of the mRNA quality by agarose gel electrophoresis. M: λDNA Hind III ladder. *Lane 1*: V5PB mRNA, *lane 2*: V5PB mRNA poly(A), *lane 3*: GFP mRNA and *lane 4*: GFP mRNA poly(A). Poly(A)-tailed RNA (2 and 4) migrates slightly slower relative to the RNA without poly(A) tail addition (1 and 3) (personal data)

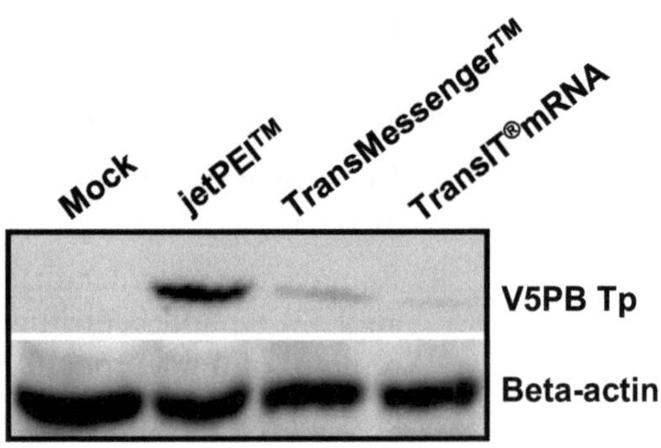

Fig. 2 Efficiency of different transfection reagents for mRNA delivery into HeLa cells. Cells were transfected with 200 ng of V5PB mRNA using jetPEI™ (N/P = 5), TransMessenger™ or TransIT®-mRNA transfection reagents. Transfection efficacy was checked by protein expression by Western-Blot analysis using an antiV5-HRP antibody 24 h post-transfection. Beta-actin is a constitutive protein used as an internal control. jetPEI gave the best result so this transfecting reagent was used for all the experiments. These data were originally published in Bire et al. [6]. The American Society for Biochemistry and Molecular Biology

3.4.1 Cells Seeding

1. Culture HeLa cells in 75 cm^2 cell culture flasks with DMEM supplemented medium at 37 °C in a 5 % CO_2 environment.
2. Split the cells 24 h prior to transfection and plate them at 80 % of confluence in 24-well plates (1.10^5 cells per well for HeLa cells) (*see* **Note 21**).
3. Plate HS-27a and HS5 stromal cells in a 24-well plate so they will be 70–85 % confluent at the time of transfection (30,000 cells/cm^2 were seeded out 1 day before the experiment).

3.4.2 Transfection Using jetPEI™ Reagent

1. Dilute 200 ng of transposase mRNA in RNase-free 150 mM NaCl to a final volume of 50 μL.
2. Dilute 0.4 μL of jetPEI™ in RNase-free 150 mM NaCl to a final volume of 50 μL (*see* **Note 22**).
3. Vortex gently each dilution and spin down briefly.
4. Add the 50 μL jetPEI™ solution to the 50 μL mRNA dilution (*see* **Note 23**).
5. Vortex immediately and spin down briefly.
6. Incubate the mix for 15–30 min at room temperature to allow the formation of mRNA-jetPEI™ complexes (*see* **Note 24**).
7. Wash the cells once with PBS (1×) and then add 500 μL of Opti-MEM.
8. Add the 100 μL jetPEI™/mRNA mix dropwise to the cells. Homogenize by gently swirling the plate.

9. Incubate the cells at 37 °C for 3 h and then replace the transfection medium by fresh medium containing serum and antibiotics.
10. Return the cell to the incubator until the next step of the experiment.

3.4.3 Transfection Using TransMessenger™ Reagent

1. Dilute 0.4 μL of Enhancer R in buffer EC-R (the final volume should be 100 μL once Enhancer R and mRNA are added).
2. Add 200 ng of transposase mRNA.
3. Mix by vortexing for 10 s.
4. Incubate for 5 min at room temperature then spin down briefly.
5. Add 0.8 μL TransMessenger transfection reagent to the mRNA–Enhancer R mixture (*see* **Note 25**).
6. Mix by pipetting up and down five times, or by vortexing for 10 s.
7. Incubate the samples for 10 min at room temperature to allow formation of the complexes.
8. Wash the cells once with PBS (1×) and then add 500 μL of Opti-MEM.
9. Add the transfection complexes dropwise onto the cells. Gently swirl the plate to ensure uniform distribution of the transfection complexes.
10. Incubate the cells at 37 °C for 3 h and then replace the transfection medium by fresh medium containing serum and antibiotics.
11. Return the cell to the incubator until the next step of the experiment.

3.4.4 Transfection Using TransIT®-mRNA Reagent

1. Wash the cells once with PBS (1×) and then add 500 μL of fresh complete DMEM (*see* **Note 26**).
2. Dilute 200 ng of transposase mRNA in 100 μL Opti-MEM and mix thoroughly by pipetting.
3. Immediately add 0.25 μL mRNA boost reagent and mix thoroughly by pipetting.
4. Immediately add 0.5 μL *Trans*IT®-mRNA reagent and mix thoroughly by pipetting.
5. Incubate the samples for 2–5 min at room temperature to allow formation of the complexes (*see* **Note 27**).
6. Add the complex mixture drop-wise onto the cells. Gently rock the plate to distribute the complexes evenly.
7. Return the cell to the incubator until the next step of the experiment.

3.4.5 Transfection Using Lipofectamine® 2000 reagent

1. Wash the cells once with PBS (1×) and then add 500 μL of Opti-MEM.
2. Dilute 2 μL of Lipofectamine® 2000 reagent in Opti-MEM medium for a final volume of 50 μL. Vortex gently.
3. Incubate the mix for 5 min (*see* **Note 27**).
4. Dilute 0.2 μg mRNA in Opti-MEM medium for a final volume of 50 μL. Vortex gently.
5. Add diluted mRNA to diluted Lipofectamine® 2000 reagent to produce a 2:1 lipid–nucleic acid charge ratio in 100 μL per well of 24-well plate. Vortex gently.
6. Incubate the samples for 15–20 min to form the complex.
7. Immediately add the nucleic acid complex to the cells.
8. Return the cell to the incubator until the next step of the experiment.

3.5 Assaying mRNA Transfection Efficiency

This step allows one to monitor the efficiency of the various transfection reagents by quantifying the amount of protein produced from the transfected mRNA (Fig. 2).

3.5.1 Total Protein Extraction

1. Wash twice the transfected cells with PBS (1×).
2. Add 200 μL of lysis buffer directly on the transfected cells in the 24-well plate and harvest the lysate in microcentrifuge tubes.
3. Sonicate twice during 10 s (*see* **Note 28**).
4. Heat the sample for 5 min at 95 °C to denaturate the proteins and cool on ice to avoid renaturation.
5. Centrifuge 5 min at 14,000 × *g* at RT at 4 °C to pellet DNA and cells debris.
6. Keep all extracts on ice if directly used or store at −80 °C.

3.5.2 SDS-PAGE and Western-Blot

1. Prepare a 4–10 % gradient SDS-PAGE gel.
2. Load 15 μg of total protein extract supernatant per lane.
3. Transfer the gel to a nitrocellulose membrane.
4. Cut the membrane in order to separate the endogenous control protein (β-actin) and the assayed protein (transposase) taking care of the sizes of the proteins. Make an incision to differentiate the 2 membranes.
5. Block the membranes with 5 % nonfat dry milk dissolved in PBS–Tween 20 0.05 % for 1 h at room temperature with gentle rocking.
6. Incubate the membranes overnight at 4 °C with gentle rocking with a mouse anti-V5 HRP antibody diluted 1/5000 (transposase protein) or Chicken polyclonal beta Actin antibody

(control) diluted 1/1000 in 5 % nonfat dry milk dissolved in PBS–Tween 20 0.05 %.

7. Wash the membranes three times for 10 min with PBS–Tween 20 0.05 % and agitation.
8. Incubate the membranes with the control protein for 1 h at room temperature with Goat Anti-Chicken IgY H&L (HRP) secondary antibody diluted 1/10,000 in 5 % nonfat dry milk dissolved in PBS–Tween 20 0.05 %.
9. Wash the control membrane three times for 10 min with PBS–Tween 20 0.05 % and agitation.
10. Detect specific *piggyBac* transposase (termed V5PB Tp) and control protein bands using chemiluminesence ECL™ Western-Blot Analysis System and LAS-4000 apparatus.
11. Quantify proteins with MultiGauge 4.0 software and normalize to the endogenous protein.

3.6 Gel Retardation Assay of jetPEI™/ mRNA Complexes

Since jetPEI gave the best results for transfection, we were interested in determining the best N/P ratio for efficient mRNA transfection (Fig. 3).

1. Dilute 1 μg of mRNA in 150 mM NaCl to a final volume of 25 μL.
2. Dilute the appropriate amount of jetPEI™ depending on the N/P ratio in 150 mM NaCl to a final volume of 25 μL (N/P ratios ranging from 0 to 5 were assayed) (*see* **Note 29**).

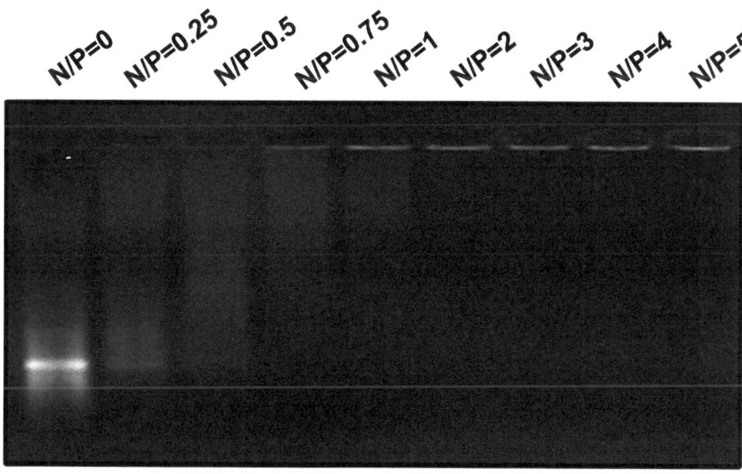

Fig. 3 Gel retardation assays. Electrophoretic migration of V5PB mRNA complexed with jetPEI™ at varying N/P ratios ranging from 0 to 5. Complexes were prepared by mixing 1 μg of V5PB mRNA with different amount of jetPEI™ according to the desired ratio. A shifted band was observed in the well so all mRNA molecules were complexed at N/P = 1. These data were originally published in Bire et al. [6]. The American Society for Biochemistry and Molecular Biology

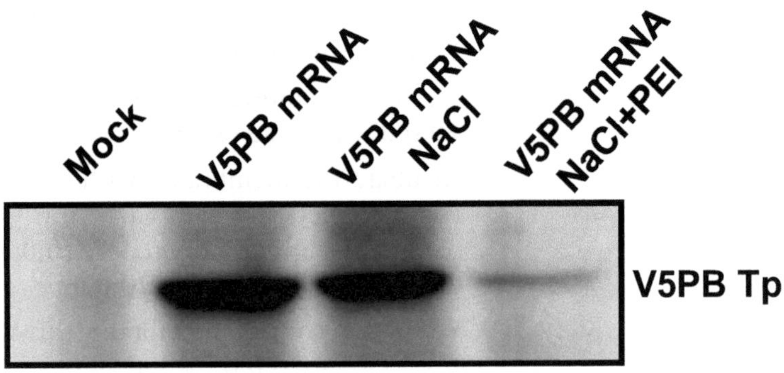

Fig. 4 Accessibility and functionality of the mRNA for in vitro translation. In vitro translation was performed on mRNA alone (V5PB mRNA), in the presence of 150 mM NaCl (V5PB mRNA-NaCl) or complexed with PEI (V5PB mRNA-NaCl+PEI). Products were separated by SDS-PAGE before chemiluminescent detection using an antiV5-HRP antibody. Mock: negative control without mRNA. The translation still occurred but with a lesser extent when mRNA was complexed with PEI. These data were originally published in Bire et al. [6]. The American society for Biochemistry and Molecular Biology

3. Add the diluted jetPEI™ to the diluted mRNA and vortex gently.
4. Incubate at room temperature for 15–30 min to allow the formation of the complexes.
5. Add 5 μL of loading buffer to the complexes solution and perform electrophoresis as described in Subheading 3.1.

3.7 mRNA In Vitro Translation

This step allows assessing the accessibility and functionality of the mRNA once complexed to jetPEI™ (Fig. 4) because in literature it was not described how much mRNA got free from the polyplexes. The question was whether mRNA could be translated when it was complexed.

3.7.1 Preparation of the mRNA Complexes

1. Dilute 2 μg of mRNA in 150 mM NaCl to a final volume of 12.5 μL.
2. Dilute 4 μL of jetPEI™ in 150 mM NaCl to a final volume of 12.5 μL.
3. Add the diluted jetPEI™ to the diluted mRNA and vortex gently.
4. Also prepare free mRNA by adding solely 25 μL of H_2O or 150 mM NaCl without jetPEI™.
5. Incubate at room temperature for 15–30 min to allow the formation of the complexes.

3.7.2 In Vitro Translation Reaction

Keep all reagents on ice while assembling the reaction.

1. Assemble 70 μL of rabbit reticulocyte lysate nuclease-treated, 1 μL amino-acid mixture minus leucine 1 mM, 1 μL amino-acid

mixture minus methionine 1 mM, 2 μL RNasin® ribonuclease inhibitor in a 0.5 mL microcentrifuge tube.

2. Add the 25 μL of mRNA either free or condensed by jetPEI™ and mix gently by pipetting.
3. Briefly centrifuge to collect the reaction sample to the bottom of the tube.
4. Incubate the translation reaction at 30 °C for 90 min.
5. Load the samples on a SDS-PAGE and performed Western-Blot analysis as described in Subheading 3.5.2.

3.8 Half-Life Calculation of mRNA Transcripts

This protocol describes the calculation of in vitro synthesized mRNA half-life following transfection in cultured cells.

3.8.1 Total RNA Extraction

1. Transfect HeLa cells with 200 ng mRNA and jetPEI™ as described in Subheading 3.4.2.
2. Perform total RNA extraction at 4 h, 8 h, 12 h, 18 h, 24 h, 36 h, and 48 h post-transfection using the Nucleospin RNA kit.
3. Completely aspirate the cell-culture medium.
4. Directly add 350 μL of lysis buffer RA1 supplemented with 3.5 μL β-mercaptoethanol to the transfected cells (*see* **Note 30**).
5. Pipette the lysate in the NucleoSpin Filter and centrifuge 1 min at 11,000 × *g* to reduce viscosity and clear the lysate.
6. Discard the filter and add 350 μL ethanol 70 %. Mix by pipetting up and down seven times.
7. Load the lysate in the NucleoSpin RNA column and centrifuge 30 s at 11,000 × *g*. Place the column in a new collection tube.
8. Add 350 μL membrane desalting buffer and centrifuge 1 min at 11,000 × *g* to dry the membrane.
9. For each isolation, add 10 μL rDNase to 90 μL reaction buffer and mix well.
10. Apply 95 μL of rDNase to the column and incubate at room temperature for 15 min.
11. Wash the membrane by adding 200 μL bBuffer RAW2 to the column and centrifuge for 30 s at 11,000 × *g*. Place the column in a new collection tube.
12. Add 600 μL buffer RA3 to the column and centrifuge for 30 s at 11,000 × *g*. Place the column in a new collection tube.
13. Add 250 μL buffer RA3 to the column and centrifuge for 2 min at 11,000 × *g* to completely dry the membrane. Place the column in a nuclease-free microcentrifuge tube.

14. Elute total RNA in 60 μL RNase-free H_2O and centrifuge for 1 min at 11,000 × *g*.
15. Quantify the RNA using a spectrophotometer. A_{260}/A_{280} should be between 1.8 and 2.1.

3.8.2 Retro-Transcription of Total RNA

Keep all the components on ice while assembling the reaction.

1. In the following order, add 1 μg of total RNA, 1 μL of random hexamer primers and adjust the volume to 12 μL with H_2O in a nuclease-free tube on ice (*see* **Note 31**).
2. Centrifuge briefly and incubate at 65 °C for 5 min. Chill on ice.
3. Add 4 μL 5× reaction Buffer, 1 μL RNAse Inhibitor, 2 μL 10 mM dNTP Mix and 1 μL retrotranscriptase. Also prepare a negative control without enzyme.
4. Centrifuge briefly and incubate for 5 min at room temperature followed by 60 min at 42 °C.
5. Terminate the reaction by heating for 5 min at 70 °C.
6. Store the cDNA at −20 °C for short-term or −80 °C for long-term storage.

3.8.3 Quantitative PCR Analysis

Keep all the components on ice. The reaction set-up can be done at room temperature.

1. Add all the following components together to prepare the reaction mix: 10 μL reaction buffer (2×), 1 μL forward primer 2 μM, 1 μL reverse primer 2 μM, and 3 μL H_2O.
2. Add 5 μL cDNA at 1 ng/μL (target gene) or 0.1 ng/μL (internal control) (*see* **Note 32**). Also prepare a negative control by adding H_2O instead of cDNA.
3. Spin down to ensure that no bubbles are present in the reaction vial since it can hamper the fluorescence reading during amplification.
4. Place the tube in the qPCR apparatus and launch the following program for 40 cycles: 3 min at 94 °C (initial denaturation), 30 s at 94 °C (dehybridization), 15 s at 55 °C (hybridization), and 30 s at 72 °C (elongation).
5. Perform the melting curve to check for specific amplification with the following program: 0.3 s from 55 °C to 94 °C (melting) and 20 s from 94 °C to 20 °C (cooling).
6. Calculate the time-point for which only half of the initial amount of mRNA remains.

3.9 Uptake and Intracellular Trafficking of mRNA Polyplexes

3.9.1 Cell Transfection, Fixation, and Permeabilization for Immunofluorescence

1. Plate 6.10^4 HeLa cells per well 24 h prior to treatment in 24-well plates containing glass coverslips.
2. Transfect the cells with 500 ng of red fluorescent-labeled mRNA using 1 μL jetPEI™ (N/P ratio of 5) as described in Subheading 3.4.2. For the mRNA/jetPEI™ co-localization experiment, use the jetPEI™-FluoF green-labeled transfection reagent.
3. Add 0.5 μL of DRAQ5® to the 500 μL transfection medium at the appropriate time points post-transfection and incubate for 5–10 min at 37 °C to allow staining of the nuclei (*see* **Note 33**).
4. Aspirate the medium and wash the transfected cells three times with 500 μL PBS (1×).
5. Add 1 mL of 5 μg/mL digitonin and incubate for 3 min at room temperature (*see* **Note 34**).
6. Wash the cells three times with 500 μL PBS (1×).
7. Add 400 μL 4 % *para*-formaldehyde diluted in PBS (1×) per well and incubate for 20 min at room temperature for cell fixation.
8. Wash the cells three times with 500 μL PBS (1×).
9. Add 400 μL 0.5 % Triton X-100 diluted in PBS (1×) and incubate for 15 min at room temperature for cell permeabilization.
10. Wash the cells three times with 500 μL 0.5 % BSA diluted in PBS (1×).
11. Incubate with the appropriate antibodies depending on the experiment (*see* following Subheadings) (*see* **Note 35**).

3.9.2 mRNA Clathrin-Dependent Endocytosis

This experiment was performed at 30 min and 2 h post-transfection (Fig. 5) to follow the internalization of mRNA/jetPEI™ complexes through endocytosis.

1. Incubate the cells for 1 h at room temperature with 200 μL Mouse anti-Clathrin HC primary antibody diluted 1/100 in 1 % BSA–PBS (1×).
2. Wash the cells three times with 500 μL 0.5 % BSA diluted in PBS (1×).
3. Incubate the cells for 45 min at room temperature in the dark with 200 μL Alexa Fluor® 488 goat anti-mouse IgG (H+L) secondary antibody diluted 1/200 in 1 % BSA–PBS (1×).
4. Wash the cells three times with 500 μL 0.5 % BSA diluted in PBS (1×) and add another 500 μL 0.5 % BSA–PBS (1×) in the wells.

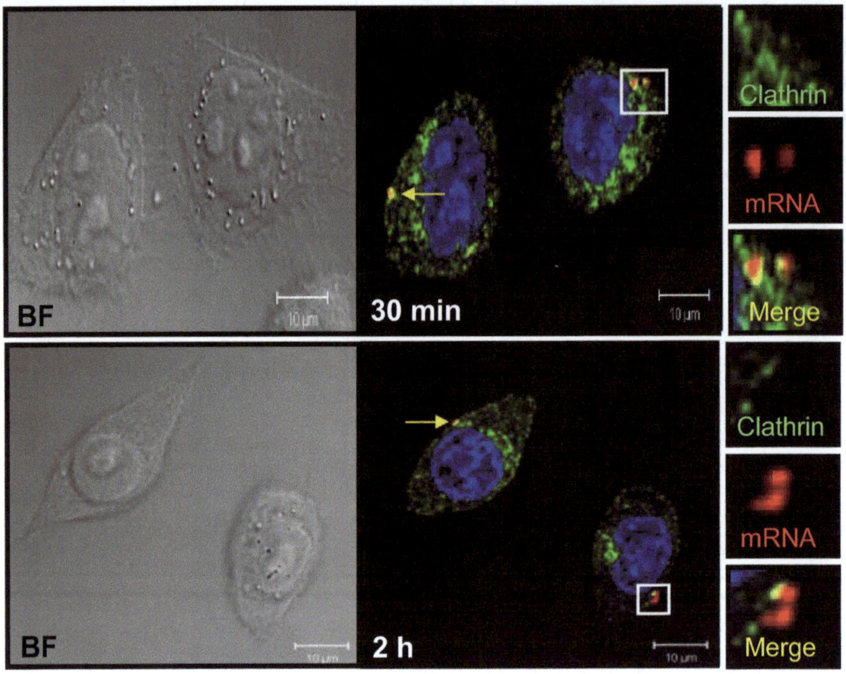

Fig. 5 Uptake pathway of mRNA polyplexes in HeLa cells through clathrin-dependent endocytosis. *Red fluorescent labeled* V5PB mRNA compacted to jetPEI™ were administered to HeLa cells. Clathrin detection was performed by immunofluorescence using an anti-clathrin primary antibody revealed with an Alexa-488 conjugated secondary antibody (*green*) at 30 min and 2 h post-transfection. *Yellow dots* correspond to colocalization of red and *green fluorescent spots*. Insets shows enlargement of boxed areas with colors separated. BF: Bright Field. Nuclei were stained in *blue* by DRAQ5®. The images shown are representative Z-sections from two experiments. mRNA polyplexes could enter the cells by clathrin-dependent endocytosis. These data were originally published in Bire et al. [6]. The American Society for Biochemistry and Molecular Biology

3.9.3 mRNA/jetPEI™ Co-Localization

This experiment is performed 3 h post-transfection (Fig. 6) to monitor the dissociation of the mRNA/jetPEI™ complexes once in the cells.

1. Incubate the cells for 1 h at room temperature with 200 μL Goat anti-Lamin A/C (N-18) primary antibody diluted 1/200 in 1 % BSA–PBS (1×).
2. Wash the cells three times with 500 μL 0.5 % BSA diluted in PBS (1×).
3. Incubate the cells for 45 min at room temperature in the dark with 200 μL Donkey anti-goat IgG-PerCP-Cy5.5 secondary antibody diluted 1/400 in 1 % BSA–PBS (1×).
4. Wash the cells three times with 500 μL 0.5 % BSA diluted in PBS (1×) and add another 500 μL 0.5 % BSA–PBS (1×) in the wells.

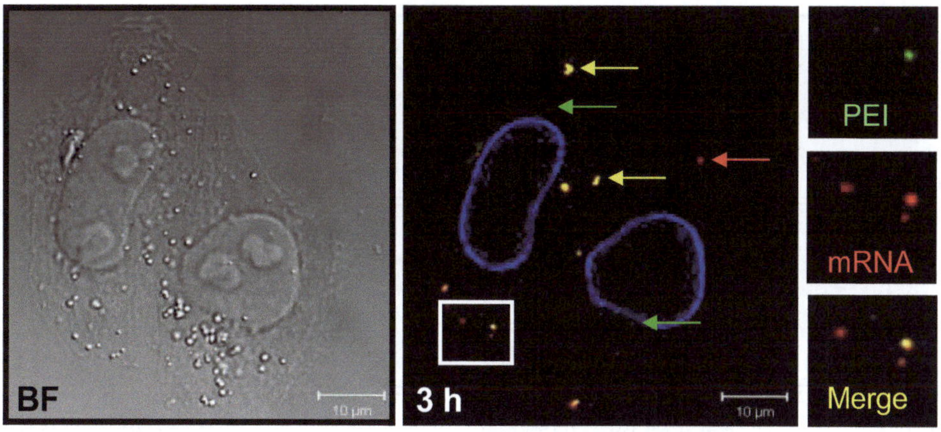

Fig. 6 Dissociation of mRNA/jetPEI™ polyplexes in HeLa cells. Confocal microscopy images of HeLa cells incubated during 3 h with Alexa Fluor® 546-labeled mRNA (*red*) condensed with jetPEI™-FluoF (*green*). Yellow dots correspond to doubly fluorescent polyplexes. *Green spots* match with jetPEI™-FluoF alone or complexed to unlabeled mRNA, and *red spots* correspond to dissociated mRNA from polyplexes. Insets show enlargements of the *boxed areas* with the colors separated. BF: Bright Field. Nuclei are stained in *blue* by lamin immunodetection. The images shown are representative Z-sections from two experiments. So mRNAs were able to dissociate from polyplexes from 3 h. These data were originally published in Bire et al. [6]. The American Society for Biochemistry and Molecular Biology

3.9.4 mRNA and SG or P Bodies Co-localizations

This experiment was performed at 3 h, 18 h, and 48 h post-transfection (Fig. 7) to monitor the fate of the exogenous mRNA in the cells. In our experiments, no P-bodies co-localization could be detected.

1. Incubate the cells for 1 h at room temperature with 200 µL Mouse anti-p70 S6 kinase monoclonal primary antibody (P bodies detection) or Goat anti-eIF3η primary antibody (stress granules detection) diluted 1/500 in 1 % BSA–PBS (1×).
2. Wash the cells three times with 500 µL 0.5 % BSA diluted in PBS (1×).
3. Incubate the cells for 45 min at room temperature in the dark with 200 µL Alexa Fluor® 488 goat anti-mouse IgG (H+L) (P bodies detection) or Donkey anti-goat IgG-FITC secondary antibody (stress granules detection) diluted 1/200 in 1 % BSA–PBS (1×).
4. Wash the cells three times with 500 µL 0.5 % BSA diluted in PBS (1×) and add another 500 µL 0.5 % BSA–PBS (1×) in the wells.

3.9.5 Confocal Microscopy

1. Put a small drop of mounting medium on a glass slide (*see* **Note 36**).

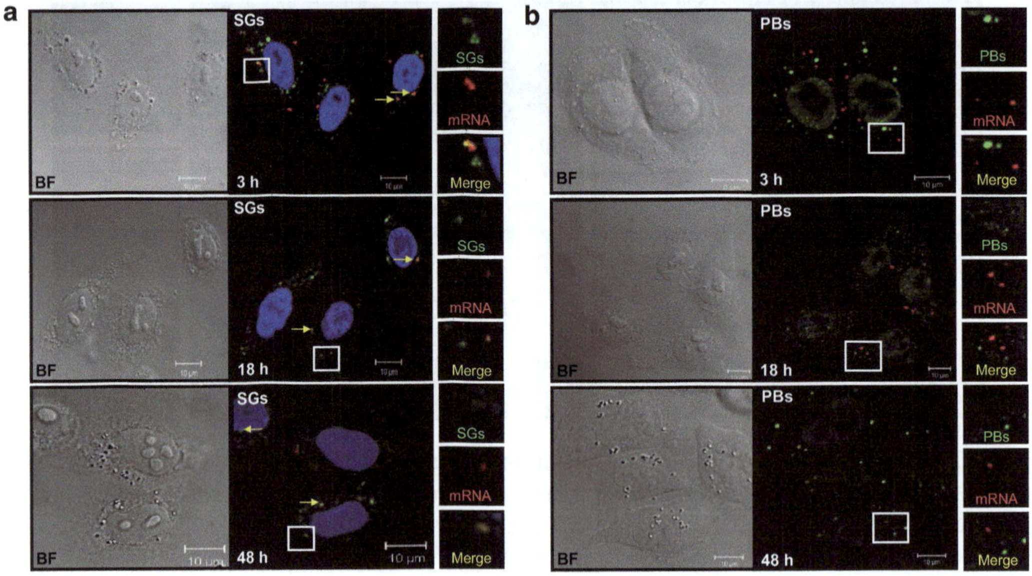

Fig. 7 mRNA localization in subcellular foci. HeLa cells were observed 3 h, 18 h and 48 h post-transfection with 500 ng of red fluorescent-labeled V5PB mRNA polyplexes. Stress granules (SGs) (**a**) and P bodies (PB) (**b**) were immunodetected using specific green fluorescent secondary antibodies. Insets show enlargements of the *boxed areas* with *separated colors. Arrows* indicate representative foci. The primary antibody directed against PBs foci exhibits nuclear staining due to nuclear p70 S6 kinase protein detection. Nuclei were stained in blue by DRAQ5® (**a**). BF: Bright Field. The images shown are representative Z-section from two experiments. There was no colocalization of mRNAs in P bodies. These data were originally published in Bire et al. [6]. The American Society for Biochemistry and Molecular Biology

2. Carefully transfer the coverslip with the stained cells onto the glass slide with fine tweezers. Make sure that the cells are in contact with the mounting medium.
3. Seal the coverslip to the slide using colorless nail polish and let it dry completely in the dark. Slides can be stored at 4 °C or −20 °C before microscopy observation.
4. Analyze the cells with confocal microscopy with the appropriate wavelengths for excitation and emission depending on the fluorophores used in the experiments.
5. Choose representative images from randomly selected areas, all analyzed using Z-sectioning.
6. Process the pictures using the routine software of the LSM 510 and ImageJ free software.

3.10 Transposition Assay in Hela Cells

The transposition assay will ensure that the transposase produced from the in vitro synthesized mRNAs is functional in cells and exhibits transposition activity (Fig. 8).

1. Transfect HeLa cells with 200 ng V5-PB mRNA and 200 ng of donor plasmid (harboring the transposon) to get a 1:1 ratio using jetPEI™ as described in Subheading 3.5.2 (*see* **Note 37**).

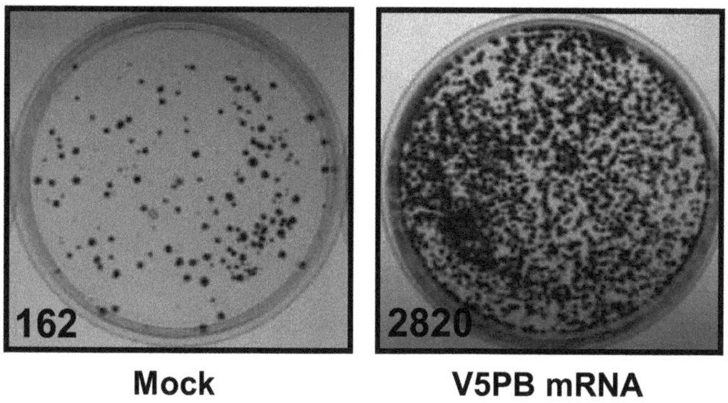

Fig. 8 mRNA functionality for transposition in HeLa cells. Cells were transfected with the mRNA encoding the V5PB transposase and the plasmid carrying the gene of neomycin resistance. Transposition events involve the integration of the NeoR gene, and thus the emergence of a resistance phenotype to G418. Positive cells were selected under antibiotic pressure for 15 days, and colonies were then stained and counted. Mock: negative control (without transposase mRNA) corresponding to recombination events. The figures indicate the number of colonies counted (mean value of three independent experiments). The transposition process occurred up to the background when transposase was supply in mRNA molecules. These data were originally published in Bire et al. [6]. The American Society for Biochemistry and Molecular Biology

Also perform a negative control without transposase using GFP mRNA.

2. Transfer the cells to 100 mm plates 48 h post-transfection.
3. Culture the cells under G418 sulfate selection (800 μg/mL) for 15 days to allow formation of resistant colonies.
4. Aspirate the culture medium and carefully rinse twice the 100 mm plates taking care not to scrap the colonies.
5. Fix and stain the cells by adding 10 mL 70 % EtOH-0.5 % methylene blue for 3 h.
6. Gently rinse the dishes with water until staining solution is completely removed.
7. Count only colonies >0.5 mm in diameter.

3.11 Transposition Assay in HS-27a and HS-5 Stromal Cells

We transfect HS-27a and HS-5 cells using Lipofectamine® 2000 and jetPEI™ (Subheadings 3.4.2 and 3.4.5). Our data show a higher transfection efficiency with Lipofectamine® 2000 in comparison to jetPEI™ whatever the molecular ratio is ($R = 1, 2, 3$, where R = transfection agent/nucleic acid charge). For these reasons, we choose to perform the remaining experiments with Lipofectamine® 2000 with $R = 2{:}1$ (Fig. 9).

In parallel, we perform a time course of in vivo translation in HS-27a and HS-5 cell lines transfected with mRNA (Subheading 3.5). As shown in Fig. 10, transposase production is detected as a plateau between 5 and 18 h post-transfection. After then, only transposase traces are detected. Overall, these data show for the first

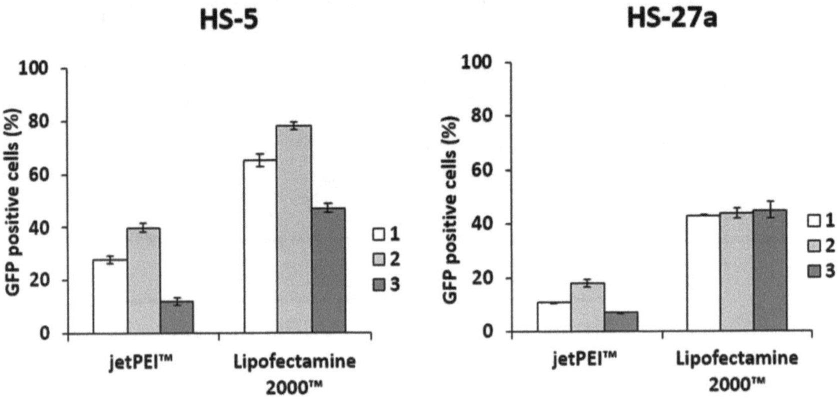

Fig. 9 Efficiency of two transfection reagents for mRNA delivery into HS-27a and HS-5 stromal cells. Cells were transfected with GFP mRNA using jetPEI™ and Lipofectamine® 2000 transfection reagents three times. Transfection efficacy was checked by percentage of GFP cells by flow cytometry analysis 18 h post-transfection. Lipofectamin ($R = 2$) gave the best results

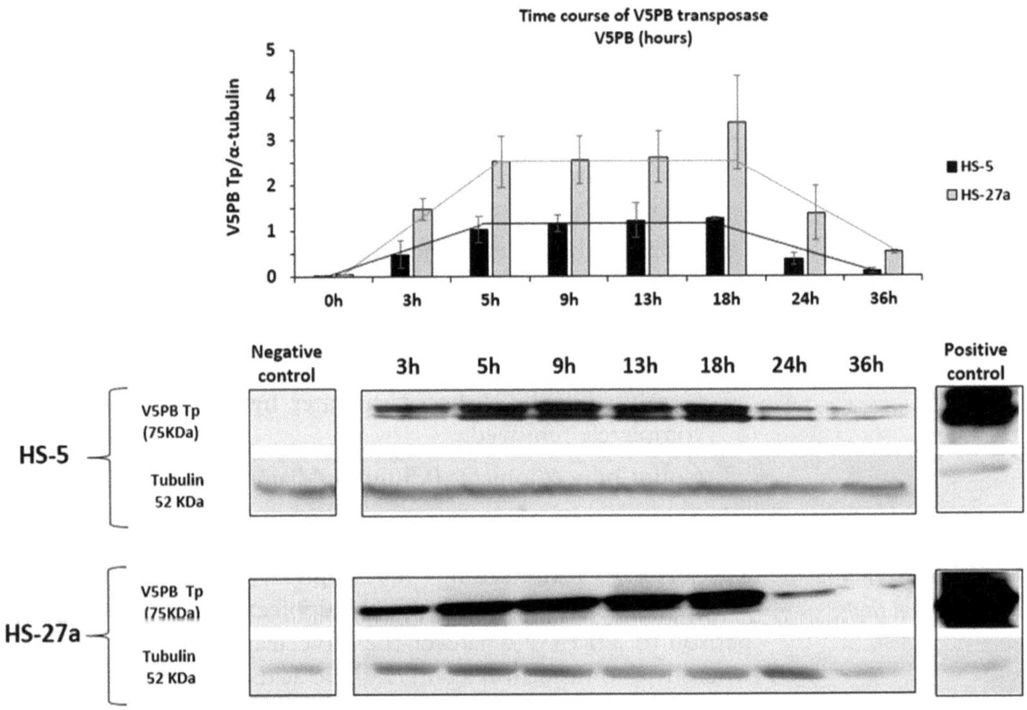

Fig. 10 Time course of in vivo production of V5PB transposase in HS-27a and HS-5 cell lines. V5PB transposase (V5PB Tp) expression was monitored by Western-Blot analysis using an antiV5-HRP antibody 24 h after mRNA transfection. Tubulin is a constitutive protein used as an internal control. Negative control: cells were transfected with transfection reagent alone, Positive control: cells were transfected with pDNA. A restricted window of transposase expression was observed between 5 h and 18 h post-transfection (personal data)

time that the mRNA transposase prototype allows a very short expression window of the *piggyBac* transposase in mesenchymal stromal cells, confirming results obtained in HeLa cells.

3.11.1 Cell Seeding

1. Plate cells in a 24-well plate so they will be 70–85 % confluent at the time of transfection (30,000 cells/cm^2 were seeded out 1 day before the experiment).
2. Calculate and prepare enough nuclei acid (mRNA and DNA) for as many wells as your experiment calls for, 1 μg of nuclei acid (0.5 μg mRNA and 0.5 μg DNA) is added to Opti-MEM for each well of the 24-well plate.

3.11.2 DNA Complex Preparation

1. Dilute 2 μL of Lipofectamine® 2000 reagent in Opti-MEM medium for a final volume of 50 μL. Vortex gently.
2. Incubate the mix for 5 min (*see* **Note 38**).
3. Dilute 0.5 μg DNA in Opti-MEM medium for a final volume of 50 μ: Vortex gently.
4. Add diluted DNA to diluted Lipofectamine® 2000 reagent to produce a 2:1 lipid–nucleic acid charge ratio in 100 μL per well of 24-well plate. Vortex gently.
5. Incubate the samples for 15–20 min to form the complexes.

3.11.3 RNA Complex Preparation

1. Dilute 2 μL of Lipofectamine® 2000 reagent in Opti-MEM medium for a final volume of 50 μL. Vortex gently.
2. Incubate the mix for 5 min.
3. Dilute 0.5 μg mRNA in Opti-MEM medium for a final volume of 50 μL. Vortex gently.
4. Add diluted mRNA to diluted Lipofectamine® 2000 reagent to produce a 2:1 lipid–nucleic acid charge ratio in 100 μL per well of 24-well plate. Vortex gently.
5. Incubate the samples for 15–20 min to form the complexes.

3.11.4 DNA and RNA Complexes Co-transfection

1. Wash the cells once with PBS (1×) and then add 500 μL of Opti-MEM.
2. Add the nucleic acid complexes separately to the cells and move gently (*see* **Note 37**).
3. Incubate the cells at 37 °C for 3 h and then replace the transfection medium by fresh medium containing serum and antibiotics.
4. Return the cell to the incubator until the next step of the experiment.
5. Transfer the cells to 100 mm plates 48 h post-transfection.
6. Culture the cells under G418 sulfate selection (800 μg/mL) for 15 days to allow establishing clone formation.

7. Seed the modified cells to confluence and coculture them with hematopoietic cells (CD34+ cells) for 5 days (*see* **Note 39**).

4 Notes

1. Plasmids quality must be checked by agarose gel electrophoresis. Plasmids must be predominantly in the superhelical form. Less nicked form and less DNA double strand cut plasmid form should be observed upon migration. The samples should be quantified by spectrophotometry, aliquoted and stored at −20 °C until manipulation.
2. Plasmid DNA must be linearized with a restriction enzyme downstream of the insert to be transcribed. Better chose a restriction enzyme leaving blunt or 5′ overhanging ends since low level transcription has been reported from linearized plasmids with 3′ extension. Moreover, this can result in double stranded RNAs as the polymerase continues to transcribe off of the opposite strand [28].
3. Linearized plasmid DNA or PCR products containing an RNA polymerase promoter site can be used for in vitro transcription of mRNA. Kits from Ambion/Life Technologies are available for T7, SP6 and T3 phage polymerase promoters. Nucleotides should be supplied separately and not premixed with any cap analog in order to allow incorporation of the ARCA or other cap variant.
4. Standard cap analog can be incorporated in the forward or the reverse orientation into the RNA. Only the forward orientation allows efficient translation, and thus half of the capped RNAs are not functional. Substitution of the traditional cap with ARCA leads to RNA transcripts capped exclusively in the correct orientation that are 100 % functional.
5. Cells in culture should be healthy and in log phase of growth, which can be achieved using pre-confluent cultures. Transfection of cells with a low passage number is recommended to ensure that the cell genotype does not become altered. For HeLa cells, less than 40 passages were used and for HS-5 and HS27-a no more than 24 passages were realized.
6. Generally, the use of serum in the culture medium enhances transfection. However, serum and antibiotics inhibit transfection complex formation when using the TransIT®-mRNA reagent. Moreover, using medium without serum or antibiotics can avoid contamination with ribonucleases (RNases). Therefore, we do not recommend using serum during transfection of RNA.

7. Lysis buffer can be prepared previously and store at room temperature without addition of dithiothreitol. Add the dithiothreitol just before use and do not keep the buffer as dithiothreitol rapidly degrades over time.
8. Electrophoresis system should be well cleaned to avoid RNAse contamination. Tris–acetate–EDTA (1×) buffer should be autoclaved before use.
9. A non-denaturing agarose gel electrophoresis is adapted for small amount RNA detection. On native gels, the samples can be loaded directly without heating. The bands should appear sharp, intact and with no smear. Sometimes a thin ladder pattern could be visualized because of RNA secondary structures. To determine the correct length of the synthesized RNA, denaturing conditions can be used.
10. The DNA template must be pure enough to be easily digested with restriction enzymes. Incomplete cut leaves circular plasmid DNA that may generate large amounts of unwanted transcripts. Routinely, 15 min are enough to cut a DNA template with a restriction enzyme. We recommend increasing the digestion time (to 2 h or overnight) to achieve 100 % plasmid linearization. High fidelity restriction enzymes allow overnight digestion without degradation of the template DNA.
11. Proteinase K treatment will remove most of the DNases and RNases and other contaminating proteins that can hamper subsequent experiments such as in vitro transcription.
12. Phenol/chloroform extraction separates nucleic acids from proteins due to solubility discrepancies. DNA will be contained in the aqueous phase whereas proteins will be stuck in the organic phase. Make sure that the pH is 7.8–8.2 since acidic pH makes the DNA go into the phenolic layer while RNA goes into the aqueous layer. To avoid losing materials during this step, increase the total volume by adding 100 μL of RNase-free water to 200 μL of DNA templates prior to extraction.
13. Make sure not touching the interphase or lower phase as persistence of chloroform may inhibit RNA transcription. Repeat **steps 4** and **5** if necessary.
14. Do not overdry the DNA pellet as it may be difficult to resuspend but be careful to ethanol traces.
15. If ARCA is incorporated during transcription, the ratio of GTP to ARCA can be varied to increase the percent of RNA transcripts containing cap, but this is done at the expense of the total amount of RNA made in the reaction. Transcription kits with ARCA use a ratio of ARCA to GTP of 4:1, which results in approximately 80 % of the transcripts capped and a 50 % reduction in the total amount of RNA made.

16. The spermidine in the reaction buffer can precipitate the template DNA if the reaction is assembled on ice. One microgram of DNA template is sufficient to generate 50–100 μg of RNA.
17. It is recommended to add the 10× reaction buffer after the water and the ribonucleotides.
18. The optimal incubation time for a given template will vary depending on the size and transcriptional efficiency of the template. Short transcripts (less than 500 nt) necessitate a longer incubation time (up to ~16 h) because more transcription initiation events are required to synthesize a given mass amount of RNA. In general, 2–4 h is sufficient but we recommend doing a time-course experiment before transcription assays.
19. Avoid long incubation time with DNase since this can lead to newly synthesized RNA degradation.
20. For single-stranded RNA, 1 A_{260} unit corresponds to 40 μg/mL. A_{260}/A_{280} should be between 1.8 and 2.1.
21. The recommended cell density (% of confluence) for transfection of adherent cells with single-stranded mRNA is 80 % for the three reagents. This ensures that the cells are in optimal physiological condition on the day of transfection. However, the optimal confluency should be determined for every new cell type to be transfected to maximize transfection efficiency. Density should be kept constant in future experiments for reproducibility by counting cells prior to seeding and by keeping the time period between seeding and transfection (24 h) constant. We described a protocol for 24-well plates, when using other plate sizes, adjust the cell seeding and volumes of reagents used for transfection.
22. The overall charge of jetPEI™/mRNA complexes is primordial for good transfection efficiency. It is determined by the number of nitrogen residues (N) in the jetPEI™ per phosphate (P) of mRNA, giving an N/P ratio. As a starting point, we use an N/P ratio of 5 in our experiments as recommended by the manufacturer's protocol.
23. Mixing the solutions in the reverse order may reduce transfection efficiency due to mRNA precipitation.
24. Once complexed to jetPEI™, mRNA is protected from degradation that can be present in the cell medium.
25. We use a Ratio of mRNA to TransMessenger transfection reagent of 1::4 as recommended by the manufacturer's protocol.
26. The *Trans*IT®-mRNA transfection reagent yields improved transfection efficiencies when transfections are performed in

complete growth medium (instead of Opti-MEM) with no media change post-transfection.

27. Do not let the complexes incubate longer than 5 min as this may decrease transfection efficiency.
28. This step will break the DNA in smaller fragment to reduce the viscosity of the sample. It will not affect the protein quality.
29. To calculate the volume of jetPEI™, one can use the following formula: $Volume_{jetPEI} = N/P \text{ ratio} \times (3^{*} \times 1\ \mu g \text{ of mRNA})$. * corresponds to the nmoles of phosphate per µg of mRNA.
30. Incomplete removal of the medium allows full lysis activity of the lysis buffer. At this point, lysate can be stored at −80 °C directly on the dish or once placed on a microcentrifuge tube. This allows collecting all the time-points, and performing total RNA extractions of the various samples all at once.
31. In this protocol, random hexamer primers were used since we chose the 18S RNA as an endogenous control to normalize the expression of V5PB mRNA for variability in the quality of RNA and the amount of cDNA input. If the internal standard is a messenger RNA (GAPDH for example), oligo$(dT)_{18}$ primers supplied in the kit should be used.
32. We consider that the retrotranscription reaction is 100 % efficient and that we obtain 1 µg of cDNA. The cDNA final concentration is 50 ng/µL.
33. Perform this step only for experiments involving mRNA co-localization with stress granules and clathrin. For jetPEI™-FluoF experiments, nuclei will be labeled in blue by Lamin immunostaining. For P bodies experiments, nuclei will be stained in green due to cross-reactivity of the primary antibody with the nuclear p70 S6 kinase.
34. Digitonin is a non-ionic detergent that solubilizes unbound proteins. Digitonin will also form some pores onto the cytoplasmic membrane for evacuation of such un-attached proteins.
35. Isotype control by treating cells with only secondary antibodies must be done to show successful specific immunostaining. No fluorescent signal should be seen in such controls.
36. Mounting medium is a glycerol-based, aqueous mountant that remains a viscous liquid on the slide. It prevents drying of the slides and has anti-fading and anti-photobleaching properties. Do not put a big drop of mounting medium as sealing will be difficult due to the viscosity of the liquid.
37. When co-transfecting mRNA and pDNA, complexes should be formed separately with the transfection reagent to avoid mRNA degradation due to RNases potentially present in the pDNA preparation. Once complexed, mRNA is protected from

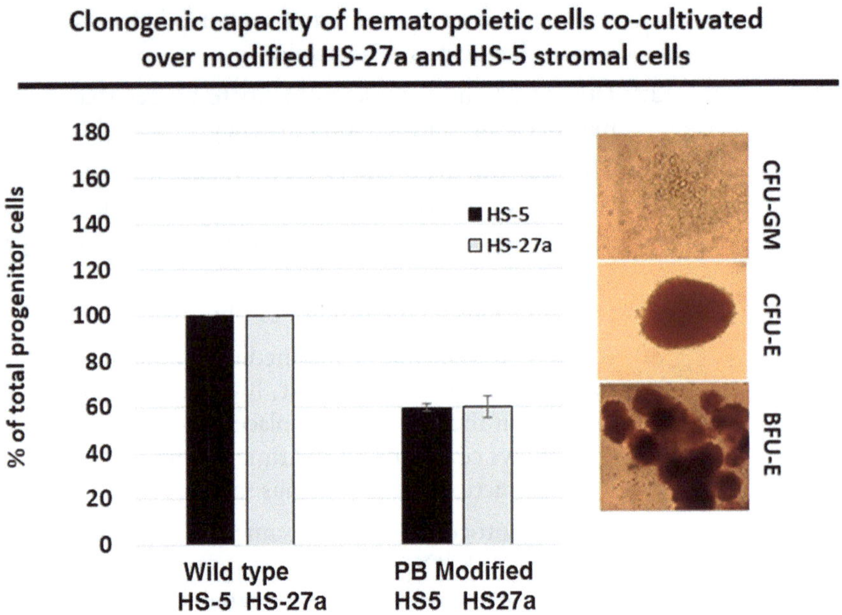

Fig. 11 Clonogenic capacity of hematopoietic cells co-cultivated over PB modified HS27-a and HS-5 stromal cells. Dexter-type short-term cultures [27] were performed by co-culturing CD34+ hematopoietic cells for 5 days over confluent layers of modified HS-27a and HS-5 cells. Modified stromal cells demonstrated a less ability to engage hematopoietic progenitors (CFU-GM, CFU-E, and BFU-E) in the differentiation process (personal data)

degradation so mRNA and pDNA complexes can be added simultaneously to the cells.

38. We found that incubating diluted lipofectamine for 5 min gives higher transfection efficiency.

39. This is a standard functional protocol to measure the capacity of stromal cells to maintain hematopoietic cells [27]. As pinpointed in Fig. 11, modified HS-27a and HS-5 cells produced less committed progenitor cells reflecting the immaturity of hematopoietic cells in comparison to control. This functional test confirms that the delivery of mRNA in mesenchymal stromal cells is as efficient as in HeLa cells to catalyze the transposition reaction. It is the first time that mRNA and DNA are simultaneously delivered in mesenchymal stromal cells to express a functional gene of interest.

References

1. Bire S, Rouleux-Bonnin F (2012) Transposable elements as tools for reshaping the genome: it is a huge world after all! Methods Mol Biol 859:1–28
2. Di Matteo M, Mátrai J, Belay E, Firdissa T, Vandendriessche T, Chuah MK (2012) PiggyBac toolbox. Methods Mol Biol 859:241–254

3. Wilson M, Coates C, George A (2007) PiggyBac transposon-mediated gene transfer in human cells. Mol Ther 15:139–145
4. Bire S, Rouleux-Bonnin F (2013) Transgene site specific integration: problems and solutions in Topics in Current Genetics edited by Renault S and Duchateau P. Springer 23:3–39
5. Palazzoli F, Bire S, Bigot Y, Bonnin-Rouleux F (2011) Landscape of chromatin control element patents: positioning effects in pharmaceutical bioproduction. Nat Biotechnol 29:593–597
6. Bire S, Gosset D, Jégot G, Midoux P, Pichon C, Rouleux-Bonnin F (2013) Exogenous mRNA delivery and bioavailability in gene transfer mediated by piggyBac transposition. BMC Biotechnol 13:75
7. Al-Dosari MS, Gao X (2009) Nonviral gene delivery: principle, limitations, and recent progress. AAPS J 11:671–681
8. Bettinger T, Carlisle R, Read M, Ogris M, Seymour L (2001) Peptide-mediated RNA delivery: a novel approach for enhanced transfection of primary and post-mitotic cells. Nucleic Acids Res 29:3882–3891
9. Mockey M, Gonçalves C, Dupuy F, Lemoine F, Pichon C, Midoux P (2006) mRNA transfection of dendritic cells: synergistic effect of ARCA mRNA capping with Poly(A) chains in cis and in trans for a high protein expression level. Biochem Biophys Res Commun 340:1062–1068
10. Reilly MJ, Larsen JD, Sullivan MO (2012) Polyplexes traffic through caveolae to the Golgi and endoplasmic reticulum en route to the nucleus. Mol Pharm 9:1280–1290
11. Duncan R, Richardson SC (2012) Endocytosis and intracellular trafficking as gateways for nanomedicine delivery: opportunities and challenges. Mol Pharm 9:2380–2402
12. Vercauteren D, Rejman J, Martens TF, Demeester J, De Smedt SC, Braeckmans K (2012) On the cellular processing of nonviral nanomedicines for nucleic acid delivery: mechanisms and methods. J Control Release 61:566–581
13. El-Sayed A, Harashima H (2013) Endocytosis of gene delivery vectors: from clathrin-dependent to lipid raft-mediated endocytosis. Mol Ther 21:1118–1130
14. Magalhães S, Duarte S, Monteiro GA, Fernandes F (2014) Quantitative evaluation of DNA dissociation from liposome carriers and DNA escape from endosomes during lipid-mediated gene delivery. Hum Gene Ther Methods 25:303–313
15. Kedersha N, Stoecklin G, Ayodele M, Yacono P, Lykke-Andersen J, Fritzler M, Scheuner D, Kaufman R, Golan D, Anderson P (2005) Stress granules and processing bodies are dynamically linked sites of mRNP remodeling. J Cell Biol 169:871–884
16. Anderson P, Kedersha N (2008) Stress granules: the Tao of RNA triage. Trends Biochem Sci 33:141–150
17. Anderson P, Kedersha N (2009) RNA granules: post-transcriptional and epigenetic modulators of gene expression. Nat Rev Mol Cell Biol 10:430–436
18. Brennecke J, Stark A, Russell RB, Cohen SM (2005) Principles of microRNA-target recognition. PLoS Biol 3(3), e85
19. Czech B, Hannon GJ (2011) Small RNA sorting: matchmaking for Argonautes. Nat Rev Genet 12:19–31
20. Tuck AC, Tollervey D (2011) RNA in pieces. Trends Genet 27(10):422–432
21. David Gerecht PS, Taylor MA, Port JD (2010) Intracellular localization and interaction of mRNA binding proteins as detected by FRET. BMC Cell Biol 11:69
22. Kuersten S, Goodwin EB (2003) The power of the 3′UTR: translational control and development. Nat Rev Genet 4:626–637
23. Elango N, Elango S, Shivshankar P, Katz M (2005) Optimized transfection of mRNA transcribed from a d(A/T)100 tail-containing vector. Biochem Biophys Res Commun 330:958–966
24. Holtkamp S, Kreiter S, Selmi A, Simon P, Koslowski M, Huber C, Türeci O, Sahin U (2006) Modification of antigen-encoding RNA increases stability, translational efficacy, and T-cell stimulatory capacity of dendritic cells. Blood 08:4009–4017
25. Roecklein BA, Torok-Storb B (1995) Functionally distinct human marrow stromal cell lines immortalized by transduction with the human papilloma virus E6/E7 genes. Blood 85:997–1005
26. Ramakrishnan A, Torok-Storb BJ (2010) The role of the marrow microenvironment in hematopoietic stem cell transplantation. Cell Ther Transplant 2:7–12
27. Dexter TM, Allen TD, Lajtha LG (1977) Conditions controlling the proliferation of haemopoietic stem cells in vitro. J Cell Physiol 91:335–344
28. Schenborn ET, Mierendorf RC Jr (1985) A novel transcription property of SP6 and T7 RNA polymerases: dependence on template structure. Nucleic Acids Res 13:6223–6236

Chapter 14

Transfection of Human Keratinocytes with Nucleoside-Modified mRNA Encoding CPD-Photolyase to Repair DNA Damage

Gábor Boros, Katalin Karikó, Hiromi Muramatsu, Edit Miko, Eszter Emri, Csaba Hegedűs, Gabriella Emri, and Éva Remenyik

Abstract

In vitro-synthesized mRNA containing nucleoside modifications has great therapeutical potential to transiently express proteins with physiological importance. One such protein is photolyase which rapidly removes UV-induced DNA damages, but this enzyme is absent in humans. Here, we apply a novel mRNA-based platform to achieve functional nonhuman photolyase production in cultured human keratinocytes. Transfection of nucleoside-modified mRNA encoding photolyase leads to accelerated repair of DNA photolesions in human keratinocytes.

Key words Pseudouridine-modified mRNA, Nonviral gene therapy, Transient transfection, Cyclobutane pyrimidine dimer-specific photolyase, Ultraviolet B, Human keratinocytes, DNA repair

1 Introduction

Expression of therapeutic proteins translated from in vitro-synthesized mRNA is a promising tool for clinical applications [1]. Nucleoside modifications and HPLC purification are big step forward in the therapeutic applicability of messenger RNA. Pseudouridine modifications and HPLC purification of RNA completely abolish immunogenicity of the RNA and makes it highly translatable [2–4]. Here, we apply this mRNA-based technology to gain insights into the pathogenesis of UV-induced DNA damages. Cyclobutane pyrimidine dimers (CPDs) are the most frequently formed deleterious DNA damages caused by UVB irradiation [5, 6]. Accumulation of the unrepaired CPD photolesions induces cytotoxic and mutagenic effects contributing primarily to UVB-induced carcinogenesis [7, 8]. The integrity of DNA can be quickly restored by a light-activated photolyase, which splits

Robert E. Rhoads (ed.), *Synthetic mRNA: Production, Introduction Into Cells, and Physiological Consequences*, Methods in Molecular Biology, vol. 1428, DOI 10.1007/978-1-4939-3625-0_14,

the CPDs using the energy of visible light in a process called "photoreactivation" [9]. This DNA repair enzyme protects most prokaryotes and eukaryotes but it is absent in placental mammals [10] that instead need to rely on the less potent nucleotide excision repair (NER) system [11]. Several studies have described reversing of the UVB effect by introduction of liposome-encapsulated photolyase enzyme into cultured cells [12, 13] and in vivo in humans as well [14–16]. However, photolyase expressed in mammalian cells following gene delivery is rarely studied [17]. We are presenting a unique tool of using nucleoside-modified mRNA encoding CPD-photolyase that is an original approach to study the molecular mechanisms of UVB-induced DNA defects in human keratinocytes. The encoded CPD-photolyase is fully functional and repairs 90 % of CPD lesions within 1 h in response to photoreactivating light. In contrast, the physiologic NER requires more than 48 h to accomplish the same level of repair. The described procedure of using in vitro-synthesized mRNA encoding CPD-specific photolyase for transforming human keratinocytes is an excellent model to investigate UVB-induced cellular mechanisms [18].

2 Materials

Prepare all solutions using analytical grade reagents. Prepare and store all reagents at room temperature unless otherwise stated.

2.1 Transient Transfection

1. Human keratinocyte cell line: HaCaT cell line (*see* **Note 1**). Media for HaCaT cells: DMEM with 4.5 g/L glucose, 2 mM L-glutamine, 0.5 % antibiotic/antimycotic solution (10,000 U/ml penicillin, 10 mg/ml streptomycin, and 25 μg/ml amphotericin B), 10 % FBS, and 0.015 g/L phenol red.
2. In vitro-synthesized mRNA encoding CPD-photolyase (*see* **Note 2**). Store at −20 °C.
3. EpiLife medium (Life Technologies) (*see* **Note 3**). Store at 4 °C.
4. Lipofectamine LTX with PLUS Reagent (Cat. number 15338-100) (Life Technologies) (*see* **Note 4**). Store at 4 °C.
5. Siliconized microcentrifuge tubes.

2.2 UVB Treatment

1. Dulbecco's phosphate-buffered saline (DPBS) (Life Technologies).
2. Hank's buffered salt solution: 137 mM NaCl, 5.4 mM KCl, 4.4 mM KH_2PO_4, 0.33 mM Na_2HPO_4, 1.3 mM $CaCl_2$, 0.81 mM $MgSO_4$, 4.2 mM $NaHCO_3$, 1 g/l glucose, pH 7.4.
3. UVB tubes (Fig. 1) (*see* **Note 5**).
4. UVX digital radiometer with UVX-31 mid-range sensor (UVP Inc.) (Fig. 1) (*see* **Note 6**).
5. Aluminum foil.

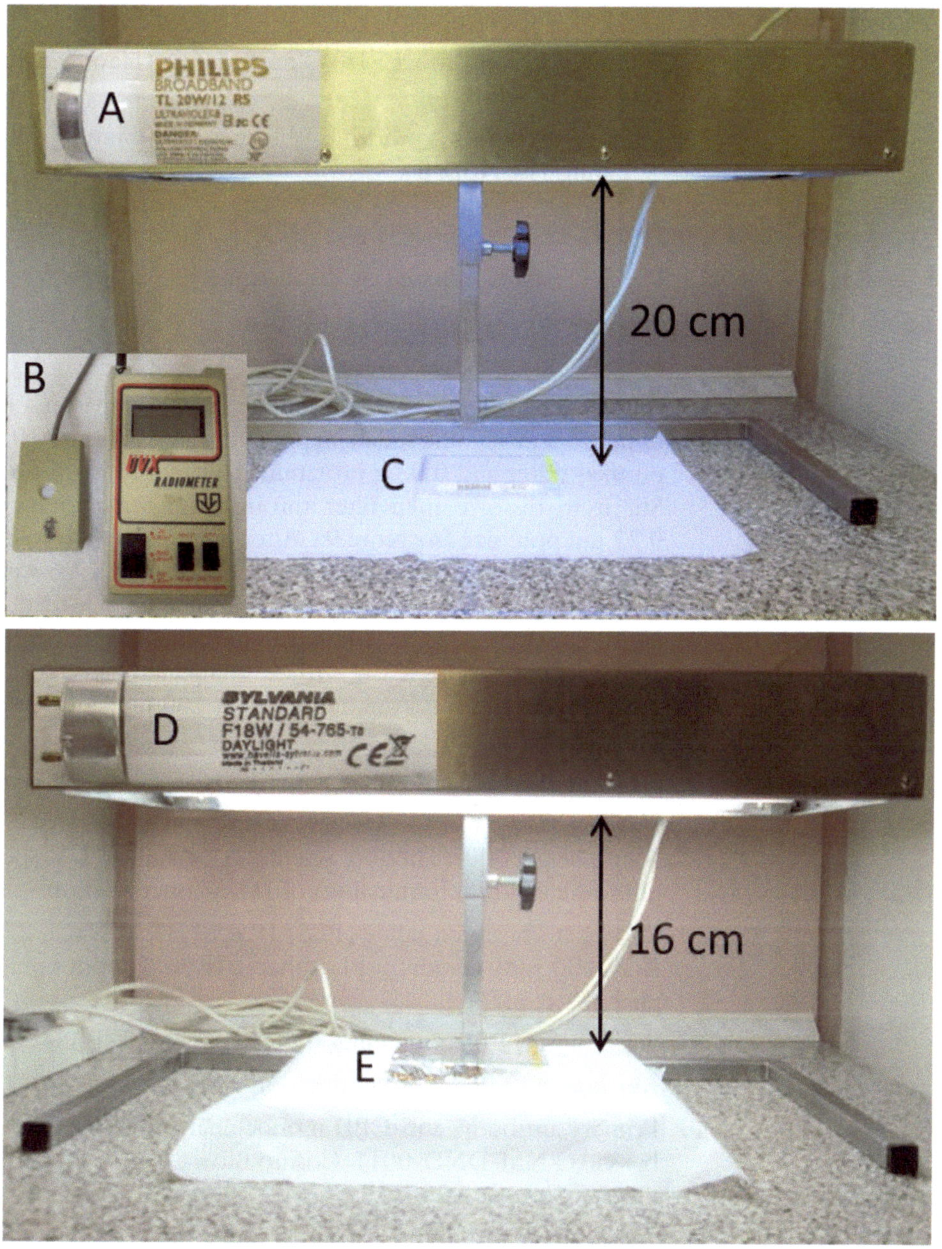

Fig. 1 Station for UVB and photoreactivating light treatments. For UVB treatment, a lamp with 2 UVB broadband tubes (**a**) and a UVX radiometer with the corresponding sensor are used (**b**). Cells cultured in a 96-well plate are placed 20 cm from the UVB source and exposed (**c**). For treatment with photoreactivating light, the UVB tubes are replaced with white fluorescent tubes (**d**). The plate with cells is covered with a glass slab and placed under the lamp (**e**). Some of the wells are covered with aluminum foil to deprive them of the energy of visible light (inactive CPD-photolyase) while others are left exposed (active CPD-photolyase)

2.3 Photoreactivation

1. DMEM without phenol red.
2. White fluorescent tubes F18W/54-765/T8 (Havells-Sylvania) (Fig. 1) (*see* **Note 7**).
3. Glass plate (length × width × thickness: 200 mm × 100 mm × 2 mm) (*see* **Note 8**).

2.4 CPD-Specific ELISA

1. DPBS.
2. Purified water (prepared by distillation).
3. Nuclease-free water.
4. Syringe filter, Filtropur S 0.22 μm.
5. A 96-well plate coated with protamine sulfate. Prepare a solution of 0.003 % protamine sulfate in purified water. Measure 10 ml of purified water and 0.03 g protamine sulfate into a glass beaker. Stir it to dissolve then filter through using membrane with 0.22 μm pore size (*see* **Note 9**). After diluting it 1000-fold with nuclease-free water aliquot 50 μl per well into a 96-well flat bottom plate. Incubate the plate without a lid at 37 °C overnight. By that time the solution evaporates and the dry protamine sulfate will coat the surface completely (*see* **Note 10**). Finally, wash the plate three times with 100 μl of nuclease-free water per well.
6. QIAamp DNA Mini Kit (Qiagen) for isolation of genomic DNA.
7. NanoDrop 2000 (Thermo Scientific) or similar UV spectrophotometer for determination of DNA concentration.
8. Phosphate-buffered saline (PBS, 10×): 1.37 M NaCl, 27 mM KCl, 100 mM disodium phosphate, 18 mM monopotassium phosphate, pH 7.4.
9. Wash buffer: PBS containing 0.05 % Tween 20 (PBS-T).
10. Blocking buffer: 2 % FBS diluted in PBS. Store at −20 °C.
11. Primary antibody: anti-CPD monoclonal antibody (Cat. number CAC-NM-DND-001) (Cosmo Bio). Clone: TDM-2, host: mouse. Store at −20 °C.
12. Secondary antibody: Goat anti-mouse IgG conjugated with horseradish peroxidase (HRP). Store at −20 °C.
13. Citrate-phosphate buffer: 51.4 mM disodium phosphate, 24.3 mM citric acid monohydrate, pH 5.0.
14. Substrate solution: 74 nM o-phenylenediamine (store at 4 °C) (*see* **Note 11**), 60 nM H_2O_2 in citrate-phosphate buffer (*see* **Note 12**).
15. Stop solution: 2 M sulfuric acid.
16. Anthos 2020 (Biochrom Ltd.) or similar microplate reader for measuring of absorbance at 492 nm.

3 Methods

In vitro transcription and HPLC purification of messenger RNA containing modified nucleosides have been reported in details [19, 20]. All procedures are performed at room temperature unless otherwise indicated.

3.1 Transfection of CPD-Photolyase mRNA into HaCaT Cells

1. Plate HaCaT cells into 96-well flat-bottom plates at a density of 2×10^4 cells per well in a final volume of 100 μl of complete DMEM 1 day prior to mRNA transfection.
2. In a siliconized tube combine CPD-photolyase-encoding mRNA (0.25 μg) with Lipofectamine PLUS reagent (0.25 μl) (*see* **Note 13**) in a final volume of 99 μl of EpiLife medium (HKGS-free, antibiotic-free) (*see* **Note 14**) and incubate for 5 min.
3. Add Lipofectamine LTX reagent (1.0 μl) into the tube containing the RNA and Lipofectamine PLUS reagent and incubate for 30 min.
4. Discard the DMEM from the cells.
5. Add 100 μl of lipofectamine-complexed RNA to a well of HaCaT cells in a 96-well plate and incubate the cells at 37 °C in a 5 % CO_2 atmosphere for 2 h.
6. After the incubation, replace the lipofectamine-RNA complex with 100 μl DMEM 10 % FBS culture medium.
7. Incubate the cells at 37 °C in a 5 % CO_2 atmosphere for 12 h (*see* **Note 15**).

3.2 Exposing HaCaT Cells Transfected With CPD-Photolyase mRNA to UVB Irradiation

1. Determine the output of the UVB tubes using radiometer. The range of measurement should be 0–2000 microW/cm^2.
2. Wash cells once with DPBS.
3. Cover cells with 50 μl DPBS (Hank's buffered salt solution can also be used) prior to UVB irradiation.
4. Expose cells to UVB at a dose of 20 mJ/cm^2 as long as it is required (*see* **Note 16**).
5. Immediately after UVB irradiation, change DPBS to 100 μl of complete DMEM without phenol red and replace UVB tubes with white fluorescent tubes in the lamp.
6. For photoreactivation, expose cells in selected wells to visible light (*see* **Note 17**) using white fluorescent tubes at a distance of 16 cm. Place a 4 mm thick glass slab on the plate to shield. Keep cells in selected wells in the dark by covering them with two layers of aluminum foil (Fig. 1) (*see* **Note 18**). Expose plate to visible light for 1 h.

7. One hour later, culture cells in complete DMEM at 37 °C in a 5 % CO_2 atmosphere for a variety of times to study DNA repair (Fig. 2). Cells treated with or without exposure to the activating visible light were incubated for 0, 5, 11, 23, and 47 h.

3.3 Isolation of Genomic DNA

1. Wash cells with 2 × 100 μl of DPBS, then collect them from the wells by trypsinization at 0, 1, 6, 12, 24, and 48 h after UVB irradiation and subsequent photoreactivation or incubation in the dark (Fig. 2).
2. Centrifuge cell suspension at 400 × *g* for 5 min and discard the supernatant.
3. Wash cell pellets with 500 μl of DPBS and centrifuge again at 400 × *g* for 5 min. Discard the supernatant and save he pellet (*see* **Note 19**).
4. Isolate genomic DNA from pellets of cells exposed to UVB irradiation and subsequent light or dark treatment using the QIAamp DNA mini kit (*see* **Note 20**) according to the manufacturer.
5. Measure the concentration of the genomic DNA isolates using spectrophotometer (e.g., NanoDrop 2000).

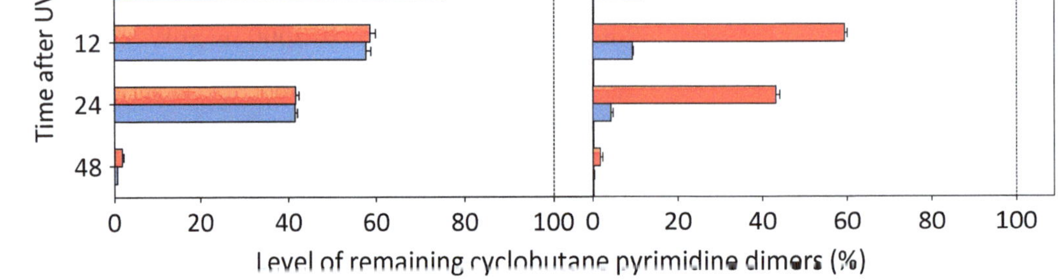

Fig. 2 Enhanced repair of UVB-induced CPDs in human keratinocytes transfected with CPD-specific photolyase mRNA. HaCaT cells were transfected with lipofectamine-complexed, pseudouridine-containing mRNA encoding CPD-photolyase or eGFP, which served as a control mRNA. Twelve hours later, cells were subjected to 20 mJ/cm^2 UVB and immediately exposed to photoreactivating light (photoreactivated) or left in the dark (non-photoreactivated) for 1 h. The remaining CPDs were measured by ELISA at the indicated times after UVB irradiation. After normalization to the basal level of cyclobutane pyrimidine dimers present in non-irradiated samples, the values were calculated relative to those obtained from cells that were harvested immediately after UVB irradiation

3.4 DNA Sample Coating

1. Dilute the isolated DNA samples in DPBS at the concentration of 0.3 ng/μl.
2. To denature the DNA, heat the samples in a PCR machine or digital dry bath at 100 °C for 10 min followed by rapid cooling on ice for 15 min.
3. Aliquot 50 μl of denatured DNA sample (15 ng) per well in triplicates to a 96-well plate precoated with protamine sulfate.
4. To coat the wells with DNA, incubate the plate without cover at 37 °C overnight until the wells are completely dry.

3.5 Measuring the Amount of UVB-Induced Cyclobutane Pyrimidine Dimers

1. Wash the DNA-coated plates three times with 100 μl per well of wash buffer (1× PBS-T).
2. Add 150 μl of blocking buffer to each well to prevent nonspecific binding of the antibody and incubate the plates at 37 °C for 30 min.
3. Wash the plates three times with 100 μl per well of wash buffer.
4. Add 100 μl of primary antibodies (anti-CPD monoclonal antibody, TDM-2) diluted in PBS (1:1000) to each well and incubate the plates at 37 °C for 30 min.
5. Wash the wells three times with 100 μl of wash buffer.
6. Add 100 μl of secondary antibody (HRP-conjugated, goat anti-mouse IgG) diluted in PBS (1:3000) to each well and incubate the plates at 37 °C for 30 min.
7. Wash the wells three times with 100 μl of wash buffer.
8. Equilibrate each sample with 150 μl of citrate-phosphate buffer per well (*see* **Note 21**).
9. Discard any citrate-phosphate buffer and add 100 μl of substrate solution (*see* **Note 22**) to each well and incubate the plates at 37 °C for 15 min.
10. Stop enzyme reaction by adding 50 μl of stop solution to each well.
11. Determine the absorbance at 492 nm using a microplate reader (e.g., an Anthos 2020) (Fig. 2).

4 Notes

1. HaCaT is a SV40-immortalized human keratinocyte cell line established by Boukamp et al. [21]. This cell line is a good substitute for normal human epidermal keratinocytes.
2. The coding sequences of CPD-photolyase gene (GenBank accession number: D26020) from *Potorous tridactylus* were codon-optimized and synthesized by Entelechon (Bad Abbach). The optimization increased the GC-content of the

encoding sequences from 51.8 % to 65.0 %, resulting in superior translation [18]. The pseudouridine-modified mRNA was generated in an in vitro transcription reaction [19] and purified by HPLC [20].

3. EpiLife is a sterile, serum-free medium for culture of normal human epidermal keratinocytes. This basal medium requires the addition of human keratinocyte growth supplement (HKGS) (Life Technologies) prior to use for culturing. However, for transfection use HKGS-free and antibiotic-free EpiLife medium.
4. Expression level of the encoded protein in HaCaT cells was highest after the transfection with Lipofectamine LTX with PLUS transfection reagent compared to other transfection reagents, including Lipofectin (Cat. number 18292-037) (Life Technologies) and TransIT-mRNA (Cat. number MIR 2225) (Mirus Bio).
5. We use a lamp with 2 Philips broadband tubes (TL 20 W/12 RS SLV) that emit radiation in the "B" bandwidth of the UV spectrum (290–315 nm).
6. Other types of radiometer can also be used. The point is to select the right sensor which corresponds to the wavelength output of the UV light source.
7. We use a standard lamp with two light tubes having neutral output. Other types of light sources with mid-range, cool white or daylight white fluorescent light can be used.
8. We use two transparent clear glass plates as a shield during photoreactivation.
9. The solution of protamine sulfate is stable at room temperature for 2 weeks, but it is the best to prepare it fresh for each application.
10. We find that at least 24 h are required to coat the 96-well plate with protamine sulfate.
11. Other organic compounds can be used as substrate to produce soluble end-products specifically for ELISA. We find that o-phenylenediamine is freely soluble in 37 °C water.
12. Prepare the substrate solution fresh each time.
13. Use a 1:1 ratio to the amount of sample.
14. Considering that RNA is quickly hydrolyzed at alkaline pH, we use EpiLife medium that has a more stable pH than DMEM.
15. Using western blotting, we detected CPD-photolyase as early as 1 h posttransfection. The level of the protein peaked between 6 and 12 h and then gradually decreased but remained detectable even 72 h after transfection with mRNA.
16. Exposure time is depend on the output of the UV light source. 1000 microW/cm^2 = 1 milliW/cm^2 = 1 milliJoule/s/cm^2.

17. Energy of visible light is required for activation of CPD-photolyase.
18. Plates were placed under the same fluorescent lamps, but the aluminum foil prevented photoreactivation, thus leaving the CPD-photolyase inactive.
19. Cell pellets can be stored at −70 °C and processed later.
20. Other DNA isolation kits are also suitable.
21. Keep the citrate-phosphate buffer on each sample at least 5 min.
22. This reaction is based on the production of yellow-orange formazan dyes.

Acknowledgements

This work was supported by National Institutes of Health (grant number R01NS029331 and R42HL87688), the Hungarian Scientific Research Fund (OTKA K105872), and the project of TÁMOP-4.2.2-A-11/1/KONV-2012-0031.

References

1. Sahin U, Kariko K, Tureci O (2014) mRNA-based therapeutics - developing a new class of drugs. Nat Rev Drug Discov 13:759–780
2. Kariko K, Buckstein M, Ni H, Weissman D (2005) Suppression of RNA recognition by Toll-like receptors: the impact of nucleoside modification and the evolutionary origin of RNA. Immunity 23:165–175
3. Kariko K, Muramatsu H, Welsh FA, Ludwig J, Kato H, Akira S, Weissman D (2008) Incorporation of pseudouridine into mRNA yields superior nonimmunogenic vector with increased translational capacity and biological stability. Mol Ther 16:1833–1840
4. Kariko K, Muramatsu H, Ludwig J, Weissman D (2011) Generating the optimal mRNA for therapy: HPLC purification eliminates immune activation and improves translation of nucleoside-modified, protein-encoding mRNA. Nucleic Acids Res 39, e142
5. Freeman SE, Hacham H, Gange RW, Maytum DJ, Sutherland JC, Sutherland BM (1989) Wavelength dependence of pyrimidine dimer formation in DNA of human skin irradiated in situ with ultraviolet light. Proc Natl Acad Sci U S A 86:5605–5609
6. Pfeifer GP (1997) Formation and processing of UV photoproducts: effects of DNA sequence and chromatin environment. Photochem Photobiol 65:270–283
7. Mitchell DL (1988) The relative cytotoxicity of (6-4) photoproducts and cyclobutane dimers in mammalian cells. Photochem Photobiol 48:51–57
8. Sage E (1993) Distribution and repair of photolesions in DNA: genetic consequences and the role of sequence context. Photochem Photobiol 57:163–174
9. Sancar A (2008) Structure and function of photolyase and in vivo enzymology: 50th anniversary. J Biol Chem 283:32153–32157
10. Li YF, Kim ST, Sancar A (1993) Evidence for lack of DNA photoreactivating enzyme in humans. Proc Natl Acad Sci U S A 90:4389–4393
11. Lehmann AR (1995) Nucleotide excision repair and the link with transcription. Trends Biochem Sci 20:402–405
12. Dong KK, Damaghi N, Picart SD, Markova NG, Obayashi K, Okano Y, Masaki H, Grether-Beck S, Krutmann J, Smiles KA, Yarosh DB (2008) UV-induced DNA damage initiates release of MMP-1 in human skin. Exp Dermatol 17:1037–1044
13. Vink AA, Moodycliffe AM, Shreedhar V, Ullrich SE, Roza L, Yarosh DB, Kripke ML (1997) The inhibition of antigen-presenting

activity of dendritic cells resulting from UV irradiation of murine skin is restored by in vitro photorepair of cyclobutane pyrimidine dimers. Proc Natl Acad Sci U S A 94:5255–5260

14. Berardesca E, Bertona M, Altabas K, Altabas V, Emanuele E (2012) Reduced ultraviolet-induced DNA damage and apoptosis in human skin with topical application of a photolyase-containing DNA repair enzyme cream: clues to skin cancer prevention. Mol Med Rep 5:570–574
15. Stege H (2001) Effect of xenogenic repair enzymes on photoimmunology and photocarcinogenesis. J Photochem Photobiol B 65:105–108
16. Stege H, Roza L, Vink AA, Grewe M, Ruzicka T, Grether-Beck S, Krutmann J (2000) Enzyme plus light therapy to repair DNA damage in ultraviolet-B-irradiated human skin. Proc Natl Acad Sci U S A 97:1790–1795
17. Asahina H, Han Z, Kawanishi M, Kato T Jr, Ayaki H, Todo T, Yagi T, Takebe H, Ikenaga M, Kimura SH (1999) Expression of a mammalian DNA photolyase confers light-dependent repair activity and reduces mutations of UV-irradiated shuttle vectors in xeroderma pigmentosum cells. Mutat Res 435:255–262
18. Boros G, Miko E, Muramatsu H, Weissman D, Emri E, Rozsa D, Nagy G, Juhasz A, Juhasz I, van der Horst G, Horkay I, Remenyik E, Kariko K, Emri G (2013) Transfection of pseudouridine-modified mRNA encoding CPD-photolyase leads to repair of DNA damage in human keratinocytes: a new approach with future therapeutic potential. J Photochem Photobiol B 129:93–99
19. Pardi N, Muramatsu H, Weissman D, Kariko K (2013) In vitro transcription of long RNA containing modified nucleosides. Methods Mol Biol 969:29–42
20. Weissman D, Pardi N, Muramatsu H, Kariko K (2013) HPLC purification of in vitro transcribed long RNA. Methods Mol Biol 969:43–54
21. Boukamp P, Petrussevska RT, Breitkreutz D, Hornung J, Markham A, Fusenig NE (1988) Normal keratinization in a spontaneously immortalized aneuploid human keratinocyte cell line. J Cell Biol 106:761–771

Part IV

Alteration of Cellular Function with Synthetic mRNA

Chapter 15

Delivery of Synthetic mRNA Encoding FOXP3 Antigen into Dendritic Cells for Inflammatory Breast Cancer Immunotherapy

Gayathri R. Devi and Sritama Nath

Abstract

Dendritic cell (DC)-based vaccines are commonly used for cancer immunotherapy. To prepare vaccines, DCs are pulsed or transfected with either: (a) defined peptides of tumor-associated antigens, (b) total protein isolated from the tumor cell, (c) autologous total RNA isolated from the tumor cell, (d) synthetic tumor-antigen-encoding mRNA, or (e) genes that encode for specific tumor-associated antigens. Introduction of tumor-associated antigen(s) and subsequent generation of mature DCs that can stimulate tumor-antigen-specific cytotoxic T lymphocytes comprise the critical steps of cancer vaccine preparation. Here, we described a method of: (a) preparing and delivering synthetic FOXP3 mRNA into human DCs, (b) generating mature DCs, (c) generating FOXP3-specific cytotoxic T lymphocytes, and (d) evaluating cytotoxicity of FOXP3-specific cytotoxic T lymphocytes against inflammatory breast cancer cells.

Key words Synthetic mRNA, Dendritic cell, Tumor-associated antigens, Cancer vaccines, Electroporation, Lipofection, Cytotoxic T lymphocyte (CTL) assay

1 Introduction

Dendritic cell (DC)-based vaccines are being extensively developed for treatment of several malignancies, some of which have successfully translated to phase II human clinical study [1, 2]. mRNAs encoding tumor antigens are introduced into bone marrow-derived DCs, a potent antigen-presenting cells capable of activating naive T cells both in vitro and in vivo, which in turn can elicit an effective antitumor response [3, 4]. Nair et al. first reported generation and use of CEA-mRNA-loaded DCs to induce CEA-specific cytotoxic T lymphocytes (CTL) in patients with metastatic malignancies expressing CEA antigen [5]. Since then several studies have demonstrated successful induction of antigen-specific CTL response using antigen-encoding synthetic mRNA-transfected DCs [6–8].

Robert E. Rhoads (ed.), *Synthetic mRNA: Production, Introduction Into Cells, and Physiological Consequences*, Methods in Molecular Biology, vol. 1428, DOI 10.1007/978-1-4939-3625-0_15,

Some of the advantages of transfecting DCs with mRNA that encodes the full-length antigen are: (a) simultaneous stimulation of a broader T cell repertoire due to presentation of both known and unknown antigenic epitopes by the mRNA [9], (b) stimulation of both CD8+ and CD4+ T cells, since the tumor-associated antigen (s) can be presented by both HLA class I and II molecules [10–12], (c) transient expression of the antigen by mRNA, which does not integrate within the host genome, making it much safer for clinical use [13], and (d) higher transfection efficiency of antigen-encoding mRNA compared to antigen-encoding plasmid DNA [14].

However, it is challenging to work with RNA as it is susceptible to degradation by ubiquitous RNAases. Various modifications are made to the 5′- and 3′-ends of synthetic mRNA to reduce its vulnerability to degradation by RNAases and increase its cellular stability. For example, incorporation of $m_2^{7,2'}$-OGppspG (β-S-ARCA) phosphorothioate cap to the 5′-end of synthetic mRNA increases its cellular stability and translational activity [15]. Further modifications of the cap include substitution of β-phosphate with BH_3, and selenium (Se), or substitution of α-β or β-γ O with NH, which improve mRNA stability, translational efficiency, and resistance to cleavage by intracellular pyrophosphatases such as Dcp2 [16]. Additionally, polyadenylation [poly(A)] on the 3′-end increases the half-life of mRNA [17].

Gilboa et al. first demonstrated that immunization against FOXP3, a member of the forkhead winged helix family of transcriptional regulators expressed in Tregs, is capable of generating FOXP3-specific CTL [18]. Soon after, our research group found that immunization of mice with FOXP3 mRNA-transfected DCs stimulates FOXP3-specific CTL, that targets FOXP3-expressing IBC cells [6], an aggressive subtype of breast cancer characterized by the formation of tumor emboli with symptoms resembling painful chest wall inflammation.

This chapter attempts to describe a method of generating FOXP3-specific DC-based vaccine, elucidating the procedure of: (a) synthesizing synthetic mRNA encoding FOXP3 antigen, (b) delivering anti-FOXP3 synthetic mRNA into DCs, (c) generating FOXP3-specific mature DCs, (d) generating FOXP3-specific CTLs, and (e) evaluating cytotoxicity of CTLs against FOXP3 expressing IBC cells.

2 Materials

2.1 IBC Cells and Culture Conditions

1. SUM149 and SUM190—human IBC cells from primary tumor (Asterand Bioscience, Detroit, MI).
2. Media for SUM149 [19, 20]: Ham's F-12 supplemented with 5 % FBS, insulin (5 μg/mL), hydrocortisone (1 μg/mL), penicillin (100 U/mL), and streptomycin (75 U/mL).

3. Media for SUM190 [19, 20]: Ham's F-12 supplemented with 0.1 % bovine serum albumin, sodium selenite, 3,3′,5-triiodo-L-thyronine, transferrin, ethanolamine, insulin (5 μg/mL), hydrocortisone (1 μg/mL), penicillin (100 U/mL), and streptomycin (75 U/mL).

2.2 Preparation of Synthetic FOXP3 mRNA [21]

1. RNase, DNase-free microcentrifuge tubes.
2. RNase, DNase-free tips.
3. RNase Zap RNAase Decontamination Solution.
4. Restriction enzymes.
5. 10× NEBuffer.
6. Nuclease-free water.
7. TE buffer: 10 mM Tris–HCl (pH 7–8), 1 mM EDTA.
8. 70 % ethanol.
9. 3 M sodium acetate.
10. 5 M ammonium acetate.
11. 0.5 M EDTA.
12. Agarose.
13. RNeasy Mini Kit (Qiagen, Valencia, CA).
14. High Capacity cDNA Reverse Transcription Kits (Life Technologies).
15. T7 mMessage mMachine kit (Ambion, Austin, TX), which provides the following reagents:
    - (a) Enzyme Mix: Buffered 50 % glycerol containing RNA polymerase, RNase Inhibitor, and other components.
    - (b) 10× Reaction Buffer.
    - (c) 2× NTP/CAP: (NTP:GTP:CAP—5:1:4).
    - (d) GTP: 30 mM (*see* **Note 1**).
    - (e) TURBO DNase (2 U/μL).
    - (f) pTRI-Xef (linearized TRIPLEscript plasmid containing *Xenopus* elongation factor 1α gene), 0.5 μg/μL (Control Template).
    - (g) Lithium Chloride Precipitation Solution: 7.5 M lithium chloride, 50 mM EDTA.
    - (h) Gel loading buffer II: 95 % formamide, 0.025 % xylene cyanol, 0.025 % bromophenol blue, 18 mM EDTA, and 0.025 % SDS.
16. 10× MOPS Buffer: 200 mM MOPS (pH 7.0), 80 mM sodium acetate, and 10 mM EDTA (pH 8.0) in nuclease-free water. Dilute the 10× MOPS buffer ten times to get 1× working solution.

17. MOPS formaldehyde agarose gel (2 %): Dissolve 2 g agarose in 100 mL of 1× MOPS Buffer. In a fume hood, add 37 % of formaldehyde to melted agarose to final concentrations of 2.2 M.
18. Tabletop microcentrifuge at 4 °C.

2.3 Generation of DCs from PBMCs

1. Falcon tube.
2. T-150 tissue culture flasks.
3. Sterile PBS.
4. Cell dissociation buffer.
5. Media for culturing DCs: 30 mL AIM-V media supplemented with human GM-CSF (800 U/mL) and human IL-4 (500 U/mL).

2.4 Transfection of DCs with Antigen Encoding Synthetic mRNA

1. Sterile polystyrene tubes.
2. RNase-free PBS.
3. Opti-MEM (Gibco/BRL, Bethesda, MD).
4. Lipid DMRIE (Vical, San Diego, CA).
5. Cationic lipid 1,2-dioleoyl-3trimethylammonium-propane/ 1,2-dioleoyl-sn-glycero-3-phosphoethanolamine DOTAP and DOPE (molar ratio 1:1; in chloroform) (Avanti Polar Lipids, Alabaster, AL).
6. Maturation media: AIM-V supplemented with maturation cytokine cocktail comprising of GM-CSF (800 U/mL), IL-4 (500 U/mL), TNF-α (5 ng/mL), IL-1β (5 ng/mL), IL-6 (150 ng/mL), and PGE2 (1 mg/mL).
7. Electro Square Porator ECM 830.

2.5 Stimulation of T Cells

1. DNase I (200 U/mL).
2. EasySep T cell Enrichment Kit (Stemcell Technologies, Vancouver, BC).
3. CTL stimulation medium: RPMI 1640 supplemented with FCS (10 %), L-glutamine (2 mM), HEPES (20 mM), Na-pyruvate (1 mM), MEM nonessential amino acids (0.1 mM), MEM amino acids (50× solution), penicillin (100 IU/mL), streptomycin (100 mg/mL), and β-mercaptoethanol (5×10^{-5} M).

2.6 CTL Assay

1. V-bottom 96-well plate.
2. Round-bottom 96-well plate.
3. Europium (Perkin-Elmer, Waltham, MA).
4. CTL stimulation medium (with no penicillin-streptomycin).

5. Delfia enhancement solution (Perkin-Elmer, Waltham, MA).
6. VICTOR3 Multilabel Counter (Perkin-Elmer, Wallac, Finland).
7. 10 % sodium dodecyl sulfate.

3 Methods

3.1 In Vitro Synthesis of FOXP3 mRNA

Large quantity of tissues are required to isolate sufficient amount of mRNA (antigens) for an effective and sustained immunization protocol. However, in most cases the tumor biopsies are not large enough to yield sufficient quantity of mRNA. Polymerase chain reaction (PCR) offers a practical solution to this problem, where inexhaustible amount of antigen-encoding mRNA can be generated from a miniscule amount of mRNA isolated from a limited tissue sample. Antigen-encoding mRNA is first reverse transcribed into cDNA, which is further cloned into a plasmid. Large quantities of the antigen-encoding mRNA are further synthesized in vitro using the mMESSAGE mMACHINE T7 Ultra Kit. The kit incorporates the Anti-Reverse Cap Analog (ARCA) [$m^7G(5')ppp(5')G$] to the 5′-end and adds a poly (A) tail to 3′-end, generating stable synthetic mRNA that yields high quantity of protein.

3.1.1 Cloning of Human FOXP3 mRNA

1. Isolate total RNA from human CD4+/CD25+ T regulatory cells.
2. Perform reverse transcription (RT) to generate single stranded cDNA of human FOXP3 mRNA using the primer pair:

   5′-TATATAAAGCTTGCCACCATGGCCCTTGGCCCAT-3′

   5′-TATATAGGATCCTCAGGGGCCAGGTGTAGGGTTG-3′
3. Setup the RT by mixing 10 μL of RNA with the 2× RT master mix consisting of 2 μL of 10× reaction buffer, 0.8 μL of 25× dNTP mix (100 mM), 2 μL of primer mix, 1 μL of reverse transcriptase, 1 μL of RNase inhibitor, and 3.2 μL of nuclease-free H_2O provided in the High Capacity cDNA Reverse Transcription Kit.
4. Setup the PCR conditions following directions in the High Capacity cDNA Reverse Transcription Kit manual.
5. Digest the human FOXP3 cDNA and pSP73-Sph/A64 plasmid using HindIII and BamHI restriction buffers.
6. Set up a ligation reaction to insert the digested human FOXP3 cDNA into pSP73-Sph/A64 plasmid.

3.1.2 Plasmid Linearization

1. Before setting up in vitro transcription reaction, linearize the pSP73-Sph/A64 plasmid DNA by digesting it with appropriate restriction enzymes (*see* **Note 2**).
2. In a 1.5 mL DNAse-free microcentrifuge tube, setup a 50 μL reaction by adding 1–3 μg of DNA, 1 μL of Restriction Enzyme (10 U/μg of DNA), 5 μL 10× NE Buffer (*see* **Note 3**), 5 μL BSA (if recommended by manufacturer), and × μL of nuclease-free water up to the total volume. Incubate the reaction for 1 h at the required temperature (*see* **Note 4**).
3. To terminate the restriction digestion, add 1/20th volume of 0.5 M EDTA, 1/10th volume of 3 M sodium acetate or 5 M ammonium acetate, and 2 volumes of ethanol. Mix the reagents well and incubate for 15 min at –20 °C.
4. Microcentrifuge at top speed (>5000 × *g*) at 4 °C for 15 min to pellet the DNA. Remove the supernatant, respin the tube for few seconds, and remove the residual fluid with a very fine-tipped pipet.
5. Resuspend the pelleted DNA in nuclease-free water or TE buffer at a concentration of 1 μg/μL (*see* **Note 5**).
6. Run the products of restriction digestion on an agarose gel to verify that the restriction digestion is complete, and most of the circular plasmid DNA has been successfully cleaved to yield linear plasmid DNA. Run the plasmid DNA in an agarose gel prepared in 1× TE Buffer.

3.1.3 In Vitro Transcription

1. Thaw the frozen reagents and place them on ice until further use.
2. Before setting up the reaction, spray RNase Zap RNA Decontamination Solution on the workbench, pipettes, and gloves to remove any contaminating RNase, which will reduce unnecessary degradation of RNA during handling.
3. Setup the transcription reactions at room temp (*see* **Note 6**). To make a 20 μL reaction, add 10 μL 2× NTP/CAP, 2 μL 10× Reaction Buffer, 1 μg linearized plasmid DNA (template), 2 μL Enzyme Mix, and 5 μL nuclease-free water in a 1.5 mL DNAse and RNAse-free microcentrifuge tube.
4. Similarly, setup another reaction using linearized pTRI-Xef (0.5 μg/μL) as template. This is positive control for in vitro transcription reaction.
5. Mix the reagents in the tubes well followed by brief centrifugation to collect the reaction mixture at the bottom of the tubes.
6. Incubate the tubes for 2 h at 37 °C for mRNA synthesis to occur. One reaction yields 25–30 μg of mRNA.
7. To terminate the reaction, add 1 μL of TURBO DNase I to the reaction tubes, mix well, and incubate for 15 min at 37 °C. DNase I treatment removes the template DNA leaving behind single stranded mRNA.

3.1.4 Transcript Purification

1. Perform Lithium chloride (LiCl) precipitation to purify the mRNA from unincorporated nucleotides and proteins (*see* **Note 7**). Add 30 μL of nuclease-free water and 30 μL LiCl Precipitation Solution. Mix the reagents thoroughly and incubate for 30 min at –20 °C.
2. Centrifuge at 4 °C for 15 min at maximum speed to pellet the RNA. Carefully remove the supernatant.
3. Wash the pellet once with ~1 mL 70 % ethanol and recentrifuge to maximize removal of unincorporated nucleotides. Carefully remove the 70 % ethanol and resuspend the RNA in TE buffer.

3.1.5 Transcript Quantitation and Storage

1. Determine the concentration of RNA using a conventional or a nanodrop spectrophotometer. Dilute the reaction 1:100 to get an absorbance reading within the linear range of spectrophotometer. Calculate the concentration of single stranded RNA using the following formulae: A260 × dilution factor × 40 = μg/mL RNA.
2. Adjust the final concentration of mRNA to 1 μg/μL. Make smaller aliquots of the mRNA and store it at −80 °C.
3. After mixing equal volumes of the mRNA sample and gel loading buffer II, heat the mixture for 3–5 min at 95 °C and run on MOPS formaldehyde agarose gel to confirm the synthesis of full-length RNA.

3.2 Generation of DCs from PBMCs

1. Thaw a cryopreserved vial of isolated human PBMC (25×10^6 cells/mL/vial) and wash them once with 1× PBS.
2. Resuspend the PBMCs in 30 mL of AIM-V media and plate them in a T-150 tissue culture flask.
3. Place the flask in an upright position and incubate the cells for 1 h at 37 °C in a 5 % CO_2 humidified chamber.
4. Bring the nonadherent cells back to suspension by gently rocking the flask from side to side.
5. Discard the media containing nonadherent cells.
6. Replenish the adherent cells in the tissue culture flask with 30 mL of AIM-V media supplemented with 800 U/mL of human GM-CSF and 500 U/mL of human IL-4. Incubate the cells at 37°C, 5 % CO_2 for 7 days to generate DCs.
7. On day 7, harvest the DCs by washing the adherent cells vigorously with ice-cold sterile PBS. Collect the cells in a falcon tube.
8. To dislodge the DCs firmly adhered to the surface of the tissue culture flask, add 10 mL Cell Dissociation Buffer to the cells and incubate at 37 °C for 5 min. Gently tap the sides of the flask to dislodge cells. After the DCs have visibly detached, add

10 mL of complete growth medium to the flask, resuspend the cells, and collect them in the falcon tube.

9. Wash the DCs with ice-cold sterile PBS and resuspend them in PBS.
10. After counting the DCs, maintain them on ice until further use.

3.3 Introduction of Synthetic FOXP3 mRNA into DCs

Below described are some of the techniques that are commonly used for transfecting DCs with antigen-encoding mRNA. In comparison to liopfection, electroporation is preferred, because it is (a) less cytotoxic for the DCs [14] and (b) GMP-approved systems are available for electroporation, which is necessary for clinical administration [22]. Internalization of mRNA liposome complex by DCs occurs preferentially by macropinocytosis and/or phagocytosis [23, 24].

3.3.1 Delivery by Electroporation

1. To transfect DCs by electroporation, use 3–4 μg of in vitro transcribed FOXP3 mRNA per 1×10^6 DCs.
2. In a 2-mm cuvette, take 5×10^6–6×10^6 DCs resuspended in 200 μL of Opti-MEM (5) and add 15–20 μg of synthetic FOXP3 mRNA (1 μg/μL). Mix the contents well.
3. In a separate cuvette, similarly load the DCs with 15–20 μg of GFP mRNA and use it as a control to measure the transfection efficiency (*see* **Note 8**).
4. Pulse the DCs and RNA at 300 V for 500 μs using an Electro Square Porator ECM 830 (*see* **Note 9**).
5. Immediately resuspend the DCs in 5–6 mL of maturation media (1×10^6 cells/mL media).
6. Plate the DCs in a 5.2 cm plate and incubate them for 6–7 h at 37 °C in a 5 % CO_2 humidified chamber.

3.3.2 Delivery by Lipofection [14]

1. To transfect DCs by lipofection, use 2.5 μg of in vitro transcribed FOXP3 mRNA or 5 μg of total RNA per 1×10^6 DCs.
2. In a sterile polystyrene tube mix 245 μL of Opti-MEM with 5 μL of FOXP3 mRNA (1 μg/μL).
3. In a separate sterile polystyrene tube mix 230 μL of Opti-MEM with 20 μL of lipid DMRIE-C (RNA to lipid ratio is 1:4).
4. Mix the two solutions (500 μL total) and incubate for 10–15 min at room temperature allowing the RNA–lipid complex to form.
5. Add the RNA–lipid complex to DCs (2×10^6 cells/mL) in Opti-MEM and incubate for 2 h at 37 °C in a 5 % CO_2 humidified chamber.
6. Wash the DCs and replenish them with maturation media. Incubate them for additional 6–7 h at 37 °C in a 5 % CO_2 humidified chamber.

3.3.3 Delivery of Synthetic mRNA with Cationic Polymers [25]

1. To make DOTAP/DOPE liposomes, transfer 100 μL of a DOTAP/DOPE chloroform solution (10 mg/mL) to a sterile glass flask.
2. Evaporate the solvent by holding it under nitrogen or argon stream. A thin film of lipid residue will be left behind on the glass surface.
3. Place the lipid residue on a vacuum pump for 10–15 min to remove any residual organic solvent.
4. Immediately dissolve the lipid film in 1 mL of nuclease-free water.
5. In the presence of sterile glass beads, sonication the solution to yield DOTAP/DOPE liposomes.
6. The total lipid concentration in these liposome dispersions will be 1 mg/mL.
7. Solution may be passed through a 0.22 μm filter to sterilize.
8. Dissolve 15 μL of the liposome dispersions in 35 μL of OptiMem.
9. Add 50 μL of the liposome containing solution to 50 μL of mRNA solution, containing 2.5 μg FOXP3 mRNA (1 μg/μL) in 47.5 μL OptiMem.
10. To allow the liposome RNA complex to form, incubate for 10 min at room temperature, followed by addition of 900 μL of OptiMem to the reaction mix.
11. Add 1 mL of the resulting solution to 5×10^5 of DCs cells and incubate for 20 h at 37 °C in a 5 % CO_2 [26].
12. Following incubation, remove the complexes, wash, and replenish the DCs with maturation media. Incubate them for additional 6–7 h at 37 °C in a 5 % CO_2 humidified chamber.

3.4 Application of FOXP3 mRNA-Transfected DCs for Anticancer Immunotherapy

3.4.1 In Vitro Stimulation of T Cells

1. Thaw a vial of cryopreserved PBMCs (10×10^6 cells/mL freezing media) and wash with PBS.
2. Treat the resuspend PBMCs in PBS with DNase I (200 U/mL) for 20 min at 37 °C in a 5 % CO_2 humidified chamber.
3. Isolate CD3+ T cells from DNase I-treated PBMCs using EasySep T cell Enrichment Kit (*see* **Note 10**).
4. To stimulate the isolated CD3+ T cells, coculture the CD3+ T cells (2–2.5 $\times 10^6$ cells/mL) with FOXP3 mRNA-transfected matured DCs (0.2–0.25 $\times 10^6$ cells/mL) in CTL stimulation medium supplemented with IL-7 (25 ng/mL). The responder (CD3+ T cells) to stimulator (DCs) ratio (*R*:*S*) being 10:1.

5. On day 3 add IL-2 (100 U/mL) to the culture.
6. On day 7 restimulate the T cells with FOXP3 mRNA-transfected DCs (*R*:*S* of 10:1) in complete RPMI supplemented with IL-2 (50–100 U/mL).
7. After 5–6 days, harvest the T cells, count, and use them as effector cells in CTL assay.

3.4.2 Evaluation of CTL Response Against FOXP3 Expressing IBC Cells

CTL assay is performed to evaluate generation of FOXP3-specific CTLs, which can effectively lyse FOXP3 positive IBC cells. In CTL assay, typically tumor cells (target cells) expressing the antigen are labeled with nonradioactive or radioactive isotopes, such as europium and chromium respectively. DCs are treated overnight in AIM-V media supplemented with GM-CSF and IL-4, on the previous day of performing the CTL assay.

1. To label, incubate 5–10 × 10^6 IBC cells with europium for 20 min at 4 °C.
2. In a V-bottom 96-well plate, seed 10,000 europium labeled IBC cells with serially diluted T cells (effector cells) in 200 μL of CTL stimulation medium with no penicillin–streptomycin, so as to get different effector to target cell ratio (e.g., *5:1, 10:1. 20:1, 40:1*).
3. Centrifuge the plate at 500 × *g* for 3 min to bring down cells from suspension and settle down onto the bottom surface of the plate.
4. Incubate cells for 4 h at 37 °C.
5. Harvest 50 μL of the supernatant from the plate and mix it with 150 μL of enhancement solution in 96-well flat-bottom plate.
6. Measure europium release using the VICTOR3 Multilabel Counter.
7. To determine maximum release, treat the target cells with 10 % sodium dodecyl sulfate that will release all the trapped europium.
8. To determine spontaneous release, incubate the target cells in media without T cells.
9. Use the following formulae to determine the specific cytotoxic activity of T lymphocytes: % specific release = [(experimental release − spontaneous release)/(maximum release − spontaneous release)] × 100. In a valid data, spontaneous release of the target cells should be less than 25 % of maximum release by detergent.

4 Notes

1. During synthesis of large transcripts (>than 5 or 6 kb), GTP availability can become rate limiting, resulting in low yield of transcript or premature abortion of the transcription. To circumvent premature termination, supplement the reaction with extra GTP.
2. The restriction enzymes are chosen based on the presence of restriction sites downstream of the insert to be transcribed.
3. Please refer the manufacture's instruction to select the type of 10× NEBuffer for the restriction digestion.
4. Please refer the manufacture's instruction to select the temperature for restriction digestion.
5. The DNA template should be relatively free of contaminating proteins and RNA to get high yields of transcript.
6. Spermidine in the 10× Reaction Buffer can co-precipitate the template DNA if the reaction is assembled on ice. Add the 10× Reaction Buffer after water and the ribonucleotides have been added to the tubes.
7. LiCl precipitation may not efficiently precipitate RNAs that are smaller than 300 nucleotides. Also, the concentration of RNA should be at least 0.1 μg/μL to assure efficient precipitation.
8. Evaluate transfection efficiency by measuring the expression of a reporter transgene such as green fluorescent protein by flow cytometry or immunofluorescence.
9. Open β-mercaptoethanol bottle in a fume hood. The reagent generates toxic fumes, which may cause skin irritation or ocular damage on contact.
10. Check the purity of the isolated CD3+ T cells by flow cytometry.

Acknowledgments

The authors would like to acknowledge the support of Department of Defense grants W81XWH-07-1-0392 (GRD) and W81XWH-13-1-0047 (GRD), members of Devi laboratory, and collaborations with Dr. Michael Morse and Dr. Smita Nair at Duke University.

References

1. Oshita C, Takikawa M, Kume A, Miyata H, Ashizawa T, Iizuka A, Kiyohara Y, Yoshikawa S, Tanosaki R, Yamazaki N, Yamamoto A, Takesako K, Yamaguch IK, Akiyama Y (2012) Dendritic cell-based vaccination in metastatic melanoma patients: phase II clinical trial. Oncol Rep 28(4):1131–1138
2. Hersey P, Menzies S, Halliday G, Nguyen T, Farrelly M, DeSilva C, Lett M (2003) Phase I/II study of treatment with dendritic cell vaccines in patients with disseminated melanoma. Cancer Immunol Immunother 53(2):125–134
3. Mitchelle D, Nair SK (2000) RNA transfected dendritic cells as cancer vaccines. Curr Opin Mol Ther 2(2):176–181
4. Gilboa E (2007) Review series DC-based cancer vaccines. J Clin Invest 117(5):1195–1203
5. Nair SK, Hull S, Coleman D, Gilboa E, Lyerly HK, Morse MA (1999) Induction of carcinoembryonic antigen (CEA)-specific cytotoxic T- lymphocyte responses in vitro using autologous dendritic cells loaded with CEA peptide or CEA RNA in patients with metastatic malignancies expressing CEA. Int J Cancer 82:121–124
6. Nair SK, Aldrich AJ, McDonnell E, Cheng Q, Aggarwal A, Patel P, Williams MM, Boczkowski D, Lyerly HK, Morse MA, Devi GR (2013) Immunologic targeting of FOXP3 in inflammatory breast cancer cells. PLoS One 8 (1):e53150
7. Heiser A, Dahm P, Yancey DR, Maurice MA, Boczkowski D, Nair SK, Gilboa E, Vieweg J (2000) Human dendritic cells transfected with RNA encoding prostate-specific antigen stimulate prostate-specific CTL responses in vitro. J Immunol 164(10):5508–5514
8. Heiser A, Maurice MA, Yancey DR, Tumors M, Coleman DM, Dahm P, Vieweg J (2001) Human dendritic cells transfected with renal tumor RNA stimulate polyclonal T-cell responses against antigens expressed by primary and metastatic tumors. Cancer Res 61 (8):3388–3393
9. Van Nuffel AM, Benteyn D, Wilgenhof S, Pierret L, Corthals J, Heirman C, van der Bruggen P, Coulie PG, Neyns B, Thielemans K, Bonehill A (2012) Dendritic cells loaded with mRNA encoding full-length tumor antigens prime CD4+ and CD8+ T cells in melanoma patients. Mol Ther 20:1063–1074
10. Bonehill A, Heirman C, Tuyaerts S, Michiels A, Breckpot K, Brasseur F, Zhang Y, Van Der Bruggen P, Thielemans K (2004) Messenger RNA-electroporated dendritic cells presenting MAGE-A3 simultaneously in HLA class I and class II molecules. J Immunol 172 (11):6649–6657
11. Su Z, Dannull J, Yang BK, Dahm P, Coleman D, Yancey D, Sichi S, Niedzwiecki D, Boczkowski D, Gilboa E, Vieweg J (2005) Telomerase mRNA-transfected dendritic cells stimulate antigen-specific CD8+ and CD4+ T cell responses in patients with metastatic prostate cancer. J Immunol 174:3798–3807
12. Bonehill A, Heirman C, Thielemans K (2005) Genetic approaches for the induction of a CD4+Tcell response in cancer immunotherapy. J Gene Med 7:686–695
13. Mitchell DA, Nair SK (2000) Nucleic acid therapeutics RNA-transfected dendritic cells in cancer immunotherapy. J Clin Invest 106 (9), 1065–1069
14. Van Tendeloo VFI, Ponsaerts P, Lardon F, Nijs G, Lenjou M, Van Broeckhoven C, Van Bockstaele DR, Berneman ZN (2001) Highly efficient gene delivery by mRNA electroporation in human hematopoietic cells: Superiority to lipofection and passive pulsing of mRNA and to electroporation of plasmid cDNA for tumor antigen loading of dendritic cells. Blood 98 (1):49–56
15. Grudzien-Nogalska E, Jemielity J, Kowalska J, Darzynkiewicz E, Rhoads RE (2007) Phosphorothioate cap analogs stabilize mRNA and increase translational efficiency in mammalian cells. RNA 13:1745–1755
16. Su W, Slepenkov S, Grudzien-Nogalska E, Kowalska J, Kulis M, Zuberek J, Lukaszewicz M, Darzynkiewicz E, Jemielity J, Rhoads RE (2011) Translation, stability, and resistance to decapping of mRNAs containing caps substituted in the triphosphate chain with BH3, Se, and NH. RNA 17:978–988
17. Drummond DR, Armstrong J, Colman A (1985) The effect of capping and polyadenylation on the stability, movement and translation of synthetic messenger RNAs in Xenopus oocytes. Nucleic Acids Res 13(20):7375–7391
18. Nair S, Boczkowski D, Fassnacht M, Pisetsky D, Gilboa E (2007) Vaccination against the forkhead family transcription factor Foxp3 enhances tumor immunity. Cancer Res 67 (1):371–380
19. Forozan F, Veldman R, Ammerman CA, Parsa NZ, Kallioniemi A, Kallioniemi OP, Ethier SP (1999) Molecular cytogenetic analysis of 11 new breast cancer cell lines. Br J Cancer 81 (8):1328–1334

20. Aird KM, Ding X, Baras A, Wei J, Morse MA, Clay T, Lyerly HK, Devi GR (2008) Trastuzumab signaling in ErbB2-overexpressing inflammatory breast cancer correlates with X-linked inhibitor of apoptosis protein expression. Mol Cancer Ther 7(1):38–47
21. Kreiter S, Selmi A, Diken M, Koslowski M, Britten CM, Huber C, Türeci Ö, Sahin U (2010) Intranodal vaccination with naked antigen-encoding RNA elicits potent prophylactic and therapeutic antitumoral immunity. Cancer Res 70(9):9031–9040
22. Linhong L, Cornell A, Shivakumar R, Peshwa MV (2012) Large volume flow electroporation of mRNA: clinical scale process. Methods Mol Biol 969:127–138
23. De Haes W, Van Greet M, Céline M, De Smedt SC, Vanham G, Rejman J (2012) Internalization of mRNA lipoplexes by dendritic cells. Mol Pharm 9(10):2942–2949
24. Diken M, Krieter S, Selmi A, Britten C, Huber C, Tureci O, Sahin U (2011) Selective uptake of naked vaccine RNA by dendritic cells is driven by macropinocytosis and abrogated upon DC maturation. Gene Ther 18:702–708
25. Rejman J, Tavernier G, Bavarsad N, Demeester J, De Smedt SC (2010) MRNA transfection of cervical carcinoma and mesenchymal stem cells mediated by cationic carriers. J Control Release 147(3):385–391
26. Pollard C, Rejman J, De Haes W, Verrier B, Van Gulck E, Naessens T, De Smedt S, Bogaert P, Grooten J, Vanham G, De Koker S (2012) Type I IFN counteracts the induction of antigen-specific immune responses by lipid-based delivery of mRNA vaccines. Mol Ther 21(1):251–259

Chapter 16

Immune Monitoring Using mRNA-Transfected Dendritic Cells

Troels Holz Borch, Inge Marie Svane, and Özcan Met

Abstract

Dendritic cells are known to be the most potent antigen presenting cell in the immune system and are used as cellular adjuvants in therapeutic anticancer vaccines using various tumor-associated antigens or their derivatives. One way of loading antigen into the dendritic cells is by mRNA electroporation, ensuring presentation of antigen through major histocompatibility complex I and potentially activating T cells, enabling them to kill the tumor cells. Despite extensive research in the field, only one dendritic cell-based vaccine has been approved. There is therefore a great need to elucidate and understand the immunological impact of dendritic cell vaccination in order to improve clinical benefit. In this chapter, we describe a method for performing immune monitoring using peripheral blood mononuclear cells and autologous dendritic cells transfected with tumor-associated antigen-encoding mRNA.

Key words Cancer immunotherapy, Immune monitoring, Vaccination, Dendritic cell, Electroporation, mRNA transfection

1 Introduction

Dendritic cells (DC) are sentinels in the human immune system and are known to be the most potent antigen presenting cells, thus being key players in orchestrating immune responses toward foreign pathogens or cancers [1]. For their survival, cancer cells often rely on overexpressed or aberrantly expressed proteins, which in turn are presented on the surface of cancer cells as antigens known as tumor-associated antigens (TAAs) [2]. Upon tumor cell death or phagocytosis, DCs take up TAAs and present them to T cells [1]. Provided proper co-stimulatory signals, T cells (CD8+ Cytotoxic T lymphocytes—CTLs) are activated and can potentially kill the tumor cells expressing the TAAs or help B cells form relevant antibodies (CD4+ T helper cells) [1].

Spontaneous responses against TAAs, such as p53, survivin, and hTERT, have been reported [3–5]. However, most anticancer

Robert E. Rhoads (ed.), *Synthetic mRNA: Production, Introduction Into Cells, and Physiological Consequences*, Methods in Molecular Biology, vol. 1428, DOI 10.1007/978-1-4939-3625-0_16,

responses are weak, and the cancer avoids immune destruction by modifying their microenvironment through upregulation of inhibitory molecules or receptors, such as PD-L1, IDO, or others [6].

At least in theory, one might overcome the cancer cell's defenses by either inducing a strong anticancer response or boosting an existing one. This has been tried in various vaccination strategies using peptides, proteins, DNA, and RNA with a wide range of adjuvants. An often tried approach is using DCs as a cellular adjuvant [1]. Usually patients undergo a leukapheresis and DCs are grown in vitro before injecting them back into the patient. In order to direct the immune response toward targeting the cancer cells, the DCs are typically loaded with the desired target TAAs by co-incubation with peptides, proteins, or tumor lysate which they take up, process and present through the major histocompatibility complex (MHC) on their surface [7]. Alternatively, antigen can be loaded directly into the DCs as mRNA encoding the full protein by electroporation [8]. The mRNA will be translated and processed, mimicking the endogenous antigen processing pathway, and will consequently be presented through the MHC class I system [9], which is desirable in a cancer setting, as it favors a CD8+ biased response. Electroporation of mRNA has been shown to be superior to other ways of transfection with a high transfection efficacy, low cytotoxicity, and high reproducibility [10]. However, transfecting with mRNA introduces the challenge predicting which epitopes are presented to the T-cell and thus complicates immune monitoring.

Despite extensive research in the field only one DC-based vaccine has been approved for treatment in metastatic cancer [11]. There is still a great need to elucidate and understand the immunological impact of DC vaccination in order to improve clinical benefit.

In this chapter, we describe a method for monitoring immune responses by two assays designed to evaluate either the proliferation of T cells or the release of IFN-γ, as a pseudo marker for cytotoxic potential, when PBMCs are stimulated with autologous DCs transfected with TAA-encoding mRNA by electroporation. An example of IFN-γ detection is showed in Fig. 1.

2 Materials

2.1 Immune Monitoring, Day Minus 1

1. 15-mL polypropylene, conical tube.
2. X-VIVO 15 (chemically defined, serum-free hematopoietic cell medium).
3. 10-mL pipet.
4. Centrifuge.

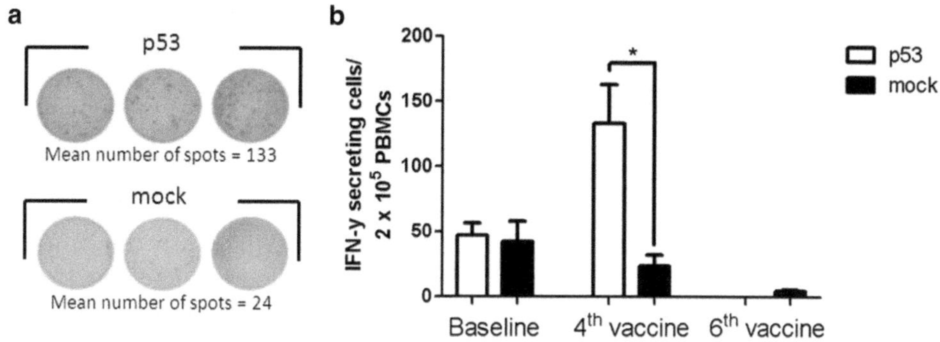

Fig. 1 Example of an induced response in a patient. (**a**) Triplicate ELISpot wells showing IFN-y producing PBMCs when stimulated with DCs transfected with p53 mRNA or mock-transfected DCs. (**b**) The number of IFN-y secreting cells per 2×10^5 PBMCs. At baseline responses are similar. At the time of fourth vaccine, a response has been induced. However, the response is completely gone at the last time point. Columns represent mean values of triplicates. Whiskers indicate SD. $^*P < 0.05$

5. Trypan blue.
6. 24-well plate.
7. Humidified CO_2 incubator.

2.2 Immune Monitoring, Day 0 (Prestimulation Culture)

1. 3-mL transfer pipet.
2. 50-mL polypropylene, conical tube.
3. X-VIVO 15.
4. Centrifuge.
5. Trypan blue.
6. 24-well plate.
7. Autologous DCs.
8. 10-mL pipet.
9. 15-mL polypropylene, conical tube.
10. Opti-MEM (reduced serum medium without phenol red).
11. 5-mL polypropylene, round-bottom tube.
12. 2-mm gap electroporation cuvette.
13. 4-mm gap electroporation cuvette.
14. In vitro synthesized mRNA—p53 [3], survivin [12], hTERT [13], EGFP [13], and CEF [14] (*see* **Note 1**).
15. BTX ECM 830 square-wave electroporator (Harvard Apparatus).
16. Humidified CO_2 incubator.

2.3 Immune Monitoring, Day 1

1. Recombinant human interleukin-2.
2. Flow cytometer (BD FACSCanto II, BD Biosciences) and analysis software (BD FACS Diva).

3. 3-mL transfer pipet.
4. 5-mL polystyrene, round-bottom tube.
5. PBS (Phosphate Buffered Saline).
6. Centrifuge.
7. Propidium iodide (PI).

2.4 Immune Monitoring, Day 6

1. ELISpot 96-well plate (Merck Millipore)—Enzyme-Linked ImmunoSpot (ELISpot) is a method for detecting the release of various cytokines from cells, in this case IFN-γ. Each spot visualized in this assay represents one single IFN-γ producing cell.
2. Anti-IFN-γ coating antibody, mAb 1-D1K (Mabtech).
3. PBS.
4. 96-well round-bottom plate.
5. Anti-CD3 coating antibody (e.g., OKT-3 clone, Ortho Biotech).

2.5 Immune Monitoring, Day 7

1. PBS.
2. X-VIVO 15.
3. 3-mL transfer pipet.
4. 50-mL polypropylene, conical tube.
5. Trypan blue.
6. 15-mL polypropylene, conical tube.
7. Diluent C and PKH dye (Sigma-Aldrich).
8. Human AB-serum.
9. Centrifuge.
10. 5-mL polystyrene, round-bottom tube.
11. Coated ELISpot 96-well plate from the day before.
12. Coated 96-well round-bottom plate from the day before.
13. 24-well plate.
14. Humidified CO_2 incubator.
15. Autologous DCs.
16. Opti-MEM.
17. 10-mL pipet.
18. 5-mL polypropylene, round-bottom tube.
19. 2-mm gap electroporation cuvette.
20. In vitro synthesized mRNA—p53, survivin, hTERT, EGFP, and CEF.
21. BTX ECM 830 square-wave electroporator.
22. Flow cytometer and analysis software.

2.6 Immune Monitoring, Day 8

1. PBS.
2. Secondary biotinylated antibody, mAb 7-B6-1-Biotin (Mabtech).
3. Diluting buffer: PBS + 1 % BSA + 0.02 % NaN_3.
4. Streptavidin-ALP (Mabtech).
5. Substrate solution, BCIP/NBT (Mabtech).
6. ELISpot reader and analysis software.
7. Flow cytometer and analysis software.
8. 3-mL transfer pipet.
9. 5-mL polystyrene, round-bottom tube.
10. PBS.
11. Centrifuge.
12. PI.

2.7 Immune Monitoring, Day 10

1. 5-mL polystyrene, round-bottom tube.
2. FACS buffer: PBS + 0.5 % BSA.
3. Centrifuge.
4. Near-infra red (NIR) live/dead marker.
5. Anti-CD3-APC conjugated antibody.
6. Anti-CD8-BV421 conjugated antibody.
7. Anti-CD4-BV500 conjugated antibody.
8. Flow cytometer and analysis software.

3 Methods

3.1 Immune Monitoring, Day Minus 1

1. Transfer four cryovials each containing 1×10^7 PBMC (*see* **Note 2**) from three different time points in patient treatment from freezer and place directly in 37 °C water bath (*see* **Note 3**).
2. Rotate cryovials in the surface of the water bath until the cell suspension is almost completely melted or a small bit of ice remains.
3. Dry off the outside of the cryovials and wipe with 70 % ethanol.
4. Transfer the cell suspension from each cryovial to a 14-mL polypropylene tube containing 9 mL 37 °C warm X-VIVO 15 medium dropwise (*see* **Note 4**).
5. Centrifuge the cells for 5 min at $400 \times g$ and room temperature.
6. Discard supernatant, resuspend in 1 mL X-VIVO 15 medium and bring all cells from the three time points together in three 14 mL polypropylene tubes, respectively. Add X-VIVO 15 medium up to 10 mL.

7. Centrifuge the cells for 5 min at 400 × *g* and room temperature.
8. Discard supernatant, resuspend and count the cells (*see* **Note 5**).
9. Adjust to 2.5–3.0 × 10^6 cells/mL and transfer them to a 24-well plate adding 2 mL cell suspension to each well (*see* **Note 6**).
10. PBMCs are rested ON in a humidified 37 °C CO_2 incubator.

3.2 Immune Monitoring, Day 0 (Prestimulation Culture)

3.2.1 Preparing PBMCs

1. Harvest PBMCs in X-VIVO 15 medium and count the cells.
2. Adjust to 5–6 × 10^6 cells/mL and transfer them to a 24-well plate adding 1 mL cell suspension to each well. Add X-VIVO 15 medium to a total volume of 1975 μL (*see* **Note 7**).
3. Prepare two wells with only 1 mL X-VIVO 15 medium without cells to be used for the transfection controls.

3.2.2 Preparing DCs

1. Transfer cryovials of DC (*see* **Note 8**) from freezer and place directly in 37 °C water bath.
2. Rotate cryovials in the surface of the water bath until the cell suspension is almost completely melted or a small bit of ice remains.
3. Dry off the outside of the cryovials and wipe with 70 % ethanol.
4. Transfer the cell suspension to a 14 mL polypropylene tube containing 9 mL 37 °C warm Opti-MEM medium dropwise (*see* **Note 9**).
5. Centrifuge the cells for 5 min at 400 × *g* and room temperature.
6. Discard supernatant, resuspend in 10 mL Opti-MEM medium.
7. Centrifuge the cells for 5 min at 400 × *g* and room temperature.
8. Resuspend DCs in Opti-MEM, transfer to 5 mL polypropylene tube and rest for 1 h (*see* **Note 10**).
9. Centrifuge the cells for 5 min at 400 × *g* and 5 °C.
10. Discard supernatant, resuspend and count cells.
11. Transfer 9 × 10^6 and 2 × 0.3 × 10^6 DCs to 3 5 mL polypropylene tubes and add Opti-MEM up to 400 μL and 2 × 100 μL, respectively, and place them on ice for 5 min (*see* **Note 11**).
12. During the incubation, setup the electroporation parameters: 500 V if you are using 4-mm cuvette or 250 V if you are using 2-mm cuvette; 2 ms pulse length, a single unipolar pulse; and 1 s interval between pulses (*see* **Note 12**).

13. In the tube containing 9 × 10^6 DCs in 400 μL, mix 5 μg of p53, survivin, or hTERT mRNA and transfer cell suspension immediately into a 4-mm electroporation cuvette (*see* **Note 13**).
14. Insert the cuvette into the electroporation chamber and trigger the pulse. Inspect that the electroporation is successful (*see* **Note 14**).
15. Transfer DCs immediately after electroporation to the prepared 24-well plate into each designated well without excessive pipetting (*see* **Note 15**). DCs are added in a DC to PBMC ratio of 1:10 (*see* **Note 16**).
16. Repeat **steps 13–15** with the second tube using EGFP mRNA and with the third tube without mRNA, electroporating in 2 mm cuvettes at 250 V (*see* **Note 17**). Transfer the DCs to the wells prepared for transfection control.
17. Incubate for 7 days at 37 °C in 5 % CO_2.

3.3 Immune Monitoring, Day 1

3.3.1 Adding IL-2 to Prestimulation Culture

1. Add 40 U/mL of IL-2 to each well.
2. Resuspend cells in each well gently and incubate for additional 6 days at 37 °C in 5 % CO_2.

3.3.2 Analyzing Transfection Control

1. Ready flow cytometer.
2. Harvest DCs from the 24-well into two FACS tubes.
3. Centrifuge the cells for 5 min at 400 × *g* and RT.
4. Discard the supernatant and leave pellet on ice.
5. Dilute 5 μg PI in 1 mL PBS (PI solution) and place on ice.
6. Resuspend pellet in 500 μL PI solution 5 min before acquiring data.
7. Acquire and analyze data (*see* **Note 18**).

3.4 Immune Monitoring, Day 6

3.4.1 Coating ELISpot Plate

1. Dilute the coating antibody (anti-IFN-γ) to 7.5 μg/mL in sterile PBS.
2. Add 70 μL of diluted antibody to 54 wells (*see* **Note 19**) of a 96-well Multiscreen ELISpot plate for detection of IFN-γ release of PBMCs stimulated with (a) p53, survivin, and hTERT mRNA-transfected DCs, (b) mock-transfected DCs (negative control of experiment), and (c) CEF mRNA-transfected DCs (positive control of ELISpot) (*see* **Note 20**). Make sure all wells are covered. Incubate ON at RT (*see* **Note 21**).

3.4.2 Coating a 96-Well Round-Bottom Plate for Proliferation Assay

1. Dilute the coating antibody (OKT-3 anti-CD3) to 0.5 μg/mL in sterile PBS.
2. Add 50 μL of diluted antibody to 9 wells of a 96-well round-bottom plate. Incubate ON at 37 °C (*see* **Note 22**).

3.5 Immune Monitoring, Day 7

3.5.1 Preparing ELISpot Plate

1. Discard primary antibody solution and wash the plate six times with 200 μL PBS (*see* **Note 23**).
2. Add 200 μL blocking solution (X-VIVO 15 medium) to each well to saturate remaining binding sites. Incubate for 2 h at room temperature (*see* **Note 24**).

3.5.2 Preparing PBMCs

1. During block of the ELISpot plate, harvest PBMCs in X-VIVO 15 medium into three corresponding 50-mL polypropylene tubes and count the cells. Adjust concentration to 3×10^6 cells/mL.
2. Transfer 3 mL PBMC from each time point to 14 mL polypropylene tubes centrifuge the cells for 5 min at $400 \times g$ (*see* **Note 25**).
3. Discard supernatant and resuspend cells in 1 mL diluent C.
4. Dilute 9 μL PKH in 3 mL diluent C, add 1 mL PKH dye solution to each tube and mix well. Incubate 2 min at room temperature (*see* **Note 26**).
5. Add 2 mL human AB-serum to stop the reaction and incubate 1 min at room temperature.
6. Add X-VIVO 15 medium up to 10 mL and centrifuge the cells for 5 min at $400 \times g$.
7. Discard supernatant and resuspend in 10 mL X-VIVO15 medium. Centrifuge the cells for 5 min at $400 \times g$.
8. Repeat **step** 7.
9. Discard supernatant, count cells and adjust to 3×10^6 cells/mL.
10. Transfer 30 μL from each time point to FACS tubes and place in the dark on ice for later FACS analysis (*see* **Note 27**).
11. Discard the blocking solution from the ELISpot plate and remove the PBS from the 96-well round-bottom plate.
12. Resuspend PBMCs well and transfer 100 μL cell suspension from either normal PBMCs or PKH-stained PBMCs to designated wells on the ELISpot plate and 96-well round-bottom plate, respectively.
13. Add X VIVO 15 medium so that final volume after adding DCs is 200 μL, *see* Table 1 for DC volumes.
14. Also add 1 mL of X-VIVO 15 medium to two wells in a 24-well plate to be used for the transfection controls.
15. Incubate at 37 °C in 5 % CO_2 while preparing DCs.

3.5.3 Preparing DCs

1. Transfer cryovials of DC from freezer and place directly in 37 °C water bath.
2. Rotate cryovials in the surface of the water bath until the cell suspension is almost completely melted or a small bit of ice remains.

3. Dry off the outside of the cryovials and wipe with 70 % ethanol.
4. Transfer the cell suspension to a 14 mL polypropylene tube containing 9 mL 37 °C warm Opti-MEM medium dropwise.
5. Centrifuge the cells for 5 min at 400 × *g* and RT.
6. Discard supernatant, resuspend in 10 mL Opti-MEM medium.
7. Centrifuge the cells for 5 min at 400 × *g* and RT.
8. Resuspend DCs and transfer to 5-mL polypropylene tube and rest for 1 h.
9. Centrifuge the cells for 5 min at 400 × *g*.
10. Discard supernatant, resuspend and count cells.
11. Transfer DCs needed according to Table 1 below to 7 × 5-mL polypropylene tubes, add Opti-MEM to a final volume of 200 μL (100 μL for CEF and EGFP) and place them on ice for 5 min.
12. During the incubation, setup the electroporation parameters: 500 V if you are using 4-mm cuvette or 250 V if you are using 2-mm cuvette; 2 ms pulse length, a single unipolar pulse; and 1 s interval between pulses.
13. In one tube, mix 5 μg of p53, survivin, or hTERT mRNA and transfer cell suspension immediately into a 2-mm electroporation cuvette.
14. Insert the cuvette into the electroporation chamber and trigger the pulse. Inspect that the electroporation is successful.

Table 1
Calculations on DC concentrations and volumes

	DC needed (round up)	Concentration of DC in cuvette	μL transferred to plates
p53, survivin, hTERT, triple transfected	6×10^5	$\frac{6 \times 10^5 \times 1000\ \mu L}{200\ \mu L} = 30 \times 10^5/mL$	$\frac{3 \times 10^4}{3 \times 10^5/mL} = 10\ \mu L$
Mock	10×10^5	$\frac{10 \times 10^5 \times 100\ \mu L}{200\ \mu L} = 50 \times 10^5/mL$	$\frac{3 \times 10^4}{50 \times 10^5/mL} = 6\ \mu L$[a]
CEF	3×10^5	$\frac{3 \times 10^5 \times 1000\ \mu L}{100\ \mu L} = 30 \times 10^5/mL$	$\frac{1.5 \times 10^4}{30 \times 10^5/mL} = 20\ \mu L$
EGFP	3×10^5	$\frac{3 \times 10^5 \times 1000\ \mu L}{100\ \mu L} = 30 \times 10^5/mL$	$\frac{3.5 \times 10^4}{30 \times 10^5/mL} = 100\ \mu L$
In total	40×10^5		

[a]60 μL is transferred to 24-well plate for transfection control analysis

15. Transfer DCs immediately after electroporation to the prepared ELISpot and 96-well round-bottom plate into each designated well without excessive pipetting. For transfection control transfer DCs to the 24-well plate.
16. Repeat **steps 28–30** with the second, third, and fourth tube using p53, survivin, or hTERT mRNA, respectively, with the fifth tube using CEF mRNA, with the sixth tube using EGFP mRNA and with the seventh tube without mRNA.
17. Incubate the ELISpot plate and 24-well plate for 18 h at 37 °C in 5 % CO_2 (*see* **Note 28**). The 96-well round-bottom plate is incubated for 3 days.

3.5.4 Baseline FACS on PKH-Stained PBMCs

1. Ready flow cytometer.
2. Add PBS up to 2-mL to FACS tubes containing PKH-stained PBMCs. Centrifuge the cells for 5 min at 400 × *g*.
3. Discard supernatant and resuspend in 100 μL PBS. Keep cells on ice until acquiring.
4. Acquire data on flow cytometer.

3.6 Immune Monitoring, Day 8

3.6.1 Processing the ELISpot Plate

1. Remove the cells and unbound cytokine by emptying the plate and wash six times with 200 μL PBS per well (*see* **Note 29**). Shake excess fluid and pat dry.
2. Dilute the biotinylated secondary antibody (7-B6-1) to 1 μg/mL in diluting buffer (PBS, 1 % BSA, 0.02 % NaN_3). Add 70 μL/well and incubate for 2 h at room temperature (*see* **Note 30**).
3. Empty the plate and wash six times with 200 μL PBS per well. Shake excess fluid and pat dry.
4. Dilute the streptavidin-ALP (1:1000) in diluting buffer and add 70 μL/well. Incubate for 1 h at room temperature.
5. Empty the plate and wash six times with 200 μL PBS per well. Shake excess fluid and pat dry.
6. Add 70 μL of substrate solution (BCIP/NBT) and incubate for 2–15 min at room temperature until blue spots develop (*see* **Note 31**).
7. Stop color development by washing extensively in distilled water.
8. Empty the plate, shake excess fluid, and pat dry.
9. Place the ELISpot plate with the membrane upwards on a paper towel and leave the plate to dry overnight in the dark.
10. When the plate is completely dry, inspect and count spots in an ELISpot reader such as the "Grand" Immunospot S6 Ultimative UV Image Analyzer.
11. Store plate in the dark at room temperature.

3.6.2 Analyzing Transfection Control

1. Ready flow cytometer.
2. Harvest DCs from the 24-well into 2 FACS tubes.
3. Centrifuge the cells for 5 min at 400 × *g* and RT.
4. Discard the supernatant and leave pellet on ice.
5. Dilute 5 μg PI in 1 mL PBS (PI solution) and place on ice.
6. Resuspend pellet in 500 μL PI solution 5 min before acquiring data.
7. Acquire and analyze data.

3.7 Immune Monitoring, Day 10

1. Harvest cells from the 96-well round-bottom plate into appropriate FACS tubes.
2. Add FACS buffer (FACS PBS + 0.5 % BSA) up to 2 mL.
3. Centrifuge the cells for 5 min at 400 × *g*.
4. Discard supernatant and resuspend the cells in 100 μL FACS buffer.
5. Add antibodies according to antibody panel and mix well (*see* **Note 32**).
6. Incubate 30 min at 5 °C in the dark.
7. During incubation, ready flow cytometer.
8. Add 2 mL FACS buffer and centrifuge the cells for 5 min at 400 × *g*.
9. Discard supernatant and resuspend cells in 2 mL FACS buffer.
10. Centrifuge the cells for 5 min at 400 × *g*.
11. Discard supernatant and resuspend cells in 250 μL FACS buffer. Keep cells on ice until acquiring.
12. Acquire data on flow cytometer.

4 Notes

1. All in vitro transcribed mRNA were generated using a linearized polyadenylated plasmid as described previously by Met and Svane [12]. The pSP73-Sph/p53/A64 plasmid is described by Met et al. [3]. The pSP73/survivin/A64 plasmid is described by Met and Svane [12]. The EGFP/pCIpA102 and hTERT/pCIpA102 plasmids were kindly provided by Dr. G. Gaudernack, The Norwegian Radium Hospital, Oslo, Norway and are described by Sæbøe-Larssen [13]. The pST1-CEF plasmid was kindly provided by Dr. JR Webb, Trev and Joyce Deeley Research Centre, British Columbia Cancer Agency, Victoria, British Columbia, Canada and described by Nielsen JS [14].
2. For the setup described here, we recommend starting with about 4 × 10^7 cells per time point of patient treatment as one

can expect to lose cells during handling and incubation because of cell fragility and inability to thrive under these conditions. Overnight a cell loss of about 30 % is expected and during the 7 day incubation another 50 % is expected. At day 7 about 14×10^6 PBMCs per time point are needed (5×10^6 for ELISpot and 9×10^6 for proliferation assay).

3. It is important for cell viability that the cells are thawed and processed quickly; thawing should take less than a minute. If PBMCs are not thawed properly, viability and cell recovery can be compromised and cells may not perform optimally in functional assays. Cells should be diluted slowly to remove DMSO. Cells with DMSO intercalated into their membranes are very fragile and must be pelleted and handled gently.
4. Prewarm medium to 37 °C by incubating one 14-mL tube containing 9 mL medium for a couple of hours or overnight.
5. We recommend resuspending the PBMCs in a fixed volume and counting the cells manually, e.g., by tryphan blue exclusion, in order to exclude dead cells.
6. In case of excess wells, these should be filled with 2 mL PBS to avoid evaporation of culture medium.
7. Total volume should be 2 mL and expectedly about 25 μL of DCs will be added after electroporation (*see* **Note 15**).
8. DCs are added to PBMCs in a DC:PBMC ratio of 1:10. Typically, 9.6×10^6 DC are needed as there should be 30×10^6 PBMC per time point and another 0.6×10^6 is needed for the transfection control.
9. Use Opti-MEM medium for electroporation as its composition ensures optimal conditions.
10. We recommend using 5-mL polypropylene tubes for resting DCs to avoid adhesion to the plastic surface.
11. It is important to incubate cells on ice prior to electroporation as the pulse might cause thermal shock to the cells and thus negatively impact cell survival. Similarly, electroporation cuvettes should be stored in a refrigerator (5 °C).
12. Both exponential decay and square-wave pulses can be used to transfect DCs. However, while exponential decay pulses are more ideal to transfect bacteria and yeast, the square-wave pulses are better tolerated by more delicate mammalian cells such as DCs. With the square-wave pulse there is a period of homeostasis to be reached in the cells before the wave is removed. As a result, there is a lower mortality rate in cells while maintaining transfection efficiencies. In our hands, transfecting using exponential decay pulses results in much higher rate of cell mortality in DCs than using square-wave pulses.

13. Add mRNA to the cell suspension just before the electroporation pulse to avoid degradation due to possible RNAses. Also make sure that no air bubbles are present in the suspension before or during delivery into the cuvette as this can cause arcing.
14. Immediately after the pulse, small air bubbles or a small layer of foam should be apparent at the top of the cell suspension. This signals that the current has passed through the solution and that the electroporation is successful.
15. After electroporation cells should be transferred immediately to 37 °C cell culture medium. Below 37 °C, the resealing of membranes after electroporation is delayed so that exchange of molecules between the cell and its environment takes place over a longer time. This causes increased stress and can reduce the survival rate. Also avoid repetitive pipetting as cells are very fragile after electroporation.
16. DCs are added to the PBMCs in a DC:PBMC ratio of 1:10. Consequently, 5–6 × 10^5 DCs are added to each well. Calculate the concentration of DCs before electroporation, e.g., 9 × 10^6 in 400 μL equals 22.5 × 10^6/mL. In that case transfer 22.2–26.6 μL DCs to each well.
17. In order to verify electroporation and transfection has been successful, DCs are electroporated with mRNA encoding enhanced green flourescent protein (EGFP) or without mRNA (mock electroporation). EGFP is measurable by FACS using the FITC channel.
18. PI binds to DNA but is generally excluded from viable cells, thus dead or dying cells will stain positive. PI can be detected in the PerCP channel.
19. Reactivity in ELISpot assay should be tested in triplicates. In this setup there is 3 time points and 6 tests: PBMCs stimulated with DCs transfected with (1) p53 mRNA, (2) survivin mRNA, (3) hTERT mRNA, (4) triple transfected, (5) CEF mRNA, or (6) mock electroporated.
20. Because cryopreservation of cells might affect their functional state, it is important to check that the effector cells to be analyzed are indeed functional. CEF mRNA codes for the proteins derived from cytomegalovirus, Epstein-Barr virus, and influenza virus. It efficiently induces IFN-γ production by $CD8^+$ T cells in more than 90 % of Caucasians and thus functions as a useful control to evaluate responsiveness of T cells in the ELISpot assay.
21. Make sure the plate is positioned in a pathogen-free environment such as inside a LAF-bench and/or wrapped in plastic. The incubation period can be altered to fit individual different

timing schedules, i.e. it can be shortened to 2 h at room temperature or prolonged to 7 days if incubated at 5 °C.

22. Anti-human CD3 efficiently binds and crosslinks the T-cell receptor and results in cellular activation and proliferation. Other kinds of T-cell activators can be used, such as staphylococcal enterotoxin B (SEB) or phytohemagglutinin (PHA). Either activating agent could also be added directly to the PBMCs on the day of the setup instead of coating the plate the day before.
23. Washing is easily accomplished using a multichannel pipette, shaking excess liquid from the plate, and patting the bottom dry on absorbent paper after the final wash.
24. Duration of the blocking step can be reduced to 30 min at 37 °C or prolonged to overnight at room temperature.
25. For the setup described 5.5×10^6 PKH-stained PBMCs are needed per time point. However, due to repeated centrifugations a cell loss of about 25 % should be expected. Consequently, we recommend starting with about 9×10^6 PBMCs.
26. PKH is a dye that is incorporated into the cell membrane. The dye is diluted on cell division and can be used for evaluating proliferation of PKH-stained cells. It is important to mix PKH dye solution just before use and the time for labeling kept shorter than 5 min, as prolonged light exposure can result in poor staining. Also, cells and PKH dye solution should be mixed thoroughly to ensure a homogenous staining. PKH intensity can be measured on a flow cytometer in the PE channel.
27. Baseline FACS on PKH-stained cells is done to evaluate the staining quality (the intensity and width of the peak), which is useful for later interpretation of analysis data on day 10.
28. Make sure the ELISpot plate is not moved or stirred while the cells are being cultured. This will lead to spots that will not be well defined. The ELISpot incubation period should be adapted to the type of antigen used and can be down to 4 h for short peptides or up to 48 h for full length proteins.
29. Sometimes cells can be sticky and washing with Tween 0.05 % is needed. However, Tween can damage the membrane and produce fluffy, uncountable spots.
30. The ELISpot incubation periods on the day of development can be varied to fit in the time schedule, i.e. 2 h incubation could be shortened to 1 h. However, it is recommended to keep these steps similar from patient to patient in an immune monitoring setting.
31. For optimal results filter the substrate solution (0.2 μm filter) before use as precipitation of the substrate can occur over time.

32. The antibody panel can be designed to fit individual needs. We are using APC αCD3, APC-Cy7 NIR (dead cell marker), BV421 αCD8, HV500 αCD4, and PE channel for detecting PKH. It should be noted that using FITC, PerCP, and PerCP-Cy5.5 takes careful compensation due to significant spillover from the PE channel (PKH).

References

1. Palucka K, Banchereau J (2012) Cancer immunotherapy via dendritic cells. Nat Rev Cancer 12(4):265–277
2. Zaffaroni N, Costa A, Pennati M, De Marco C, Affini E, Madeo M et al (2007) Survivin is highly expressed and promotes cell survival in malignant peritoneal mesothelioma. Cell Oncol 29(6):453–466
3. Met Ö, Balslev E, Flyger H, Svane IM (2011) High immunogenic potential of p53 mRNA-transfected dendritic cells in patients with primary breast cancer. Breast Cancer Res Treat 125(2):395–406
4. Vonderheide RH, Schultze JL, Anderson KS, Maecker B, Butler MO, Xia Z et al (2001) Equivalent induction of telomerase-specific cytotoxic T lymphocytes from tumor-bearing patients and healthy individuals advances in brief bearing patients and healthy individuals 1. Cancer Res 61:8366–8370
5. Andersen MH, Pedersen LØ, Capeller B, Class MHC, Epitopes IT, Bro E (2001) Spontaneous cytotoxic T-cell responses against survivin-derived MHC Class I-restricted T-cell epitopes in situ as well as ex vivo in cancer patients advances in brief as well as ex vivo in cancer patients 1. Cancer Res 61:5964–5968
6. Mittal D, Gubin MM, Schreiber RD, Smyth MJ (2014) New insights into cancer immunoediting and its three component phases-elimination, equilibrium and escape. Curr Opin Immunol 27(1):16–25
7. Figdor CG, de Vries IJM, Lesterhuis WJ, Melief CJM (2004) Dendritic cell immunotherapy: mapping the way. Nat Med 10 (5):475–480
8. Met Ö, Eriksen J, Svane IM (2008) Studies on mRNA electroporation of immature and mature dendritic cells: effects on their immunogenic potential. Mol Biotechnol 40 (2):151–160
9. Gilboa E (2007) Review series DC-based cancer vaccines. J Clin Invest 117 (5):1195–1203
10. Van Tendeloo VFI, Ponsaerts P, Lardon F, Nijs G, Lenjou M, Van Broeckhoven C et al (2001) Highly efficient gene delivery by mRNA electroporation in human hematopoietic cells: Superiority to lipofection and passive pulsing of mRNA and to electroporation of plasmid cDNA for tumor antigen loading of dendritic cells. Blood 98(1):49–56
11. Kantoff P, Higano C (2010) Sipuleucel-T immunotherapy for castration-resistant prostate cancer. N Engl J Med 363(5):411–422
12. Met Ö, Svane IM (2013) Synthetic messenger RNA and cell metabolism modulation. Methods Mol Biol 969:275–292
13. Sæbøe-Larssen S, Fossberg E, Gaudernack G (2002) mRNA-based electrotransfection of human dendritic cells and induction of cytotoxic T lymphocyte responses against the telomerase catalytic subunit (hTERT). J Immunol Methods 259(1-2):191–203
14. Nielsen JS, Wick DA, Tran E, Nelson BH, Webb JR (2010) An in vitro-transcribed-mRNA polyepitope construct encoding 32 distinct HLA class I-restricted epitopes from CMV, EBV, and Influenza for use as a functional control in human immune monitoring studies. J Immunol Methods 360(1-2):149–156

Chapter 17

Transfection of Tumor-Infiltrating T Cells with mRNA Encoding CXCR2

Manja Idorn, Per thor Straten, Inge Marie Svane, and Özcan Met

Abstract

Adoptive T-cell therapy based on the infusion of patient's own immune cells after ex vivo culturing is among the most potent forms of personalized treatment among recent clinical developments for the treatment of cancer. However, despite high rates of successful initial clinical responses, only about 20 % of patients with metastatic melanoma treated with tumor-infiltrating lymphocytes (TILs) enter complete and long-term regression, with the majority either relapsing after initial partial regression or not benefiting at all. Previous studies have shown a positive correlation between the number infused T cells migrating to the tumor and the clinical response, but also that only a small fraction of adoptively transferred T cells reach the tumor site. In this chapter, we describe a protocol for transfection of TILs with mRNA encoding the chemokine receptor CXCR2 transiently redirecting and improving TILs migration toward tumor-secreted chemokines in vitro.

Key words Tumor-infiltrating lymphocytes (TILs), mRNA transfection, Cancer immunotherapy, Tumor homing, Chemokine receptor

1 Introduction

Adoptive cell transfer (ACT) of tumor-infiltrating lymphocytes (TILs) after nonmyeloablating chemotherapy represents a breakthrough in the treatment of disseminated malignant melanoma. However, despite its high degree of efficacy over current therapeutic options, the majority of treated patients either relapse after an initial partial regression (~30 %) or do not appear to benefit at all (~50 %) [1, 2]. TIL treatment takes advantage of the enriched numbers of tumor specific T cells at the tumor site, which are mostly lacking in the peripheral blood under normal conditions [3]. However, although high numbers of T cells (up to 1.99×10^{11} cells) [4] are infused in patients, only a small fraction of these cells are shown to recognize tumors and mediate tumor regression [5].

Robert E. Rhoads (ed.), *Synthetic mRNA: Production, Introduction Into Cells, and Physiological Consequences*, Methods in Molecular Biology, vol. 1428, DOI 10.1007/978-1-4939-3625-0_17,

The approaches so far proposed to improve TIL efficacy have relied mostly on combination with synergistic agents [6, 7] or improvement of phenotypic or functional characteristics of the infused cells [8, 9]. In this regard, one limiting factor for ACT is the inefficient homing of TILs to the tumor site. Early studies in which TILs were labeled with ^{111}indium showed that only a small fraction of intravenously infused T cells were able to home into tumor tissue [10]. Therefore, strategies which focus on improving the migratory capacity of T cells into tumor sites are likely to improve TIL therapy and thus potentiate durable clinical response rates.

Current research efforts in this area have focused on chemokines, produced by tumor cells and tumor-associated stromal cells, to enhance TIL migration into the tumor site [11, 12]. Many tumors express chemokines that divert homing of antitumor T cells and rather promote homing of cells with suppressive action [13]. Interleukine-8 (IL-8), a chemokine widely expressed by several cancers, is an important player in immunological escape, progression, and metastatic behavior of the cancer. Importantly, cells that are more prone to antitumor activity do not express the IL-8 receptor, CXCR2, and are thus not attracted to the tumor site [14]. This includes TILs used for ACT. The T cells therefore may never find the target.

Although genetic modification of T cells using viral vectors holds great promise for improving ACT, i.e., affinity-enhanced TCRs and chimeric antigen receptors (CARs) [15], the practical application of this strategy remains significantly challenging. Regulatory concerns such as insertional mutagenesis and genotoxicity as well as severe and irreversible immunologically adverse events [16, 17] following viral transduction of T cells have prompted the development of gene transfer methods that can circumvent these issues and at the same time archive meaningful modifications of T cell function. In this regard, the use of mRNA electroporation as means for gene transfer into cells of the immune system has been of particular interest [18, 19]. High reproducibility and ease of performance are the inherent advantages of this method which can be readily applied clinically.

In this chapter, we present a protocol for the transfection of TILs by electroporation using in vitro synthesized mRNA encoding CXCR2, and the subsequent analyses of CXCR2 expression kinetics and signaling by flow cytometry-based methods. Furthermore, the chapter presents a protocol to assess and quantify the improved migration of CXCR2-transfected TILs toward tumor-secreted IL-8 by an in vitro transwell migration assay. This protocol can be applied to other chemokine receptors of interest which are suitable targets for directing specific T cells toward tumor.

2 Materials

2.1 In Vitro Synthesis of CXCR2 mRNA

1. pSP73/CXCR2/A64 serves as template DNA vector for in vitro transcription (*see* Fig. 1; *see* **Note 1**).
2. Qiagen Plasmid *plus* Maxi Kit (Qiagen, Hilden, Germany).
3. Restriction enzyme *Spe*I kit incl. BSA and NEB buffer 4 (New England Biolabs, Ipswich MA, USA).
4. Nuclease-free water.
5. RNAse *Zap*® (Life Technologies).
6. Heating block.
7. Wizard DNA Clean-Up System (Promega, Madison WI, USA).
8. mMESSAGE mMACHINE T7 Ultra Kit (Ambion, Austin TX, USA).
9. MEGAclear Kit (Ambion).
10. Agilent 2100 Bioanalyzer (Agilent Technologies, Palo Alto CA, USA).
11. RNA 6000 Nano LabChip Kit (Agilent Technologies).

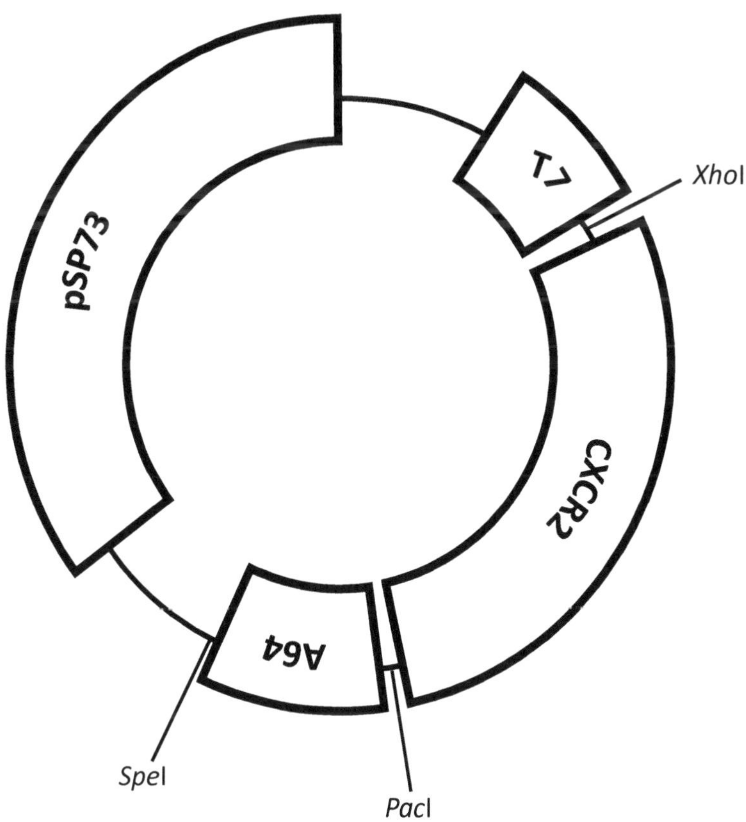

Fig. 1 Schematic of plasmid vector pSP73 containing the coding sequence for CXCR2, a T7 promoter necessary for in vitro transcription of the chemokine receptor and coding sequences for a 64 nt Poly(A) chain for increased stability of the transcribed mRNA

2.2 Isolation and Expansion of Human TILs

1. Tumor biopsy (*see* **Note 2**).
2. 9-cm petri dish.
3. 50-mL conical tube.
4. 24-well cell culture plate.
5. Sterile forceps and scalpel.
6. TIL medium: RPMI 1640; 10 % Human AB serum; 6000 U/mL IL-2; 1 % penicillin/streptomycin; 1.25 μg/mL Fungizone.
7. IL-2 (Proleukine, Novartis).
8. Heat-inactivated human AB serum (Sigma-Aldrich, St. Louis MO, USA) (*see* **Note 3**).
9. Humidified CO_2 incubator.
10. Hemocytometer.
11. Trypan blue.
12. Antibiotics: 1 % 10,000 U/mL Penicillin, 10,000 μg/mL Streptomycin.
13. Fungizone (Amphotericin B).
14. Functional grade anti-CD3 (Clone OKT3).

2.3 Electroporation of TILs

1. 6-well cell culture plate.
2. Opti-MEM medium (Life Technologies).
3. 1.8-mL nunc cryotubes.
4. BTX ECM 830 square-wave electroporator (BTX Harvard Apparatus, Holliston MA, USA).
5. 2-mm gap cuvettes.
6. TILs isolated from tumor biopsy.
7. 14- and 50-mL conical tubes.
8. PBS.
9. X-VIVO 15 medium (Lonza, Basel, Switzerland).
10. RNAse *Zap*®
11. Hemocytometer.
12. Trypan blue.
13. Hereus mulitfuge centrifuge.
14. Humidified CO_2 incubator.

2.4 Expression and Signaling of CXCR2 by Flow Cytometry

1. Transfected TILs.
2. PBS.
3. EDTA.
4. Bovine serum albumin (BSA).

5. Fluo-4-AM (Life Technologies).
6. Fluorochrome conjugated monoclonal antibodies toward PE-Cy7 CD3, PB CD4, FITC CD8, APC CXCR2, APC IgG1 (Biolegend, San Diego, CA, USA).
7. Live/Dead® Fixable Near-Infrared dead cell stain (DCM) (Life technologies).
8. Hemocytometer.
9. Trypan blue.
10. Hereus mulitfuge centrifuge.
11. Humidified CO_2 incubator.
12. Flow Cytometer (BD FACScanto II, BD Biosciences, Franklin Lakes NJ, USA).

2.5 Specific Migration of CXCR2-Transfected TILs

1. TILs isolated from tumor biopsy, transfected with CXCR2.
2. 6.5-mm transwell with 3-μm polycarbonate insert (Corning, NY, USA).
3. Sterile forceps.
4. Recombinant human CXCL8/IL-8 (R&D systems, Minneapolis MN, USA).
5. RPMI 1640.
6. 14-mL conical tubes.
7. Hemocytometer.
8. Trypan blue.
9. Hereus mulitfuge centrifuge.
10. Humidified CO_2 incubator.
11. Automated multipipette.
12. 7-Amino-actinomycin D (BD Biosciences).
13. Flow Cytometer (BD FACScanto II, BD Biosciences).

3 Methods

3.1 In Vitro Synthesis of CXCR2 mRNA

1. Purify the pSP73/CXCR2/A64 plasmid (*see* **Note 1**) with the Qiagen Plasmid *plus* Maxi Kit according to the manufacturer's protocol.
2. For linearization of the pSP73/CXCR2/A64 plasmid, prepare the following in one Eppendorf tube: 29.5 μL of Nuclease-free water (prewarmed to 37 °C), 5.0 μL of NEB buffer 4 (10×) (prewarmed to 37 °C), 5.0 μL BSA (10×) (prewarmed to 37 °C), 10.0 μL of pSP73/CXCR2/A64 plasmid (2 μg/μL), 0.5 μL of *Spe* I restriction enzyme (10 U/μL) (*see* **Note 4**).
3. Incubate for 1 h in a heating block at 37 °C.

4. Purify linearized pSP73/CXCR2/A64 plasmid with Wizard DNA Clean-Up System according to the manufacturer's protocol.
5. Produce CXCR2 mRNA with mMessage mMACHINE Ultra T7 Ultra Kit according to the manufacturer's protocol.
6. Purify CXCR2 mRNA transcripts with the MEGAclear Kit according to the manufacturer's protocol.
7. Evaluate the mRNA length, concentration and purity with the Agilent 2100 Bioanalyzer and the RNA 6000 Nano LabChip Kit according to the manufacturer's protocol.

3.2 Isolation and Expansion of Human TILs

1. After informed consent, obtain tumor biopsy from surgeons, and transport it in a sterile 50-mL conical tube containing 20 mL of RPMI 1640 + 1 % penicillin/streptomycin.
2. All procedures are subsequently performed under sterile conditions in a flow cabinet and with sterile tools and reagents.
3. Prepare 24-well plate, by adding 1 mL of TIL medium (RPMI 1640 supplemented with 10 % Human AB serum, 6000 U/mL IL-2, 1 % penicillin/streptomycin and 1.25 μg/mL Fungizone) to each plate, and prewarm in a humidified CO_2 incubator at 37 °C.
4. Transfer tumor biopsy and transportation medium to a 9-cm petri dish and cut it into 1–2 mm^2 fragments using sterile forceps and scalpel.
5. Transfer 1–2 fragments to each well containing 1 mL prewarmed medium (*see* **Note 5**). Add an additional 1 mL prewarmed TIL medium for a total volume of 2 mL per well.
6. Incubate plates in humidified CO_2 incubator at 37 °C for 5 days (*see* **Note 6**).
7. After 5 days of incubation, refresh medium by removing 1 mL—do not resuspend the cells/fragments—and adding 1 mL of newly made and prewarmed TIL medium (*see* **Note 7**).
8. Refresh medium as in **step 7**, two to three times per week.
9. Split wells into two wells when the cells at the bottom of the well form a confluent layer (*see* **Note 8**). Add fresh TIL medium for a total volume of 2 mL per well.
10. Culture up to 50×10^6 cells. Cells may be pooled, cryopreserved (*see* **Note 9**), or put into a Rapid Expansion Protocol (REP) (*see* **Note 10**).

3.3 Electroporation of TILs

1. Prior to electroporation, prepare a 6-well plate by adding 4 mL of fresh TIL medium to each well, and prewarm it in humidified CO_2 incubator at 37 °C.
2. Harvest pooled TIL culture by thorough pipetting, avoiding bubbles.

3. Transfer the cells to a 50-mL tube.
4. Add PBS to a final volume of 40 mL and centrifuge it for 5 min at 400 × *g*.
5. Discard supernatant by carefully pouring it into a waste bottle.
6. Wash the cells with 10 mL of OPTI-MEM medium by centrifugation for 5 min at 400 × *g*.
7. Discard supernatant by carefully pouring it into a waste bottle and resuspend in OPTI-MEM.
8. Count the cells and adjust the volume for a concentration of 50×10^6 cells/mL.
9. Transfer 200 μL of cell suspension (10×10^6 cells) to two sterile cryotubes.
10. Place the cells on ice for at least 5 min (*see* **Note 11**).
11. Setup the RNase-free electroporation station and parameters (*see* **Note 12**) as follows: Voltage at 250 V, pulse length at 2 ms, pulse as 01, interval at 0.01 s, polarity as unipolar.
12. Mix 10 μg of mRNA encoding CXCR2 into one of the cryotubes containing 10×10^6 TILs by carefully stirring (*see* **Note 13**).
13. Immediately transfer cell suspension into a precooled 2-mm electroporation cuvette (*see* **Note 14**).
14. Insert the cuvette into the electroporation chamber and apply the pulse. Make sure to note that the electroporation is successful (*see* **Note 15**).
15. Use the sterile cuvette-pipette to transfer TILs into the 6-well plate containing prewarmed TIL medium immediately after electroporation. Avoid excessive pipetting and bubbles (*see* **Note 16**).
16. Repeat **steps 9–11** with the second cryotube, either without mRNA or including 10 μg of control mRNA, e.g., EGFP, for a mock transfection control cells.
17. Transfer the 6-well plate to a humidified CO_2 incubator at 37 °C (*see* **Note 17**).

3.4 Expression and Signaling of CXCR2 by Flow Cytometry

1. Harvest transfected TILs post-electroporation (*see* **Notes 18** and **19**) by pipetting and transfer to a 14-mL tube.
2. Count the cells and transfer 1×10^6 cells (CXCR2 and mock-transfected TILs, respectively) to labeled Falcon tubes including the appropriate controls (*see* **Note 20**).
3. Wash the cells in 10 mL of PBS, 0.5 % BSA, 2 mM EDTA.
4. Centrifuge FACS tubes 5 min at 400 × *g*.
5. Repeat **steps 3** and ®.

6. For labeling of cells for expression kinetics, cells are stained in 50 μL of mAb cocktail (CD8 FITC, CD3 PE-Cy7, CD4 PB, DCM-NiR, and CXCR2 APC or isotype IgG1 APC) and incubated for 30 min at 4 °C in the dark.
7. After incubation, wash the cells twice in 2 mL of PBS, 0.5 % BSA, 2 mM EDTA, by centrifugation at 400 × *g* for 5 min.
8. For labeling of cells for Ca^{2+}-influx assay, cells are labeled according to manufacturer's recommendations, in 1 mL of PBS, 0.5 % BSA, 2 mM EDTA.
9. Add 5 μM of Fluo-4-AM to the cells and mix by vortexing.
10. Incubate the cells at 37 °C for 30 min (*see* **Note 21**).
11. Wash cells at least thrice in 2 mL of PBS, 0.5 % BSA, 2 mM EDTA by centrifugation for 5 min at 400 × *g*, to ensure that all dye nonspecifically associated to the cells are removed.
12. Resuspend the cells in appropriate volume (*see* **Note 22**) of PBS, 0.5 % BSA, 2 mM EDTA, and acquire on FACScantoII or equal flow cytometer (*see* **Note 23**).

3.5 Specific Migration of CXCR2-Transfected TILs

1. Harvest transfected TILs 2 h post-electroporation by pipetting and transfer to 14-mL tube.
2. Add RPMI 1640 to a total volume of 10 mL and centrifuge it for 5 min at 400 × *g*.
3. Discard the supernatant and repeat **step 2**.
4. Resuspend cells in RPMI 1640 (*see* **Note 24**).
5. Count cells and adjust the volume to a concentration of 1.5×10^6 cells/mL.
6. Prepare transwell setup (*see* Fig. 4).
7. Add chemoattractant medium to the lower wells as follows: Control "Max" and "Min" wells are added 500 μL of RPMI 1640; "Supernatant" wells are added 500 μL of RPMI 1640 + 50 ng/mL of hrIL-8.
8. Carefully place transwell inserts into all wells except the "max" wells.
9. Add 200 μL of cell suspension to the upper chambers immediately (*see* **Note 25**). "Max" wells are added 200 μL of cell suspension directly into the lower chamber.
10. Incubate the plate for 5 h in a humidified CO_2 incubator at 37 °C.
11. Following incubation, carefully remove and discard the transwell inserts.
12. Thoroughly resuspend the cells in the lower chamber by pipetting.
13. Transfer the cells directly to prelabeled FACS tubes.

14. Wash the empty wells with PBS and transfer to corresponding FACS tubes.
15. Centrifuge FACS tubes 5 min at 400 × *g*.
16. Carefully discard the supernatant. Make sure to remove all supernatant!
17. Resuspend the cells by carefully running the tubes over the rack.
18. Use an automated multipipette to add exactly 100 μL of PBS to all tubes.
19. Immediately prior to flow acquisition, add 2 μL of 7-AAD to all tubes for dead cell exclusion.
20. Acquire all tubes on flow cytometer for exactly 60 s, using a standard acquisition setup and a high flow rate (*see* **Note 26**).

4 Notes

1. The cDNA encoding CXCR2 (NM_001557.3) was synthesized and cloned into pSP73-SphA64 (kindly provided by Professor E. Gilboa, Duke University Medical Center, Durham, NC) using 5′XhoI/3′PacI restriction sites (Geneart/Life Technologies, Regensburg, Germany).
2. It is also possible to use other sources for chemoattaction of T cells, such as cancer cell line supernatant. Cancer cell line supernatant is obtained by harvesting medium 2 days post seeding 0.5–1 × 10^6 tumor cells in 6-well plate containing 2 mL of RPMI and 10 % FCS. The cell line supernatant should be centrifuged for 5 min at 400 × *g* to sediment any cells that may have been removed along with the medium. The Cell line supernatant can be aliquoted and stored at −20 °C until use.
3. Aliquot human AB serum, as well as FBS, to 50-mL tubes and heat-inactivate in a 56 °C water bath for 30 min to inactivate proteases and complement proteins. A centrifugation step at 800 × *g* for 10 min will sediment any precipitates. Following centrifugation, pipette the serum into new tubes and keep at −20 °C until use.
4. One U of restriction enzyme cleaves 1 μg of DNA in 1 h. We add an excess of restriction enzyme (5 U) to cleave 2 μg of DNA to be sure of a complete digestion.
5. To make sure that tumor fragments do not dry out, only dissect the tumor biopsy in the transportation medium-filled petri dish and transfer fragments to 24-well plate containing prewarmed medium. The prewarmed media is to accommodate a more favorable environment for the cells.
6. As it takes some time for the TILs to respond to the high IL-2 concentration and start proliferating, it is not necessary to change medium for the first 5 days.

7. Fresh medium must be prepared at every medium change, as the half-live of the supplements varies. Most importantly, Proleukine (IL-2) is the most sensitive to storage. The manufacturers can document that reconstituted IL-2 retains its chemical and physical properties for up to 48 h when stored at refrigerated temperatures. Prewarming the medium prior to use helps increase the yield of the TIL isolation.
8. TILs will grow out from the tumor fragment. Depending on how much resuspension occurs during medium change, early TIL outgrowth will be noted as a ring of single-layered cells around the fragment or around the edges of the wells. It is important not to stir the cells too much in the beginning of the outgrowth. As the cell count increases, the cell concentration should be kept at around 1–1.5 × 10^6 cells/mL. When a well is confluent and/or there are more than 1–1.5 × 10^6 cells/mL, the well content needs to be divided into two wells after thorough resuspension of the cells. If there are wells with little or no TIL outgrowth, it is preferable to split wells into these.
9. Cryopreservation of cells should be done using FBS supplemented with 10 % of DMSO as freezing medium. Cell viability is sensitive to handling and DMSO exposure, which is why the cells should be placed in the freezer within 3–5 min of resuspension in freezing medium. Additionally, when thawing the cells, the use of 37 °C prewarmed medium is recommended. The cells should be frozen using a gradual freezing device such as a CoolCell® with a freezing rate of 1 °C/min in an −80 °C freezer to obtain the highest viability of cryopreserved cells. [20].
10. The Rapid Expansion Protocol is a method in which T cells can be expanded >1000-fold within 14 days of culture. We adapted the protocol from Dudley et al. [21, 22]. In short, TILs isolated from tumor biopsies are cultured in TIL medium in a T25 culture flask (Research scale) with irradiated (40 Gy) feeder cells at a 1:200 ratio, along with anti-CD3 and high dose IL-2. In order to prevent the dilution of anti-CD3 in the initiation phase, cultures are left for 5 days, after which half the medium is replaced every second day.
11. During the electroporation procedure, the electrical current will generate heat. To protect the cells from heat shock during electroporation, the cells are kept on ice until use, as a heat shock will severely affect viability of the cells.
12. This chapter describes the mRNA transfection of TILs; however, naive T cells have been shown to be successfully transfected with mRNA using square-wave electroporation [23, 24]. We have found that electroporation settings for optimal mRNA transfection of naive T cells are: a single pulse of 250 V, 5 ms in a 2 mm cuvette containing 10 × 10^6 T cells/100 μL.

13. The expression of mRNA products after transfection is dose dependent. Therefore, a higher surface expression of the chemokine receptor (or protein products in general) can be achieved by increasing the mRNA concentration in the electroporation cocktail. However, keep in mind that to maintain the optimal conductivity and transfection efficacy, the mRNA volume added should not exceed 10 % of the total volume of the electroporation cocktail.
14. 2-mm electroporation cuvettes are kept refrigerated until use, as this will decrease a potential heat development during electroporation. Be aware that cooling the cuvettes may result in condensation collecting on the exterior surface of the electroporation cuvette.
15. Looking at the surface of your cell suspension during electroporation, you should note a change in appearance, as current is applied. This is most commonly described as a slight puffing or foaming at the surface. This indicates that a current went through your cell suspension. Also note from the apparatus the exact voltage, pulses and duration applied, as this may vary.
16. Carefully transfer the cell suspension from cuvette to prewarmed medium without resuspension. Aspirate all the cells from the cuvette, and slowly distribute the cell suspension to the well, drop by drop. The cell membrane is porous due to the electroporation, and cell viability may be affected by the physical stress of resuspension.
17. Electroporated cells should theoretically recover in 5–10 min when cultured subsequently at 37 °C [25], although this will depend on the applied pulse.
18. Successful experiments using mRNA, transfected TILs is dependent on using the TILs within the optimal "expression window." The expression kinetics of a given mRNA is dependent on the construct, the stability of the mRNA and the amount of mRNA that has entered the cells. Thus, it is recommended to establish the window of expression every time a new construct or cell type is used. This can be done by analysis of the surface expression at different time points by flow cytometry in the case of chemokine receptor encoding mRNAs (*see* Fig. 2). The first 24 h, testing 2, 8 and 24 h will tell something about the immediate expression of the mRNA. If expression is detected after 24 h, testing surface expression every 24 h until it is negative will give insights to the stability of the mRNA.
19. Analyses of signal transduction of the transfected chemokine receptor using Ca^{2+}-influx assay depend on the surface expression. Thus, it is important to know the optimal expression window of the receptor of interest before starting.

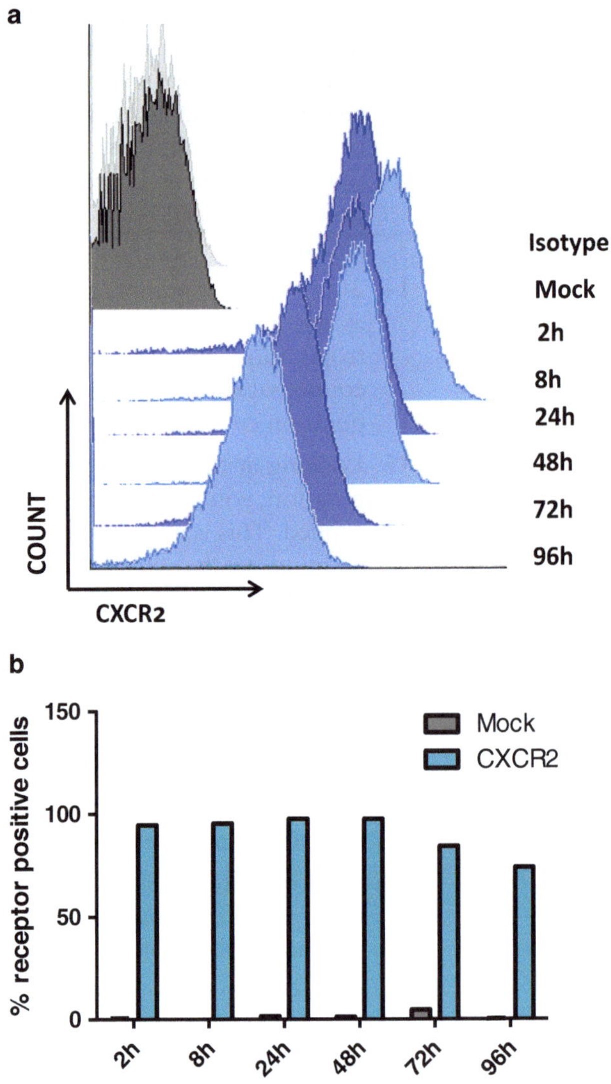

Fig. 2 Time-dependent expression of CXCR2 on the surface of mRNA-electroporated TILs. (**a**) MFI histograms of expression, including an isotype control and the mock-transfected TIL control, and (**b**) the percentage of CXCR2$^+$ TILs at time points; 2, 8, 24, 48, 72, and 96 h posttransfection

20. The mock-transfected cells are the most important control for chemokine receptor transfection efficacy and signal transduction; however, it is important to include other controls. An isotype IgG mAb labeled with the same fluorochrome as the chemokine receptor mAb is recommended for guiding the gating of CXCR2$^+$ cells. Similarly, it is important to include a positive control for the assessment of Fluo-4 AM labeling of your cells. Thus Ionomycin, a Ca^{2+} ionophore, can be used to mimic the receptor-dependent influx of calcium (*see* Fig. 3).

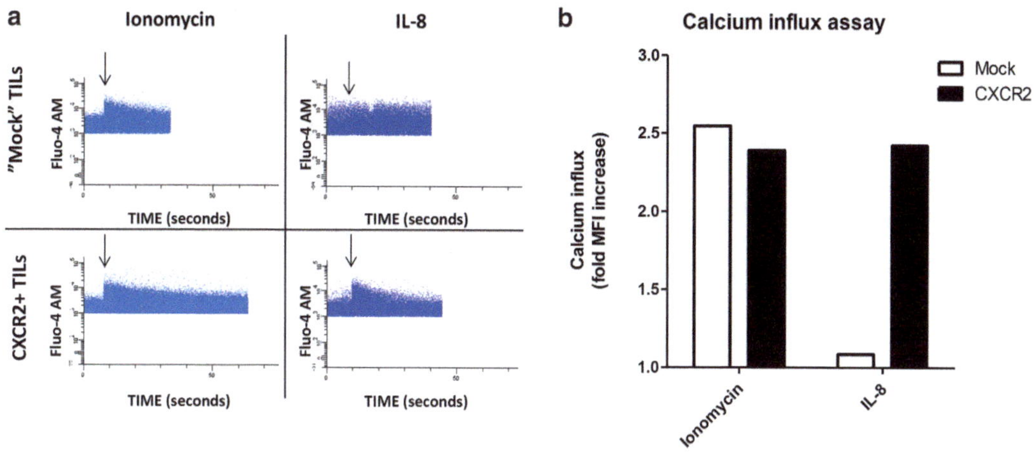

Fig. 3 Calcium influx can be measured by excitation by the 488 nm blue laser of a flow cytometer. (**a**) Calcium-mediated excitation of intracellular Fluo-4-AM upon addition of ionomycin or IL-8 upon binding of the IL-8 receptor CXCR2. Mock-transfected cells do not excite Fluo-4-AM upon addition of IL-8. (**b**) Calculated fold increase in Fluo-4-AM MFI after addition of ionomycin or IL-8

21. Fluo-4-AM is a Ca^{2+} indicator, increasing fluorescence excitation at 488 nm upon binding of Ca^{2+}, which allows for detection by the blue laser on a flow cytometer. The AM-ester is easily loaded into cells by dilution into cell-containing medium. Labeling is done by using an adaptation of manufacturer's recommendations.
22. The resuspension volume is dependent on what application the cells are used for on the flow cytometer. For analysis of surface marker expression, a flow rate of 3500–5000 events/s is recommended to ensure that all events are recorded. However, to analyze signal transduction, one must measure the change in fluorescence over time post Ca^{2+}-influx, and thus a very low flow rate is recommended. Flow rate can be adjusted on the flow cytometer prior to acquisition.
23. The FACScantoII allows for 8-color flow cytometry. Make sure to adjust the mAb staining panel to accommodate the flow cytometer of use, as many flow cytometers have only 4-colors. Additionally, it is of great importance to have the correct compensation settings and run controls before use.
24. It is important to consider what medium to use for the cell suspension applied the upper well (the transwell insert). Serum—both human and fetal bovine—contains many factors that act as chemoattractants. Therefore, it is the medium of the lower well that should determine the medium of the upper well. In this case, we use conditioned medium from a tumor cell line supernatant in the lower wells. In this setup, tumor cell lines are cultivated in R10. Thus, to minimize the background,

we use the same medium in the upper well, which would cancel out any FBS-mediated migration.

25. It is recommended to use sterile forceps for translocation of the transwell inserts during setup and conclusion of the experiment. The inserts should be placed and removed carefully, as the system works by creating a gradient between the medium in the upper and lower wells. This may be affected by too forceful an application of the transwell insert, as this may cause a spill-over of medium from the lower to the upper well.

26. Migration can be measured in a number of ways. We use flow-cytometry-based quantification, where cells from the lower wells are resuspended in a volume of 100 μL and counted on FACScanto. By including a "maximal" and a "minimal" migration control, the percentage of specific migration can be calculated (Sample − minimal)/(maximal − minimal) × 100% (*see* Fig. 4).

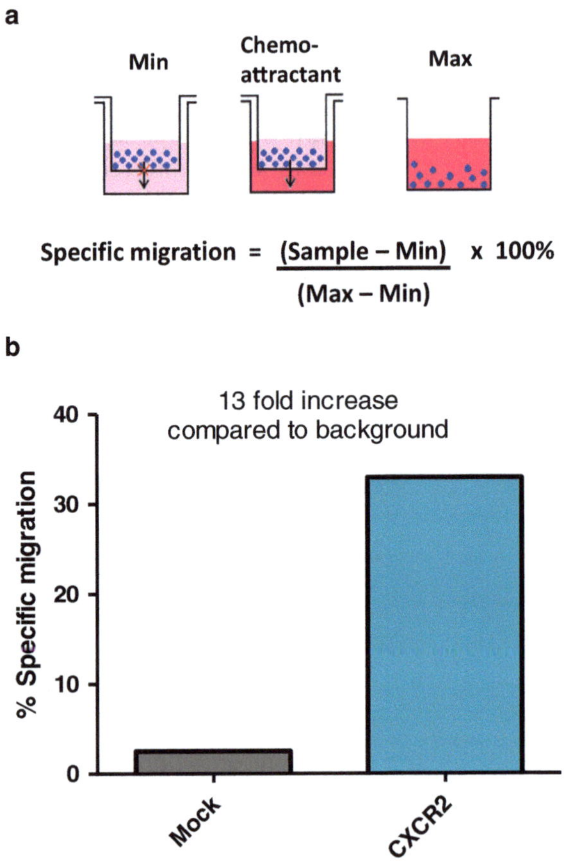

Fig. 4 (**a**) Transwell migration assay setup and calculation. (**b**) Graph depicting the specific migration of mock- and CXCR2-transfected TILs toward an IL-8 source

References

1. Rosenberg SA, Yang JC, Sherry RM et al (2011) Durable complete responses in heavily pretreated patients with metastatic melanoma using T-cell transfer immunotherapy. Clin Cancer Res 17:4550–4557
2. Besser MJ, Shapira-Frommer R, Itzhaki O et al (2013) Adoptive transfer of tumor-infiltrating lymphocytes in patients with metastatic melanoma: Intent-to-treat analysis and efficacy after failure to prior immunotherapies. Clin Cancer Res 19:4792–4800
3. Rosenberg SA, Spiess P, Lafreniere R (1986) A new approach to the adoptive immunotherapy of cancer with tumor-infiltrating lymphocytes. Science 233:1318–1321
4. Donia M, Larsen SM, Met Ö, Svane IM (2014) Simplified protocol for clinical-grade tumor-infiltrating lymphocyte manufacturing with use of the Wave bioreactor. Cytotherapy 16:1117–1120
5. Rosenberg SA, Dudley ME (2004) Cancer regression in patients with metastatic melanoma after the transfer of autologous antitumor lymphocytes. Proc Natl Acad Sci U S A 101(Suppl):14639–14645
6. Donia M, Hansen M, Sendrup SL et al (2013) Methods to improve adoptive T-cell therapy for melanoma: IFN-γ enhances anticancer responses of cell products for infusion. J Invest Dermatol 133:545–552
7. Koya RC, Mok S, Otte N et al (2012) BRAF inhibitor vemurafenib improves the antitumor activity of adoptive cell immunotherapy. Cancer Res 72:3928–3937
8. Li Y, Liu S, Hernandez J et al (2010) MART-1-specific melanoma tumor-infiltrating lymphocytes maintaining CD28 expression have improved survival and expansion capability following antigenic restimulation in vitro. J Immunol 184:452–465
9. Chacon JA, Wu RC, Sukhumalchandra P et al (2013) Co-stimulation through 4-1BB/CD137 improves the expansion and function of CD8+ melanoma tumor-infiltrating lymphocytes for adoptive T-cell therapy. PLoS One. doi:10.1371/journal.pone.0060031
10. Pockaj BA, Sherry RM, Wei JP et al (1994) Localization of 111indium-labeled tumor infiltrating lymphocytes to tumor in patients receiving adoptive immunotherapy. Augmentation with cyclophosphamide and correlation with response. Cancer 73:1731–1737
11. Peng W, Ye Y, Rabinovich BA et al (2010) Transduction of tumor-specific T cells with CXCR2 chemokine receptor improves migration to tumor and antitumor immune responses. Clin Cancer Res 16:5458–5468
12. Moon EK, Carpenito C, Sun J et al (2011) Expression of a functional CCR2 receptor enhances tumor localization and tumor eradication by retargeted human T cells expressing a mesothelin-specific chimeric antibody receptor. Clin Cancer Res 17:4719–4730
13. Curiel TJ, Coukos G, Zou L et al (2004) Specific recruitment of regulatory T cells in ovarian carcinoma fosters immune privilege and predicts reduced survival. Nat Med 10:942–949
14. Waugh DJJ, Wilson C (2008) The interleukin-8 pathway in cancer. Clin Cancer Res 14:6735–6741
15. June CH, Blazar BR, Riley JL (2009) Engineering lymphocyte subsets: tools, trials and tribulations. Nat Rev Immunol 9:704–716
16. Porter DL, Levine BL, Kalos M et al (2011) Chimeric antigen receptor-modified T cells in chronic lymphoid leukemia. N Engl J Med 365:725–733
17. Cameron BJ, Gerry AB, Dukes J, et al. (2013) Identification of a Titin-derived HLA-A1-presented peptide as a cross-reactive target for engineered MAGE A3-directed T cells. Sci Transl Med 5:197ra103.
18. Rabinovich PM, Weissman SM (2013) Cell engineering with synthetic messenger RNA. Methods Mol Biol 969:3–28
19. Met O, Balslev E, Flyger H, Svane IM (2011) High immunogenic potential of p53 mRNA-transfected dendritic cells in patients with primary breast cancer. Breast Cancer Res Treat 125:395–406
20. Yokoyama WM, Thompson ML, Ehrhardt RO (2012) Cryopreservation and thawing of cells. Curr Protoc Immunol. doi:10.1002/0471142735.ima03gs99
21. Dudley ME, Wunderlich JR, Shelton TE et al (2008) Generation of tumor-infiltrating lymphocyte cultures for use in adoptive transfer therapy for melanoma patients. J Immunother 26:332–342
22. Donia M, Junker N, Ellebaek E et al (2011) Characterization and comparison of "Standard" and "Young" tumor infiltrating lymphocytes for adoptive cell therapy at a Danish Translational Research Institution. Scand J Immunol 157–167
23. Schaft N, Dörrie J, Müller I et al (2006) A new way to generate cytolytic tumor-specific T cells: Electroporation of RNA coding for a T cell receptor into T lymphocytes. Cancer Immunol Immunother 55:1132–1141

24. Zhao Y, Zheng Z, Cohen CJ et al (2006) High-efficiency transfection of primary human and mouse T lymphocytes using RNA electroporation. Mol Ther 13:151–159

25. Teissié J, Rols MP (1993) An experimental evaluation of the critical potential difference inducing cell membrane electropermeabilization. Biophys J 65:409–413

Chapter 18

mRNA Electroporation of Dendritic Cells with WT1, Survivin, and TriMix (a Mixture of caTLR4, CD40L, and CD70)

An Coosemans, Sandra Tuyaerts, Kim Morias, Jurgen Corthals, Carlo Heirman, Kris Thielemans, Stefaan W. Van Gool, Ignace Vergote, and Frédéric Amant

Abstract

The immune system is a crucial player in the development of cancer. Once it is in imbalance and immunosuppressive mechanisms supporting tumor growth take over control, dendritic cell immunotherapy might offer a solution to restore the balance. There are several methods to manufacture dendritic cells but none of them has yet proven to be superior to others. In this chapter, we discuss the methodology using electroporation of mRNA encoding Wilms' tumor gene 1, survivin, and TriMix (mixture of caTLR4, CD40L, and CD70) to simultaneously load and mature dendritic cells.

Key words mRNA, Electroporation, Dendritic cell, WT1, Survivin, TriMix

1 Introduction

It becomes increasingly clear that the immune system is important in the development of cancer. Once initial neoplastic cells arise, the host immune system strives to control their progression mainly through the presence of effector T cells (Teff) and NK cells. This stage is called the cancer immune surveillance. One of the different innate immune cells controlling tumor development during the immune surveillance stage is the dendritic cell (DC). These are professional antigen presenting cells, that can capture tumor-associated antigens (TAA), expressed by cancer cells, in their immature state (DCi) in a tumor milieu. Upon ingestion of the TAA and in the presence of an appropriate "danger" signal, for example, inflammatory cytokines, DC start to mature (DCm) and migrate to the lymph nodes, where they make contact with and activate T cells that

Robert E. Rhoads (ed.), *Synthetic mRNA: Production, Introduction Into Cells, and Physiological Consequences*, Methods in Molecular Biology, vol. 1428, DOI 10.1007/978-1-4939-3625-0_18,

will then traffic to the target organ (the tumor) to selectively destroy the neoplastic cells. If this process is successful, the neoplastic cells are either eliminated or come to reside in a dormant state (elimination versus equilibrium). However, the problem is that tumor cells might also be capable to escape from the control of the immune system, start proliferating, and cause the tumor to become clinically apparent [1]. Tumor immune escape is facilitated by two main pathways: (1) mechanisms that eventually cause antigen loss and (2) the infiltration of immunosuppressive cells, which are attracted to the tumor microenvironment due to the sustained secretion of chemokines and cytokines (TGF-β (tumor growth factor beta), IL-10 (interleukin-10), VEGF (vascular endothelial growth factor), galectin-1, etc.,) produced by the tumor and immune cells present in and around the tumor. These immunosuppressive cells will in turn also start producing several mediators that will strengthen their effects and will cause more negative feedback on other cells, leading to a downward spiral of effects all contributing to the escape of tumor cells from the immune surveillance. There are several players within this group of immunosuppressive cells. However, the most currently relevant cells in pelvic gynecological tumors are the regulatory T cells (Treg), the myeloid-derived suppressor cells (MDSC), and the tumor-associated macrophages (TAM) [2, 3].

DC immunotherapy is an attempt to increase the number of potent DCm (and consequently tumor-specific T cells) in order to shift the balance from immunosuppression toward cancer immune surveillance [4]. At this level, there is room for improvement. In a recent review, we summarized in a nonexhaustive list possible strategies that can be used to improve the effectiveness of DC [5]. At present, no method has proven to be superior to another. Therefore, we opted for a combination of three different routes of improvement: (1) the use of two predefined TAA, Wilms' tumor gene 1 (WT1) and survivin, both linked to a DC-LAMP targeting signal to obtain both $CD4^+$ and $CD8^+$ T cell induction; (2) the use of TAA at the mRNA level for DC electroporation; (3) the use of TriMix-mRNA to simultaneously mature DCi.

The choice of two TAA is explained by the rational that the risk of immune escape is reduced and the group of tumors that can be targeted is broadened [6]. The in vitro work of several research groups has suggested that RNA transfection of TAA is an effective method to generate immunostimulatory DCs, regardless of the patients HLA-type [4, 7]. Further modifications of the coding sequence by adding lysosomal targeting sequences (e.g., DC-LAMP) lead to the presentation of antigenic peptides in the context of both HLA class I and class II [8–10]. Our research group has proven the safety and feasibility of DC immunotherapy in uterine and ovarian tumors (*WT1* mRNA-loaded DCs) [11, 12]. In a recent review by our research group and the group of Kris

Thielemans, the use of TriMix was explained. In brief, TriMix is constituted of three different mRNAs: caTLR4 [constitutively active Toll-like receptor 4], CD40L, and CD70. Transfection with TriMix leads to mature, cytokine/chemokine-secreting DC that are capable of stimulating naive T cells and support their proliferation. The technique is fast and results in semimature DC 3.5 h after electroporation. As a result, TriMix DC will mature and secrete most of their immunostimulatory cytokines and chemokines in vivo after human injection [6, 13].

This chapter will especially outline the procedure of the electroporation. Details will be given on how to handle electroporation cuvettes and mRNA, what steps should be followed consecutively to obtain an optimal result and what should be avoided. The protocol will also give you an insight in how to work with large volume culture material (ULA).

2 Materials

2.1 Reagents

1. Supplemented RPMI: RPMI, 1 % human autologous plasma (heat-inactivated 30 min at 56 °C), IL-4 (500 IU/mL), GM-CSF (1000 IU/mL).
2. DPBS + EDTA: Dulbecco's phosphate-buffered saline with ethylenediaminetetraacetic acid (460 mg/L).
3. OptiMEM, Reduced Serum Medium, no phenol red.
4. Rapamycin.
5. *WT1.DC-LAMP* mRNA.
6. *survivin.DC-LAMP* mRNA.
7. *TriMix*-mRNA.

2.2 Equipment

1. Warm water bath (56 °C).
2. Tabletop centrifuge consisting of a rotor and buckets capable of holding 50 mL tubes and reaching at least 300 × *g*.
3. 5 % CO_2 humidified incubator.
4. 5 and 50 mL conical tubes.
5. 2 ml cryovials.
6. 4 mm electroporation cuvettes.
7. RNase-free tips of 40–200 μL and of 0.5–10 μL (*see* **Note 1**).
8. Corning® CellSTACK® cell culture chambers.
9. Electroporation device.
10. Bürcker chamber (microscope counting).

3 Methods

3.1 mRNA Electroporation of Immature Dendritic Cells

The procedure is for 400×10^6 to 600×10^6 DCi in total (*see* **Note 2**). **Steps 1–6** are carried out in advance of **step** 7 (*see* **Note 3**):

1. Fill the ULA with 200 mL supplemented RPMI (*see* **Notes 4** and **5**).
2. Fill a 50 ml conical tube with supplemented RPMI.
3. Centrifuge the *WT1.DC-LAMP*-mRNA, *survivin.DC-LAMP*-mRNA, and *TriMix*-mRNA at maximal speed during 5 s.
4. Take out 60 μg *WT1.DC-LAMP*-mRNA + 60 μg *TriMix*-mRNA and transfer into cryovial A; take out 60 μg *survivin.DC-LAMP*-mRNA + 60 μg *TriMix*-mRNA and transfer into cryovial B.
5. Complete cryovial A and B with OptiMEM until a final volume of 200 μL has been reached.
6. The process described in **steps 4** and **5** is sufficient to electroporate 50×10^6 DCi. Repeat the procedure for cryovial A for half of the total amount of DCi. Repeat the procedure for cryovial B for half of the total amount of DCi.
7. Dissolve 50×10^6 immature DC (DCi) in 10 ml OptiMEM in a 50 ml conical tube.
8. Repeat this step until all DCi are used.
9. Centrifuge DCi at $300 \times g$ during 10 min at room temperature.
10. Discard the supernatant.
11. Dissolve each DCi pellet in 400 μl OptiMEM in the 50 mL conical tube.
12. Transfer the content of one 50 mL conical tube into one electroporation cuvette.
13. Rinse each 50 mL conical tube with one mRNA mix (200 μL volume) to collect all remaining DCi (**step 5**).
14. Transfer the 200 μL into the electroporation cuvette. Final volume in electroporation cuvette is now 600 μL (*see* **Note 6**).
15. Mix contents very gently by repeated pipetting (*see* **Note** 7).
16. Electroporate at 300 V and 450 μF, using exponential decay.
17. Transfer the contents of the electroporation cuvette into the ULA (*see* **Note 8**).
18. Rinse the cuvette twice with 800 μL RPMI medium from the 50 mL conical tube prepared in **step 2**.
19. Repeat **step** 7 until 13 for all 50 mL conical tubes.

20. Gently shake the ULA to homogenate the cells.
21. Add 100 nM rapamycin to the ULA.
22. Place the ULA at 37 °C with 5 % CO_2 during 3, 5 h.

3.2 Preparation of Mature Dendritic Cells

1. Gently shake the ULA.
2. Empty the ULA with a 25 mL pipet into 50 mL conical tubes.
3. Add 75 mL DPBS + EDTA to the ULA (*see* **Note 9**).
4. Place the ULA at 37 °C with 5 % CO_2 during 15 min.
5. Gently shake the ULA.
6. Empty the ULA with a 25 mL pipet into 50 mL conical tubes.
7. Rinse the ULA twice with 75 mL supplemented RPMI.
8. Empty the ULA with a 25 mL pipet into 50 mL conical tubes.
9. Centrifuge all 50 mL conical tubes at 300 × *g* during 10 min at room temperature.
10. Discard the supernatant.
11. Pool the content of all 50 mL conical tubes in a total volume of 40 mL supplemented RPMI.
12. Count the amount of mature DC (DCm) in the Bürcker chamber.
13. Centrifuge DCm at 300 × *g* during 10 min at room temperature.
14. Discard the supernatant.
15. DCm can now be used in further experiments (*see* **Note 10**).

4 Notes

1. RNA is fragile and subject to the activity of nucleases, particularly RNases. Therefore, use RNase-free material (indicated on the packing of disposable material) when handling pure mRNA (before electroporation) and work with sterile technique. Always use gloves and never touch anything with bare hands, since they contain nucleases. Once the mRNA is electroporated into the DCi, non-RNase-free material can be used.
2. DCi can be obtained from culturing monocytes into DC. In our hands, peripheral blood mononuclear cells are cultured in cell factories (plastic adherence) over 6 days. IL-4, GM-CSF, and autologous plasma are added to the cell factories on day 0, 2, and 4. At day 6, the cells have been differentiated into DCi. Several other protocols have been described to generate DCi [5].

3. Before you start the procedure, make sure that your filter tips can reach the bottom of the electroporation cuvette.
4. RPMI should be at room temperature before use.
5. Supplemented RPMI can be kept at 4 °C after filling the ULA.
6. When filling the electroporation cuvette, try not to overfill. The fluid level cannot overreach the electrode plates.
7. When working in the electroporation cuvettes, try not the create air bubbles because they will prevent effective electroporation.
8. After electroporation, cells can clot. Make sure you do not use a filter tip with a small diameter to empty the electroporation cuvette—tips from 0.5 to 10 μL are too small. Preferably use a filter tip from 200 to 1000 μL and use it at 800 μL. After rinsing, the electroporation cuvette, you can use a small pipette filter tip of 0.5–10 μL to empty the cuvette.
9. Never use DPBS + EDTA longer than 15 min, because it will become toxic for the DC.
10. All work should be performed in a biological safety cabinet in order to work sterile.

Acknowledgments

An Coosemans is supported by the Fund for Scientific Research Flanders (FWO-V). Sandra Tuyaerts is supported by the Anticancer Fund. Frédéric Amant and Stefaan Van Gool are senior researchers for the Research Fund Flanders (FWO-V). The research was supported by the Anticancer Fund.

References

1. Schreiber RD, Old LJ, Smyth MJ (2011) Cancer immunoediting: integrating immunity's roles in cancer suppression and promotion. Science 331:1565–1570
2. Baert T, Timmerman D, Vergote I et al (2015) Immunological parameters as a new lead in the diagnosis of ovarian cancer. Facts Views Vis Obgyn 7:67–72
3. Vanderstraeten A, Luyten C, Verbist G et al (2014) Mapping the immunosuppressive environment in uterine tumors: implications for immunotherapy. Cancer Immunol Immunother 63:545–557
4. Gilboa E (2007) DC-based cancer vaccines. J Clin Invest 117:1195–1203
5. Coosemans A, Baert T, Vergote I (2015) A view on dendritic cell immunotherapy in ovarian cancer: how far have we come? Facts Views Vis Obgyn 7:73–78
6. Coosemans A, Tuyaerts S, Vanderstraeten A et al (2014) Dendritic cell immunotherapy in uterine cancer. Hum Vaccin Immunother 10:1822–1827
7. Michiels A, Tuyaerts S, Bonehill A et al (2005) Electroporation of immature and mature dendritic cells: implications for dendritic cell-based vaccines. Gene Ther 12:772–782
8. Bonehill A, Heirman C, Thielemans K (2005) Genetic approaches for the induction of a CD4+ T cell response in cancer immunotherapy. J Gene Med 7:686–695
9. Bonehill A, Heirman C, Tuyaerts S et al (2004) Messenger RNA-electroporated dendritic cells presenting MAGE-A3 simultaneously in HLA

class I and class II molecules. J Immunol 172:6649–6657
10. Su Z, Dannull J, Yang BK et al (2005) Telomerase mRNA-transfected dendritic cells stimulate antigen-specific CD8+ and CD4+ T cell responses in patients with metastatic prostate cancer. J Immunol 174:3798–3807
11. Coosemans A, Vanderstraeten A, Tuyaerts S et al (2013) Immunological response after WT1 mRNA-loaded dendritic cell immunotherapy in ovarian carcinoma and carcinosarcoma. Anticancer Res 33:3855–3859
12. Coosemans A, Vanderstraeten A, Tuyaerts S et al (2013) Wilms' Tumor Gene 1 (WT1)-loaded dendritic cell immunotherapy in patients with uterine tumors: a phase I/II clinical trial. Anticancer Res 33:5495–5500
13. Van Lint S, Wilgenhof S, Heirman C et al (2014) Optimized dendritic cell-based immunotherapy for melanoma: the TriMix-formula. Cancer Immunol Immunother 63:959–967

Chapter 19

Redirecting T Cell Specificity Using T Cell Receptor Messenger RNA Electroporation

Sarene Koh, Noriko Shimasaki, and Antonio Bertoletti

Abstract

Autologous T lymphocytes genetically modified to express T cell receptors or chimeric antigen receptors have shown great promise in the treatment of several cancers, including melanoma and leukemia. In addition to tumor-associated antigens and tumor-specific neoantigens, tumors expressing viral peptides can also be recognized by specific T cells and are attractive targets for cell therapy. Hepatocellular carcinoma cells often have hepatitis B virus DNA integration and can be targeted by hepatitis B virus-specific T cells. Here, we describe a method to engineer hepatitis B virus-specific T cell receptors in primary human T lymphocytes based on electroporation of hepatitis B virus T cell receptor messenger RNA. This method can be extended to a large scale therapeutic T cell production following current good manufacturing practice compliance and is applicable to the redirection of T lymphocytes with T cell receptors of other virus specificities such as Epstein-Barr virus, cytomegalovirus, and chimeric receptors specific for other antigens expressed on cancer cells.

Key words Messenger RNA, Electroporation, T cells, Cell therapy, T cell receptor, Hepatitis B virus, Hepatocellular carcinoma

1 Introduction

Cytotoxic T lymphocytes, our body's natural serial killers, play an important role in killing cancerous and infected cells. They express T cell receptors that recognize a specific antigen that is often produced intracelluarly by cancer cells or viruses [1]. The intracellular viral or tumor proteins are processed and presented as peptide fragments on class I MHC molecules that are recognized by a particular T cell receptor (TCR) specific for that antigen. Once cytotoxic T cells recognize their target, they promote cell death of the target cell through a combination of cytotoxins and receptor-mediated mechanisms [2]. The potent killing ability of cytotoxic T lymphocytes can be harnessed as mediators of antitumor immunity.

Robert E. Rhoads (ed.), *Synthetic mRNA: Production, Introduction Into Cells, and Physiological Consequences*, Methods in Molecular Biology, vol. 1428, DOI 10.1007/978-1-4939-3625-0_19,

Early immunotherapy using autologous cytokine activated killer cells showed some contribution to tumor regression or patient survival [3, 4]. Later on, in vitro studies showed that human tumor-infiltrating lymphocytes obtained from resected melanomas contained specific T cells capable of recognition of autologous tumors in vitro [5], and the first successful adoptive cell therapy in patients with metastatic melanoma achieved objective regression of cancer [6]. However, the isolation and expansion of preexisting tumor-specific T cells from most cancer patients is extremely difficult and time consuming. To overcome these limitations, gene transfer-based strategies have been developed to retarget patients' lymphocytes to a chosen tumor antigen or tumor neoantigen expressed by the cancer cells. This is achieved by the transfer of a chimeric antigen receptor composed of antibody binding domains fused to T cell signaling domains or transfer of TCR αβ heterodimers. Adoptive T cell therapy using chimeric antigen receptor- [7, 8] or TCR-engineered [9–11] T cells have shown highly promising results in the treatment of several cancers, including melanoma and leukemia, and may offer a lifeline to patients with certain cancers who have exhausted all other treatment options.

In addition to redirecting T cell specificity towards tumor-associated antigens and tumor neoantigens, tumors expressing viral peptides, for example, hepatitis B virus (HBV)-related hepatocellular carcinoma (HCC), are also attractive targets for cell therapies because their expression is limited to infected cells. A high frequency of HBV integrations has been observed in HBV-related HCC [12, 13] that might result in the expression of HBV antigens in tumor cells. Targeting nonself viral antigens expressed on HCC cells present several advantages over the targeting of classical tumor-associated antigens (NY-ESO-1, glypican-3, alpha fetoprotein, MAGE-A3) [14]. The use of TCR-engineered T cells specific for classical tumor-associated antigens can lead to severe side effects due to the targeting of normal tissues [15, 16] or induce a state of tolerance [17] in tumor-specific T cells. Moreover, TCRs specific for tumor-associated antigens often have low affinities and require modifications in order to efficiently recognize tumor cells. By contrast, TCRs recognizing HBV antigens that were cloned from patients who resolved acute HBV infection have high affinities and can be activated at low concentration of peptide in the femtogram per milliliter range [18]. We have also demonstrated that transfer of these HBV-specific TCRs can be used to produce therapeutic T cells that lyse not only HBV infected cells but also natural HCC lines with HBV–DNA integration [18]. Furthermore, these TCR-redirected T cells are not inhibited by the large quantity of circulating antigens (HBeAg and HBsAg) present in the patients' sera that might instead interfere with recognition of infected cells by HBV-specific chimeric antigen receptor-T cells [19]. Indeed, we recently demonstrated that adoptive transfer of

HBV-TCR-redirected T cells caused a profound inhibition of HBsAg production [20] in a patient with extrahepatic HCC metastases with HBV–DNA integration producing only HBsAg.

The use of HBV antigen as a target for immunotherapy, however, has one considerable drawback: in patients with chronic HBV infection who developed HCC, HBV antigens are expressed not only by HCC cells with HBV–DNA integration but also by HBV infected hepatocytes. Thus, there is a risk of inducing a severe liver damage due to the targeting of normal infected hepatocytes. For this reason, we developed an alternative and practical production method of HBV-TCR-redirected T cells through direct electroporation of TCR mRNA into lymphocytes [21]. We found that electroporation of mRNA encoding HBV-TCR with the method described here resulted 24 h after electroporation in a median receptor expression of 80.0 % in activated T cells ($n = 20$ donors) with excellent cell viability (Fig. 1a). Expression was detectable 6 h after electroporation, reaching maximum levels at 24 h (Fig. 1b). Stimulation of HBV-TCR-electroporated T cells with T2 cells loaded with HBV envelope peptide resulted in antigen-specific production of IFN-γ, TNF-α, and IL-2 (Fig. 1c), indicating that functional T cells were engineered. We can produce large quantity of T cells redirected with HBV-specific TCRs in less than a month that are functionally efficient in vitro and in vivo and can be developed for clinical use. TCR mRNA production and direct electroporation of lymphocytes is easier, faster, and considerably less expensive than viral vector production. In addition to these practical criteria, TCR expression in T cells electroporated with TCR mRNA is transient, which offers a safety advantage in the initial testing of TCR-modified T cells in HBV infected HCC patients.

In this chapter, we describe the methods for redirecting the specificity of primary human T lymphocytes to target HBV viral antigens expressed on HCC tumor cells with HBV–DNA integrations based on electroporation of HBV-TCR mRNA. We provide the protocols to activate T cells for efficient expression of introduced TCR, produce mRNA encoding the TCR based on in vitro transcription, electroporation procedures, and analysis of TCR expression and function using flow cytometry. This method can produce 5×10^6 to 2×10^7 T cells per electroporation reaction for research purposes and can also be extended to a large scale current good manufacturing practice compliant therapeutic T cell production of up to 1×10^{10} T cells. In addition, this mRNA electroporation technology is applicable to the redirection of T lymphocytes with TCRs of other virus specificities such as Epstein-Barr virus, cytomegalovirus, and chimeric receptors specific for other tumor antigens expressed on cancer cells.

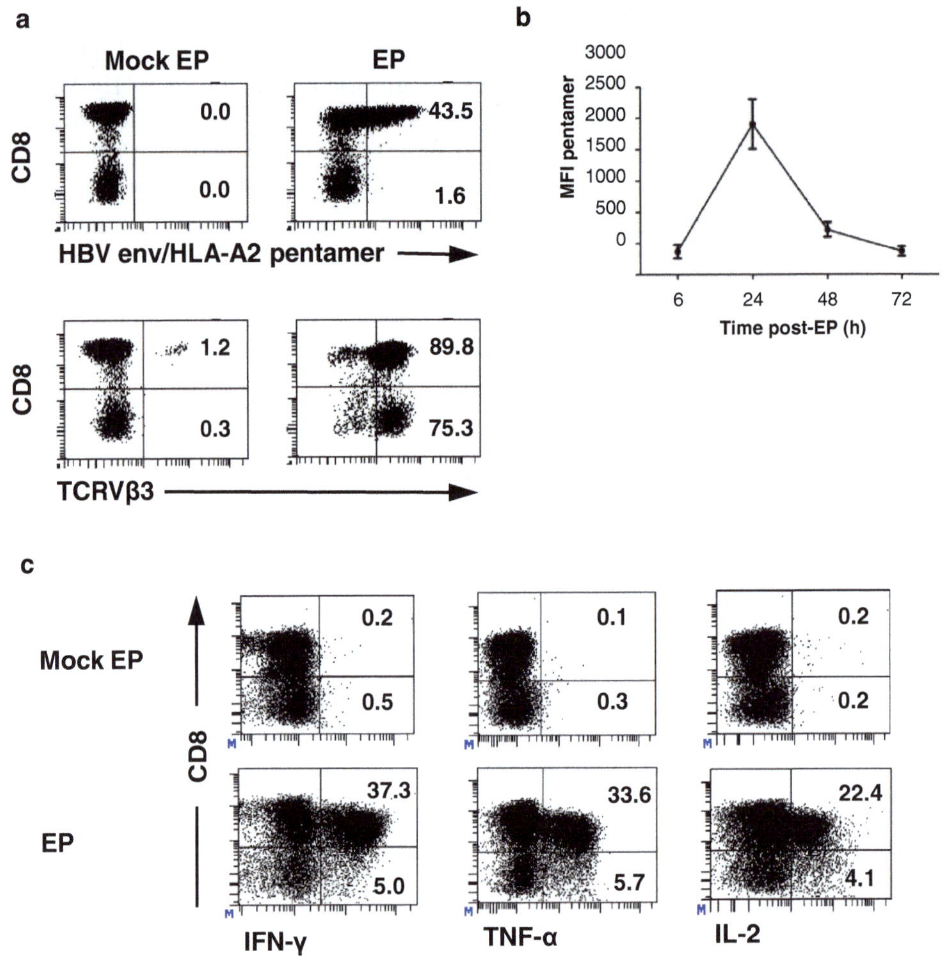

Fig. 1 TCR expression and functionality of HBV-TCR mRNA electroporated T cells. (**a**) *Dot plots* from a representative HBV envelope/HLA-A2 pentamer (*top*) and TCR β-chain variable region (*bottom*) staining in HBV envelope TCR mRNA electroporated T cells at 24 h postelectroporation. Mock electroporated T cells served as negative control. The percentages of pentamer + or TCRVβ3 + cells out of CD8+ or CD8− cells are indicated. (**b**) Expression of TCR on electroporated CD8+ T cells expressed as mean fluorescence intensity (MFI) of pentamer expressing cells at 6, 24, 48, and 72 h postelectroporation. (**c**) *Dot plots* from a representative HBV-TCR mRNA electroporated T cells, after overnight coculture with HBV envelope peptide-loaded HLA-A2+ T2 cells and stained for CD8 and IFN-γ (*first column*), TNF-α (*second column*), and IL-2 (*third column*). Mock electroporated T cells cocultured with peptide-loaded T2 cells served as negative control. The percentages of cytokine-producing cells out of total T cells are indicated

2 Materials

2.1 Expansion of T Cells

1. Peripheral blood mononuclear cells (PBMC) obtained by density-gradient separation from peripheral blood.
2. Anti-CD3 antibody.
3. Anti-CD28 antibody.

4. TexMACS medium (Miltenyi Biotec) or equivalent T cell culture medium.
5. Recombinant human interleukin-2 (IL-2).

2.2 Electroporation of Expanded T Cells with mRNA

2.2.1 Linearization of Plasmid

1. Vector plasmid containing a gene of anti-HBV-TCR and a T7 promoter site [21].
2. Restriction enzyme and suitable digestion buffer.

2.2.2 In Vitro Transcription

1. Transcription kit (T7 mScript Standard mRNA Production System, Cellscript, Madison, WI) containing 10× mScript T7 transcription buffer, NTP solution (25 mM each GTP, ATP, UTP, CTP), 100 mM dithiothreitol, ScriptGuard RNase inhibitor, mScript T7 enzyme solution, RNase-free DNase I, RNase-free water, and 5 M ammonium acetate.

2.2.3 Synthesis of Capped and Tailed RNA

1. Transcription kit (T7 mScript Standard mRNA Production System, Cellscript, Madison, WI) containing 10× ScriptCap capping buffer, 20 mM GTP, 20 mM *S*-adenosyl-methionine, ScriptGuard RNase inhibitor, ScriptCap 2′-*O*-methyltransferase, ScriptCap capping enzyme, 10× A-Plus tailing buffer, 20 mM ATP, A-Plus Poly(A) polymerase, RNase-free water, 5 M ammonium acetate, and lithium chloride.

2.2.4 Purification of Capped and Tailed mRNA

1. Transcription kit (T7 mScript Standard mRNA Production System, Cellscript, Madison, WI) containing 70 % ethanol, RNase-free water, $T_{10}E_1$ buffer (10 mM Tris–HCl, pH 7.5, 1 mM EDTA), and 5 M ammonium acetate.

2.3 Electroporation

1. Electroporator (Nucleofector 2b Device, Lonza, Swizerland) or other equivalent electroporator.
2. Cell Line Nucleofector Kit V (Lonza) containing Cell Line Nucleofector Solution V, Supplement, certified cuvettes, and plastic pipettes.
3. TexMACS medium (Miltenyi Biotec) or equivalent T culture medium.
4. IL-2.
5. Phosphate buffered saline.
6. Antibody against specific TCR (such as HLA class I pentamer and antibody against TCR β-chain variable region).

2.4 Assay for TCR Function

1. HBV peptide.
2. Antigen presenting cells, e.g., T2 cells for HLA-A2 restricted TCRs or HLA-matched EBV-transformed B cell lines.

3. Brefeldin A.
4. Antibodies against cytokines (such as interferon (IFN)-γ, tumor necrosis factor (TNF)-α, and IL-2).

3 Methods

3.1 Expansion of T Cells

1. Resuspend PBMC in TexMACS medium at a concentration of 1×10^6 cells/ml in a conical tube.
2. Add 10 μg/ml of anti-CD3 antibody, 1 μg/ml of anti-CD28 antibody, and 100 IU/ml of IL-2 to the tube (*see* **Notes 1** and **2**) [22].
3. Transfer the cells with antibodies and IL-2 to a tissue culture plate (*see* **Note 3**).
4. Incubate at 37 °C in 5 % CO_2 for 3 days.
5. On day 3, add the same volume of TexMACS medium with 200 IU/ml IL-2.
6. Every 2 days, add the same volume of TexMACS medium with 200 IU/ml IL-2 until use (*see* **Notes 4** and **5**).

3.2 Electroporation of Expanded T Cells with mRNA

3.2.1 Linearization of Plasmid

1. Linearize the vector plasmid with a restriction enzyme downstream of the anti-HBV-TCR gene (*see* **Note 6**).
2. Purify the linearized plasmid and resuspend the linearized plasmid in RNase-free water at a concentration of 0.2–1 mg/ml (*see* **Notes** 7 and **8**).

3.2.2 In Vitro Transcription

1. Mix 1 μg of linearized plasmid, 2 μl of 10× mScript T7 transcription buffer, 7.2 μl of NTP solution, 2 μl of 100 mM dithiothreitol, 0.5 μl of ScriptGuard RNase inhibitor, 2 μl of mScript T7 enzyme solution, and RNase-free water to reach a total of 20 μl in a RNase-free tube (*see* **Notes 9–11**).
2. Incubate at 37 °C for 1 h.
3. Add 1 μl of RNase-free DNase I to the reaction.
4. Incubate at 37 °C for 30 min.
5. Add 179 μl of RNase free water and 200 μl of 5 M ammonium acetate.
6. Mix thoroughly and incubate for 15 min on ice.
7. Pellet the transcribed mRNA by centrifugation at 12,000 × *g* for 15 min at 4 °C.
8. Discard the supernatant with a pipette.
9. Add 1 ml of 70 % ethanol.
10. Centrifuge at 12,000 × *g* for 15 min at 4 °C.
11. Discard the supernatant with a pipette.

12. Centrifuge at 12,000 × *g* for 1 min.
13. Discard the remaining supernatant with a pipette.
14. Air-dry the pellet of the transcribed mRNA (*see* **Note 12**).
15. Dissolve the mRNA in 75 μl of RNase-free water.
16. Determine the concentration of the mRNA by spectrophotometry.
17. Store the mRNA at −20 °C until use.

3.2.3 Synthesis of Capped and Tailed RNA

1. Dilute 55 μg of transcribed mRNA with 72 μl of RNase-free water.
2. Incubate the diluted mRNA at 65 °C for 10 min and cool the heat-denatured mRNA on ice.
3. Mix 10 μl of 10× ScriptCap capping buffer, 5 μl of 20 mM GTP, 2.5 μl of 20 mM *S*-adenosyl-methionine, 2.5 μl of Script-Guard RNase inhibitor, and 4 μl of ScriptCap 2′-*O*-methyl-transferase in a RNase-free tube thoroughly.
4. Add 72 μl of heat-denatured RNA from **step 2** and 4 μl of ScriptCap capping enzyme to the tube from **step 3** and mix thoroughly.
5. Incubate at 37 °C for 1 h.
6. Add 0.5 μl of ScriptGuard RNase inhibitor, 12 μl of 10× A-Plus tailing buffer, 6 μl of 20 mM ATP, and 5 μl of A-Plus Poly (A) polymerase to 100 μl of the capped mRNA from **step 5** and mix thoroughly (*see* **Note 13**).
7. Incubate at 37 °C for 1 h.
8. Add 62 μl of lithium chloride and mix thoroughly (*see* **Note 14**).
9. Incubate at −20 °C for 16 h.

3.2.4 Purification of Capped and Tailed mRNA

1. Pellet the capped and tailed mRNA by centrifugation at 12,000 × *g* for 15 min at 4 °C.
2. Discard the supernatant with a pipette.
3. Add 1 ml of 70 % ethanol.
4. Centrifuge at 12,000 × *g* for 15 min at 4 °C.
5. Discard the supernatant with a pipette.
6. Centrifuge at 12,000 × *g* for 1 min.
7. Discard the remaining supernatant with a pipette.
8. Air-dry the pellet of the mRNA for 3 min.
9. Dissolve the mRNA with 100 μl of RNase-free water.
10. Add 100 μl of 5 M ammonium acetate and mix thoroughly.
11. Incubate for 15 min on ice.

12. Pellet the mRNA by centrifugation at 12,000 × *g* for 15 min at 4 °C.
13. Discard the supernatant with a pipette and add 1 ml of 70 % ethanol.
14. Pellet the mRNA by centrifugation at 12,000 × *g* for 15 min at 4 °C.
15. Discard the supernatant with a pipette.
16. Centrifuge the sample at 12,000 × *g* for 1 min.
17. Discard the supernatant with a pipette.
18. Air-dry the pellet of the mRNA for 3 min.
19. Dissolve the mRNA with $T_{10}E_1$ buffer (10 mM Tris–HCl, pH 7.5, 1 mM EDTA) and store the mRNA at −80 °C (*see* **Notes 15** and **16**).
20. Determine the concentration and the size of the mRNA by spectrophotometry and agarose gel electrophoresis, respectively.

3.3 Electroporation Procedure

1. Mix 12 ml of TexMACS medium and 100 IU/ml of IL-2 in a conical tube.
2. Transfer 8 ml of TexMACS medium with IL-2 to a culture flask and incubate at 37 °C in 5 % CO_2.
3. Prewarm the remaining 4 ml of culture medium at 37 °C.
4. Add 0.5 ml of supplement provided in Cell Line Nucleofector Kit V to 2.25 ml of Nucleofector Solution V and prewarm the supplemented Nucleofector Solution V at ambient temperature (*see* **Note 17**).
5. Turn on Nucleofector 2b device and select program "X-001" (*see* **Notes 18** and **19**).
6. Collect expanded T cells in a conical tube (*see* **Note 20**).
7. Determine the cell number and transfer the required number of cells to another conical tube (*see* **Note 21**).
8. Centrifuge at 300 × *g* for 5 min.
9. Discard the supernatant.
10. Resuspend the cells in PBS and transfer to an RNase-free tube.
11. Centrifuge at 300 × *g* for 5 min.
12. Discard the supernatant with a pipette (*see* **Note 22**).
13. Centrifuge at 300 × *g* for 1 min.
14. Discard the supernatant with a pipette completely.
15. Resuspend the cells in warmed Nucleofector Solution from **step 4** at 5×10^7 to 2×10^8 cells/ml (*see* **Note 23**).

16. Transfer 100 μl of resuspended cells to a tube containing 10–20 μg of mRNA and mix gently by pipetting (*see* **Notes 24** and **25**).
17. Transfer the cells with mRNA to a cuvette for electroporation and close the cuvette with the cap (*see* **Note 26**).
18. Place the cuvette in Nucleofector 2b device.
19. Press the "X" bottom to start the program.
20. Take the cuvette out of the device once the program is finished.
21. Add 500 μl of warmed TexMACS medium with IL-2 from **step 3** to the cuvette in a dropwise manner.
22. Transfer the diluted cells gently into the prepared flask containing culture medium from **step 2** using a supplied plastic pipette (*see* **Note 27**).
23. Repeat **steps 21** and **22** using the same plastic pipette three times.
24. Incubate the cells at 37 °C in 5 % CO_2 until use.
25. Examine expression of the anti-HBV-TCR by staining cells with the specific pentamer and antibody against TCR β-chain variable region (*see* **Notes 28** and **29**).

3.4 Assay for TCR Function

1. Pulse antigen presenting cells expressing HLA class I molecules matching the restriction of the anti-HBV-TCR at 1×10^6 cells/ml with HBV peptide at 1–10 μg/ml (*see* **Note 30**) for at least 1 h.
2. Wash the pulsed antigen presenting cells twice with phosphate buffered saline. Resuspend cells in TexMACS medium.
3. Coculture T cells expressing the anti-HBV-TCR with the target cells from **step 2** at an effector:target ratio of 1:1 in the presence of 10 μg/ml of brefeldin A for 5 h at 37 °C in 5 % CO_2 incubator.
4. Examine intracellular cytokines secreted by T cells by staining cells with the antibodies (*see* **Note 31**).

4 Notes

1. T cells can be expanded using magnetic beads coated with antibodies against CD3 and CD28 cell surface molecules (Dynabeads human T-activator CD3/CD28, Invitrogen, Waltham, MA). It is not necessary to remove the beads before electroporation.
2. IL-2 described in this protocol is aldesleukin. Several types of recombinant IL-2 are available. The biological effects are variable. The dose of IL-2 is recommended to be optimized [23].

3. The recommended cell density is 1.0×10^6 cells/ml/cm^2.
4. Only T cells can be expanded in this culture condition. On day 5, the percentage of T cells should be higher than 90 %.
5. The cells should be monitored daily. If the medium is acidic, the medium must be changed. If the cells appear to be too crowded, the cells must be split.
6. It is recommended to choose a restriction enzyme leaving 5′ protruding ends.
7. Various kits for purification of DNA are available from several manufacturers. We usually purify the linearized DNA using NucleoSpin Gel and PCR Clean-up (Macherey-Nagel, Germany).
8. The complete digestion should be confirmed by agarose gel electrophoresis.
9. Use RNase-free tubes and pipette tips and perform the following procedures in RNase-free work environment.
10. In this protocol, methods using T7 mScript Standard mRNA Production System (Cellscript) is described. Several manufacturers provide various kits for synthesis of capped and tailed mRNA. In our experiences, the mRNA synthesized using mMESSAGE mMACHINE T7 Ultra Kit (Ambion, Austin, TX) has the equivalent electroporation efficiency to the mRNA synthesized using T7 mScript Standard mRNA Production System (Cellscript) [24].
11. For higher yields of mRNA, reactions can be scaled up.
12. Avoid drying the mRNA pellet completely as it becomes insoluble.
13. This reaction can produce 100–200 bp of poly (A)—tails on the capped RNA.
14. In this protocol, precipitation method using lithium chloride is described. Several methods have been established to purify the mRNA, for example, phenol:chloroform extraction and isopropanol precipitation. The method can be chosen according to resources and the application.
15. The expected yield of the mRNA is 55 μg. The desired concentration is at 1.5–5.0 mg/ml.
16. It is recommended to aliquot 20 μg of mRNA into a RNase-free tube before freezing.
17. Alternatively, Human T Cell Nucleofector Kit (Lonza) can be used.
18. Alternatively, expanded T cells and unstimualated T cells resuspended in Human T Cell Nucleofector Kit (Lonza) can be electroporated with program "T-023" and "V-024", respectively

19. In this protocol, electroporation using Nucleofector 2b device (Lonza) is described. Several manufacturers provide electroporator, for example, MaxCyte GT (MaxCyte, Gaithersburg, MD). Use appropriate buffer and the program for the electroporator [24].
20. Unstimulated T cells can be electroporated, but electroporation efficiency in expanded T cells is much higher than that in unstimulated T cells.
21. The recommended cell density per one electroporation reaction is 5×10^6 to 2×10^7 T cells in 100 μl of supplemented Nucleofector Solution.
22. Use RNase-free tubes and pipette tips and perform the following procedures in RNase-free work environment.
23. Electroporate cells in 15 min after resuspending in Nucleofector Solution, because Nucleofector Solution is relatively toxic.
24. The recommended final concentration of mRNA is 150–200 μg/ml.
25. DNA instead of mRNA can be used, but electroporation efficiency with capped and tailed mRNA is much higher than that with DNA.
26. Ensure that the sample covers the bottom of the cuvette.
27. Do not pipet up and down, as it decreases the cell viability.
28. Expression of TCR can be detectable 4 h after electroporation, reaching maximum levels at 24 h. The expression declines gradually after 2 days and become undetectable after 5–7 days (*see* Fig. 1a, b).
29. Pentamer can bind specific TCR determined by the HLA allele and the peptide combination.
30. We routinely use Epstein-Barr virus (EBV)-transformed B lymphocytes expressing HLA matching the restriction of the anti-HBV-TCR.
31. We routinely examine the secretion of IFN-γ, TNF-α, and IL-2 from T cells (*see* Fig. 1c).

References

1. Janeway CA, Travers P, Walport M, Shlomchik MJ (2001) Immunobiology: the immune system in health and disease, 5th edn. Garland Science, New York
2. Barry M, Bleackley RC (2002) Cytotoxic T lymphocytes: all roads lead to death. Nat Rev Immunol 2(6):401–409
3. Rosenberg SA et al (1993) Prospective randomized trial of high-dose interleukin-2 alone or in conjunction with lymphokine-activated killer cells for the treatment of patients with advanced cancer. J Natl Cancer Inst 85(8): 622–632
4. Olioso P et al (2007) Immunotherapy with cytokine induced killer cells in solid and hematopoietic tumours: a pilot clinical trial. Hematol Oncol 25(3):127–131
5. Muul LM, Spiess PJ, Director EP, Rosenberg SA (1987) Identification of specific cytolytic immune responses against autologous tumor

in humans bearing malignant melanoma. J Immunol 138(3):989–995
6. Rosenberg SA et al (1988) Use of tumor-infiltrating lymphocytes and interleukin-2 in the immunotherapy of patients with metastatic melanoma. N Engl J Med 319(25):1676–1680
7. Kochenderfer JN et al (2010) Eradication of B-lineage cells and regression of lymphoma in a patient treated with autologous T cells genetically engineered to recognize CD19. Blood 116(20):4099–4102
8. Lee DW et al (2015) T cells expressing CD19 chimeric antigen receptors for acute lymphoblastic leukaemia in children and young adults: a phase 1 dose-escalation trial. Lancet 385 (9967):517–528
9. Morgan RA et al (2006) Cancer regression in patients after transfer of genetically engineered lymphocytes. Science 314(5796):126–129
10. Johnson LA et al (2009) Gene therapy with human and mouse T-cell receptors mediates cancer regression and targets normal tissues expressing cognate antigen. Blood 114 (3):535–546
11. Robbins PF et al (2011) Tumor regression in patients with metastatic synovial cell sarcoma and melanoma using genetically engineered lymphocytes reactive with NY-ESO-1. J Clin Oncol 29(7):917–924
12. Sung WK et al (2012) Genome-wide survey of recurrent HBV integration in hepatocellular carcinoma. Nat Genet 44(7):765–769
13. Amaddeo G, et al. (2014) Integration of tumour and viral genomic characterisations in HBV-related hepatocellular carcinomas. Gut 1–10
14. Gehring AJ et al (2009) Profile of tumor antigen-specific CD8 T cells in patients with hepatitis B virus-related hepatocellular carcinoma. Gastroenterology 137(2):682–690
15. Morgan RA et al (2010) Case report of a serious adverse event following the administration of T cells transduced with a chimeric antigen receptor recognizing ERBB2. Mol Ther 18 (4):843–851
16. Parkhurst MR et al (2011) T cells targeting carcinoembryonic antigen can mediate regression of metastatic colorectal cancer but induce severe transient colitis. Mol Ther 19 (3):620–626
17. Ghorashian S et al (2014) CD8 T cell tolerance to a tumor-associated self-antigen is reversed by CD4 T cells engineered to express the same T cell receptor. J Immunol 194 (3):1080–1089
18. Gehring AJ et al (2011) Engineering virus-specific T cells that target HBV infected hepatocytes and hepatocellular carcinoma cell lines. J Hepatol 55(1):103–110
19. Krebs K et al (2013) T cells expressing a chimeric antigen receptor that binds hepatitis B virus envelope proteins control virus replication in mice. Gastroenterology 145 (2):456–465
20. Qasim W et al (2015) Immunotherapy of HCC metastases with autologous T cell receptor redirected T cells, targeting HBsAg in a liver transplant patient. J Hepatol 62(2):486–491
21. Koh S et al (2013) A practical approach to immunotherapy of hepatocellular carcinoma using T cells redirected against hepatitis B virus. Mol Ther Nucleic Acids 2(August), e114
22. Hombach AA, Kofler D, Hombach A, Rappl G, Abken H (2007) Effective proliferation of human regulatory T cells requires a strong costimulatory CD28 signal that cannot be substituted by IL-2. J Immunol 179 (11):7924–7931
23. Hank JA et al (1999) Distinct clinical and laboratory activity of two recombinant interleukin-2 preparations. Clin Cancer Res 5 (2):281–289
24. Shimasaki N, Campana D (2013) Natural killer cell reprogramming with chimeric immune receptors. Methods Mol Biol 969:203–220

Chapter 20

Measuring Hematocrit in Mice Injected with In Vitro-Transcribed Erythropoietin mRNA

Azita Josefine Mahiny and Katalin Karikó

Abstract

In vitro-transcribed (IVT) mRNA encoding therapeutic protein has the potential to treat a variety of diseases by serving as template for translation in the patient. To optimize conditions for such therapy, reporter protein-encoding mRNAs are usually used. One preferred reporter is erythropoietin (EPO), which stimulates erythropoiesis and leads to an increase in hematocrit. Measurement of hematocrit is a fast and reliable method to determine the potency of the in vitro-transcribed EPO mRNA. However, frequent blood draw from mice can increase hematocrit due to blood loss. Therefore, instead of using conventional hematocrit capillary tubes, we adapted glass microcapillaries for hematocrit measurement. Daily monitoring of mice can be accomplished by drawing less than 20 μL of blood, thus avoiding blood loss-related hematocrit increase. Due to the small volume of the withdrawn blood the hematocrit remains the same for mice injected with control mRNA, whereas significant hematocrit increase is measured between day 4 and 20 postinjection for those injected with pseudouridine-modified EPO mRNA. Following hematocrit measurement the microcapillaries are snapped easily to recover plasma for further analyses, including EPO measurement by ELISA.

Key words Erythropoietin (EPO), Hematocrit, Erythropoiesis, In vitro-transcribed mRNA, mRNA therapy, Protein replacement therapy, Modified nucleosides

1 Introduction

The pleiotropic erythropoietin (EPO) is a hematopoietic cytokine that is produced by the kidney and the fetal liver [1]. EPO is essential for terminal proliferation and differentiation of progenitor cells into erythrocytes, and thus it is used to treat anemia caused by kidney failure [2]. Erythropoiesis in humans occurs exclusively in the bone marrow, but in mice ~10 % of the red blood cells are produced by the spleen [3]. To determine the volume percentage of red blood cell fraction in the blood hematocrit is measured.

Increased hematocrits have been measured in model animals following administration of EPO-encoding viral constructs [4],

Robert E. Rhoads (ed.), *Synthetic mRNA: Production, Introduction Into Cells, and Physiological Consequences*,
Methods in Molecular Biology, vol. 1428, DOI 10.1007/978-1-4939-3625-0_20,

plasmids [5, 6], and in vitro-transcribed (IVT) mRNAs [7–9]. To determine optimal translation, IVT mRNAs containing a variety of modified nucleosides and different structural elements were evaluated in combination with diverse complexing agents.

Here, we describe a fast method to evaluate the performance of erythropoietin-encoding IVT mRNA administered into mice. The injected EPO mRNA is translated into functional protein, which stimulates red blood cell production resulting in increased hematocrit. Reproducible results for hematocrit and EPO measurements can be achieved by drawing as little as 18 μL of blood. This amount is sufficient to determine hematocrit after centrifugation as well as to measure EPO protein in the plasma by ELISA.

2 Materials

2.1 Treatment of Mice

1. BALB/c mice.
2. Both pseudouridine (Ψ)-modified and unmodified murine-specific EPO mRNA(s) and firefly luciferase mRNA can be purchased (TriLink, San Diego, CA) or synthesized [8].
3. Narcotics, depending on the route of administration; 3.5 % halothane in a mixture of N_2O and O_2 (70 %:30 %) or Ketamin.
4. Transfection reagent (e.g., TransIT-mRNA transfection kit, Mirus Bio, Madison, WI), store at +4 °C.
5. Syringes 1 mL.
6. Nuclease-free Eppendorf tubes 1.5 mL.
7. Serum-free DMEM with 4.5 g/L glucose store at +4 °C.

2.2 Blood Draw and Hematocrit Measurement

1. Needle 27-G.
2. Microcentrifuge tubes, 0.2 mL.
3. Micropipettes and RNase-free tips.
4. Drummond microcaps glass microcapillary tube.
5. Sealing putty, Cha-seal.
6. Rubber bulb (usually included into the package of microcapillaries).
7. Phosphate-buffered saline (PBS).
8. EDTA (0.2 mol/L) to inhibit blood clotting.
9. Lamp with infrared light bulb or small water bath to warm up mice before blood draw.
10. Hematocrit centrifuge.
11. Sticky tape that is easy to manually tear apart.
12. Computer and scanner.

13. Image processing computer program to measure the length of the erythrocyte column in the scanned image of the hematocrit capillaries (e.g., ImageJ).
14. ELISA kit for erythropoietin.
15. Bovine serum albumin (BSA) to prepared diluents for ELISA.

3 Methods

3.1 Experimental Planning

1. First select the EPO mRNA to be tested and the formulations that are suitable to protect the mRNA from degradation. Then choose the application route, e.g., intravenous (i.v.) or intraperitoneal (i.p.) to deliver the complexed mRNA. If a certain mouse model does not limit the options, the best is to select a BALB/c background, simply because it is easier to detect the tail vein on the light-skinned tail.
2. Calculate the amount of EPO mRNA per animal to be tested. In the case of purified, Ψ-modified EPO mRNA, usually 1 μg of TransIT-formulated mRNA is sufficient to increase the hematocrit significantly [8].
3. Formulate IVT mRNA with transIT following the manufacturer's instructions. The use of DMEM from a bottle that is freshly opened is critical to avoid alkaline lysis of the mRNA. DMEM is CO_2-buffered with and its solubility is temperature and pressure dependent. Prepare the formulation directly before administration into the animals.
4. Schedule of blood draw from BALB/c mice for hematocrit measurements is shown in Fig. 1. Due to the short, 2 h half-life of EPO it can only be measured between 6 h and 96 h (day 4) following injection of the encoding mRNA. Due to the continuous presence of high levels of EPO, significant hematocrit increases can be first measured at day 4, then values reach maximum between day 7 and 10 postinjection.

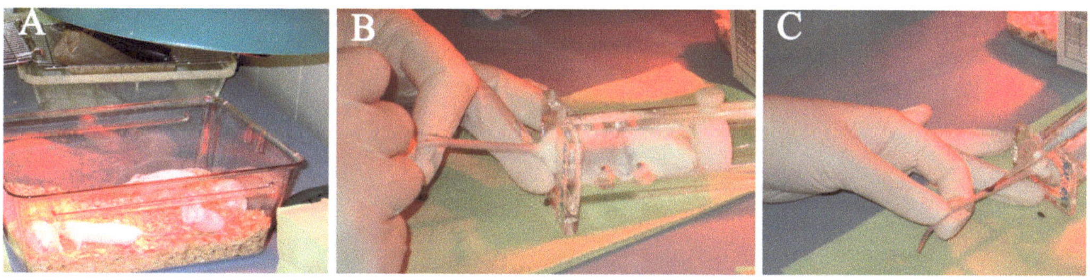

Fig. 1 Blood drawing. To induce vasodilation and facilitate blood draw mice are prewarmed under lamp with infrared light for a couple of minutes (**A**) followed by immobilization in a restrainer to puncture the tail vein (**B**). An 18 μL of blood is be taken by pipetting (**C**). The tip of the pipette has been rinsed with 0.2 M EDTA prior to collecting the blood. The blood is then transferred into a 0.2 mL microcentrifuge tube and mixed with 2 μL 0.2 M EDTA

3.2 Experimental Procedure

1. Intraperitoneal injection is the preferred route for administration of EPO mRNA, considering the ease of injection, which does not require anesthesia.
2. For blood collection, measure 2 μL aliquots of 0.2 M EDTA into 0.2 mL microcentrifuge tubes, label sufficient amount of capillaries and put together according to the animal groups in the experiment (*see* **Notes 1–3**).
3. Before drawing blood prewarm the animals by placing their cage under a lamp with infrared light (Fig. 1A) (*see* **Note 4**). At different time points (6 h, day 1–5, day 7, day 14, day 19) draw blood by tail vein puncture (Fig. 1B). First, set the pipette for 18 μL which is sufficient to fill up the microcapillary and provide enough plasma sample for EPO measurement by ELISA. Rinse the pipette tip with 0.2 M EDTA before aspiration of blood then mix the blood quickly with the 2 μL of EDTA to prevent it from clotting. Leave the blood in the provided 0.2 mL reaction tubes and collect blood similarly from all the animals.
4. In the next step, fill hematocrit microcapillaries with blood by capillary force (Fig. 2A) and close at one end with sealing putty (Fig. 2B), by gently punching the microcapillary into the sealing putty several times (*see* **Notes 5–7**).
5. Put the microcapillaries into a centrifuge equipped with a hematocrit rotor (Fig. 3A). Place the putty-sealed end pointing away from the center of the centrifuge. Close the inner lid, which keeps the microcapillaries in the grooves. Centrifuge the samples at 13,000 × *g* for 3 min (Fig. 3B).
6. Centrifugation separates the plasma from red blood cells (Fig. 3C) and the length of the red blood cell column can be measured. For this purpose make groups of five capillaries and

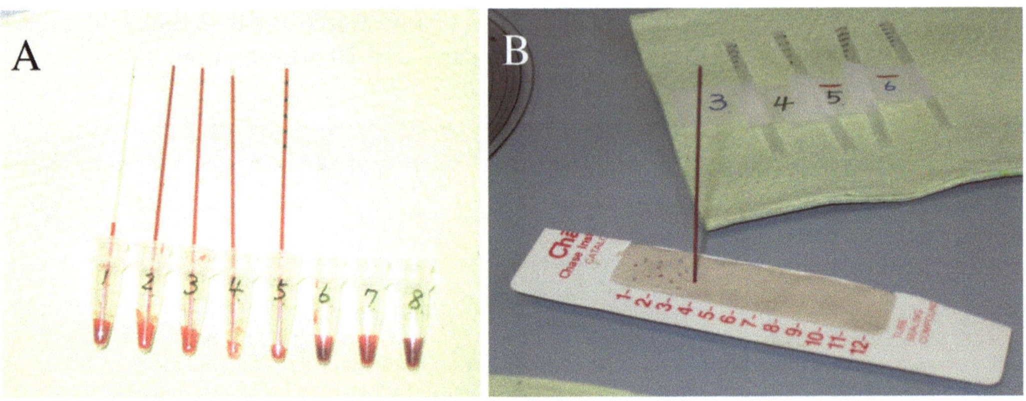

Fig. 2 Filling the capillaries with blood. Glass microcapillaries are filled up with blood by capillary action (**A**). Gently turning the capillaries around and/or holding them upside-down can instigate this process. Subsequently, the capillaries are closed with sealing putty on the end that is not marked (**B**)

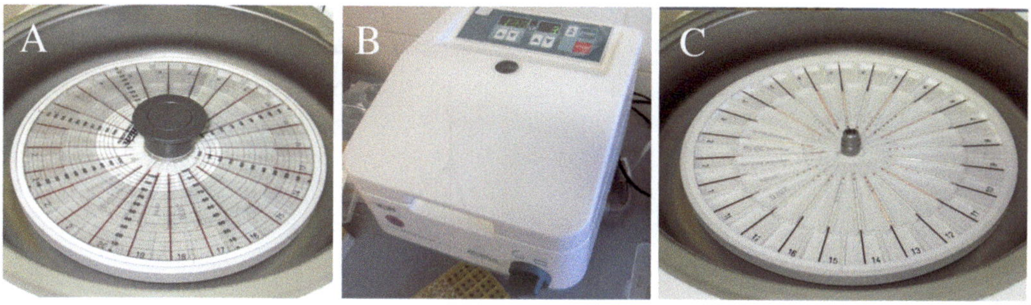

Fig. 3 Centrifugation. The grooves of the hematocrit rotor are filled with capillaries, the sealed ends of the microcapillary tubes are pointing away from the center of the centrifuge and the inner lid is locked in (**A**). The capillary tubes are centrifuged at 13,000 $\times$ *g* 3 min (**B**). After centrifugation the erythrocytes and plasma are separated (**C**)

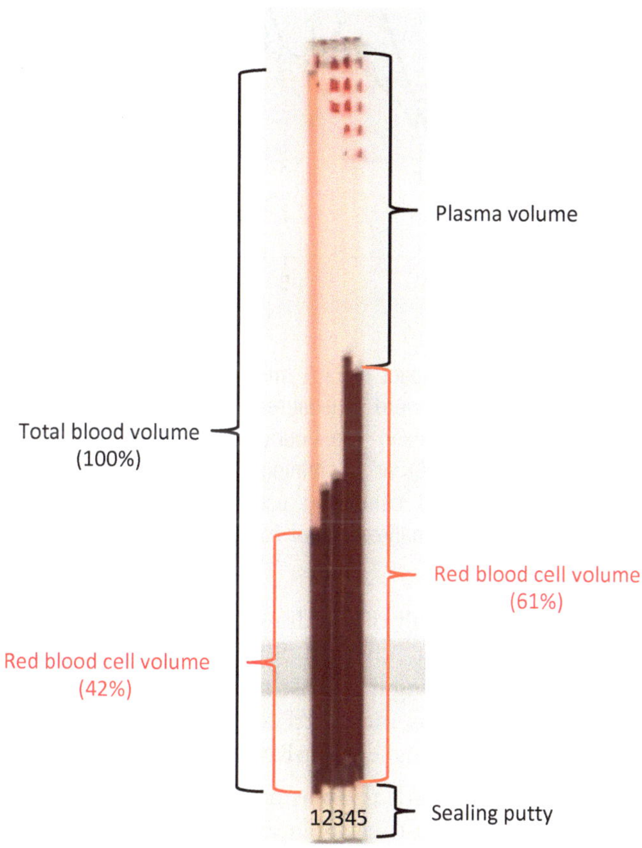

Fig. 4 Hematocrit determination. Capillaries filled with blood collected from five animals were grouped together, taped onto the bed of a scanner and scanned. The distance corresponding to the whole blood, the plasma, and the red blood cells are indicated on the image

scan (Fig. 4) (*see* **Note 8**). Any suitable computer program for image processing fits to measure the length corresponding to the volumes of total blood and the red blood cells (Fig. 4) (*see* **Note 9**).

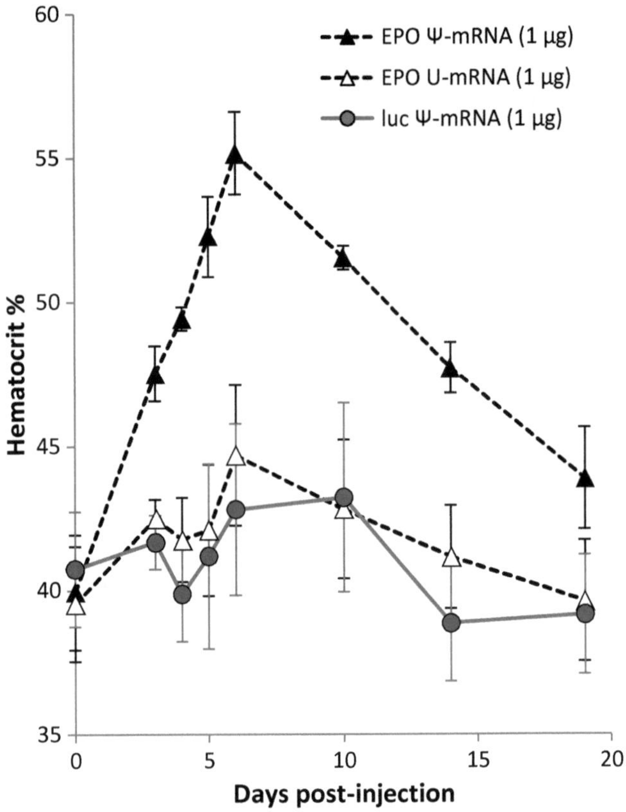

Fig. 5 The impact of EPO mRNA injection on the hematocrit levels in mice. Hematocrits were measured in mice injected intraperitoneally with TransIT-complexed pseudouridine-containing mRNA coding for murine EPO (EPO Ψ-mRNA) or containing uridine (EPO U-mRNA) or coding for firefly luciferase and containing pseudouridine (luc Ψ-mRNA). Five animals per groups were analyzed. *Error bars* are SEM

7. As a result of the drawn small volume of blood the hematocrit did not change for mice injected with pseudouridine-containing luciferase mRNA, but significant increase of hematocrit was measured for animals injected with pseudouridine-containing EPO mRNA (Fig. 5).
8. Afterwards the plasma can be collected for further analysis. By breaking the capillaries the plasma can be discharged (Fig. 6A) into a clean reaction tube (Fig. 6B) and stored at −20 °C until measuring EPO by ELISA (*see* **Note 10**). For this purpose the plasma samples should be diluted depending on the time point the blood was drawn. Treatment with Ψ-modified EPO mRNA results in high EPO levels (Fig. 7), therefore dilute samples collected at 6 h, day 1 and day 2 postinjection 1:100 or 1:50 in general. Dilute samples collected at day 3 and day 4 1:20.

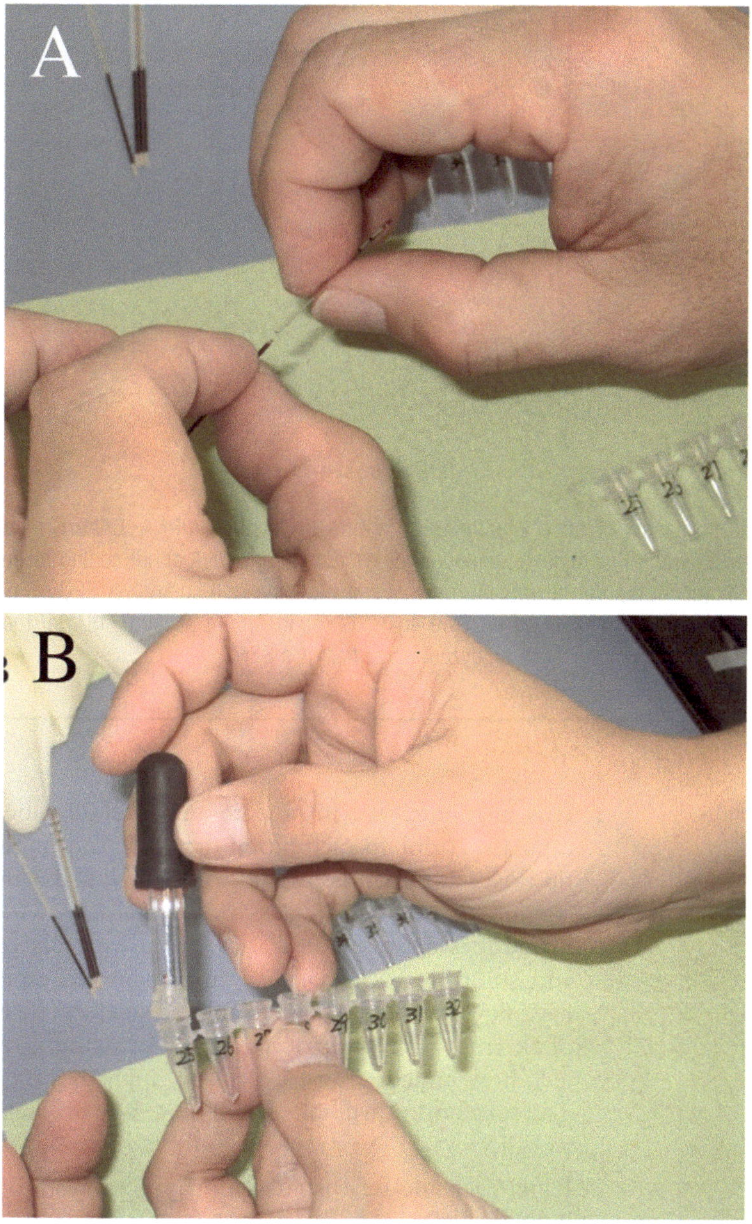

Fig. 6 Plasma collection. The microcapillaries are broken manually to separate red blood cells from the plasma (**A**). The portion containing the red blood cells is discarded, while the plasma discharged into a clean tube using rubber bulb (**B**)

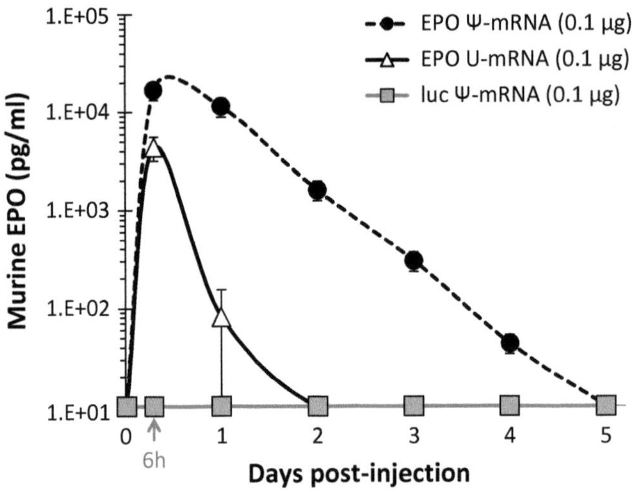

Fig. 7 EPO levels in plasma of mice injected with in vitro-transcribed mRNA. BALB/C mice were injected intraperitoneally with TransIT-complexed pseudouridine-containing mRNA either coding for murine EPO (EPO Ψ-mRNA) or firefly luciferase (luc Ψ-mRNA), or containing uridine and coding for EPO (EPO U-mRNA). EPO levels were measured in the plasma by enzyme-linked immunosorbent assay (ELISA) at the indicated time points. Five animals per condition were analyzed. *Error bars* are SEM

4 Notes

1. It is recommended to prepare and label all tubes and microcapillaries prior to each blood drawing. You might use pipette tip boxes and their lids or lids of 96-well plates for short-term storage of the fragile glass microcapillaries. Assemble some extra tubes and microcapillaries, because mistakes could happen and some microcapillary might also break, especially, if you have never used glass microcapillaries before. Test the fragility of an empty glass microcapillary by breaking it on purpose to feel how much pressure can be applied, before it breaks. The preparation will take some time, but will help to keep the workflow.
2. Label each mouse within the groups to monitor the animals individually.
3. Label capillaries accordingly (e.g., draw rings with colored marker).
4. A water bath is an alternative to lamp with infrared light. Either place the mouse tail directly into the warm water at 37 °C or place a small container such as an empty pipette tip box on a support (for example, a rack for reaction tubes) into the water bath in a way that the container is not in direct contact with the water. Furnish the container with tissue paper and prewarm the mice no longer than 2 min. The prewarming will cause vasodilatation of the tail veins and ease the blood collection.

5. Mix the blood well with the EDTA supplied in the tubes, but avoid creating bubbles. EDTA is used to inhibit blood clotting because according to our experience heparin and citrate interfere with ELISA measurements of EPO. Heparin can interfere with some antibody–antigen-binding reactions. EDTA chelates magnesium and calcium ions. These ions often functions as coenzymes for some proteases. Hence, plasma samples may be more stable in EDTA compared to heparin.
6. If the capillaries do not fill easily with blood, try to rotate the capillaries a bit and turn them upside-down together with the reaction tube containing the blood. Try to avoid air bubbles entering into the microcapillary.
7. Be careful, when sealing the capillaries with wax, as they break upon "just a tiny bit more pressure." Practice sealing of microcapillary with empty tubes.
8. Keep filled capillaries of each group in a separate box to avoid mix-up with others, especially before and right after centrifugation by paying additional attention to the putative easy workflow. It is the best to group and tape the centrifuged capillaries immediately after centrifugation onto the bed of the scanner. Use soft sticky tape to avoid glue residues on the screen of the scanner. You may want to try to break the microcapillaries with bare hands instead of wearing gloves, to have a better feeling for the material, but this is up to personal preference. Put finger nails closely together in order to fix the right point to break.
9. When measuring the length of the hematocrit column pay attention to the transition of sealing putty to blood and to the sinus of the plasma, meaning the end of the liquid column of the plasma, which is close but not at the very end of the microcapillary.
10. Although the procedure can be done by one person, the help of two colleagues makes the work very effective. We found the following distribution of work is the most effective: one person handles mice and does tail vein puncture, the other person collects the blood by pipetting. And the third person fills the capillaries and takes care of centrifugation. Meanwhile this person can also exchange cages. After scanning everybody helps to collect the plasma. Processing 50 mice takes approximately 2.5 h.
11. **Do not** draw too much blood, 18 μL is sufficient. **Do not** collect blood daily for more than a week. Drawing too much, or smaller volume, but daily for extended period will increase the hematocrit independently from injection of Ψ-modified EPO mRNA.

References

1. Vogel J, Gassmann M (2011) Erythropoietic and non-erythropoietic functions of erythropoietin in mouse models. J Physiol 589:1259–1264
2. Sankaran VG, Weiss MJ (2015) Anemia: progress in molecular mechanisms and therapies. Nat Med 21:221–230
3. Richmond TD, Chohan M, Barber DL (2005) Turning cells red: signal transduction mediated by erythropoietin. Trends Cell Biol 15:146–155
4. Rivera VM, Gao GP, Grant RL, Schnell MA, Zoltick PW, Rozamus LW, Clackson T, Wilson JM (2005) Long-term pharmacologically regulated expression of erythropoietin in primates following AAV-mediated gene transfer. Blood 105:1424–1430
5. Pitard B, Bello-Roufai M, Lambert O, Richard P, Desigaux L, Fernandes S, Lanctin C, Pollard H, Zeghal M, Rescan PY, Escande D (2004) Negatively charged self-assembling DNA/poloxamine nanospheres for in vivo gene transfer. Nucleic Acids Res 32:e159
6. Sebestyén MG, Hegge JO, Noble MA, Lewis DL, Herweijer H, Wolff JA (2007) Progress toward a nonviral gene therapy protocol for the treatment of anemia. Hum Gene Ther 18:269–285
7. Kormann MS, Hasenpusch G, Aneja MK, Nica G, Flemmer AW, Herber-Jonat S, Huppmann M, Mays LE, Illenyi M, Schams A, Griese M, Bittmann I, Handgretinger R, Hartl D, Rosenecker J, Rudolph C (2011) Expression of therapeutic proteins after delivery of chemically modified mRNA in mice. Nat Biotechnol 29:154–157
8. Karikó K, Muramatsu H, Keller JM, Weissman D (2012) Increased erythropoiesis in mice injected with submicrogram quantities of pseudouridine-containing mRNA encoding erythropoietin. Mol Ther 20:948–953
9. Thess A, Grund S, Mui BL, Hope MJ, Baumhof P, Fotin-Mleczek M, Schlake T (2015) Sequence-engineered mRNA without chemical nucleoside modifications enables an effective protein therapy in large animals. Mol Ther 23:1456–1464

Chapter 21

Genetic Modification of Human Pancreatic Progenitor Cells Through Modified mRNA

Song Lu, Christie C. Chow, Junwei Zhou, Po Sing Leung, Stephen K. Tsui, and Kathy O. Lui

Abstract

In this chapter, we describe a highly efficient genetic modification strategy for human pancreatic progenitor cells using modified mRNA-encoding GFP and Neurogenin-3. The properties of modified mRNA offer an invaluable platform to drive protein expression, which has broad applicability in pathway regulation, directed differentiation, and lineage specification. This approach can also be used to regulate expression of other pivotal transcription factors during pancreas development and might have potential therapeutic values in regenerative medicine.

Key words Modified mRNA, Pancreatic progenitors, Transcription factors, Human pancreas development, Gene therapy, Regenerative medicine

1 Introduction

Chemically modified mRNA (modRNA) is a novel nonintegrating gene delivery strategy that allows one to regulate expression levels of target proteins in a temporal and spatial manner. Compared to conventional DNA-based gene delivery, modRNA enables transient yet highly efficient expression of target proteins without the risk of genomic integration. Unlike virus-mediated gene delivery, the chemical modifications employed in synthesis of modRNA minimize innate immune responses against viral RNAs and unmodified RNAs [1]. To date, modRNA has been successfully applied in stem cell reprogramming [1], directed progenitor cell differentiation [2–4], and therapeutic gene delivery in various mouse models [5–7].

In both type 1 and type 2 diabetes, the gradual loss of functional pancreatic beta cells (β cells) is a common pathological hallmark. Despite the appeal of pancreatic islet transplantation is the hope to replenish β cell mass for diabetic patients [8], more than

Robert E. Rhoads (ed.), *Synthetic mRNA: Production, Introduction Into Cells, and Physiological Consequences*, Methods in Molecular Biology, vol. 1428, DOI 10.1007/978-1-4939-3625-0_21,

one cadaveric donor is often required to collect sufficient islet cells for a single transplantation. To overcome the shortage of donors, recent advances in generation of pancreatic β cells from pluripotent stem cells [9] or pancreatic progenitor cells [10, 11] offer alternative sources of β cells for transplantation. It has been reported that Neuorgenin-3 (Ngn3), a basic helix–loop–helix transcription factor, is essential for endocrine-lineage commitment in human fetal pancreas [12]. Around gestational week 8, the exocrine and endocrine lineages begin to separate, and the endocrine progenitor cells, characterized by transient expression of Ngn3, eventually differentiate into various endocrine cells including β cells in the mature islet [13]. Ngn3 initiates expression of many downstream transcription factors crucial for maintenance of the endocrine lineage, of which Neurogenic differentiation 1 (NeuroD1) is one [14]. The significance of Ngn3 in pancreas development has been demonstrated in mice and other animal models [15, 16], and recent studies have also revealed that Ngn3 plays an important role in β cell regeneration [17–19].

Previously, we successfully utilized modRNA to drive vascular regeneration in the heart by virtue of its transient yet localized and highly efficient properties [4, 7, 20, 21]. During development of human pancreas, transient expression of transcription factors including Ngn3 is found to facilitate commitment of progenitor cells to the endocrine lineage. Therefore, the modRNA technique might promote maturation and functions of human fetal PPCs for transplantation.

In this chapter, we provide detailed procedures of modRNA synthesis using Ngn3 and GFP reporter gene as examples. We utilize human pancreatic progenitor cells (hPPCs) to illustrate the application of modRNA in efficient gene delivery. hPPCs were derived from isolated islets of the human fetal pancreas (gestational weeks 11–13). The cells can be maintained in the proliferative stage for at least 10 passages and are able to form islet-like clusters in differentiation medium [10, 11, 22]. Using the protocol described here, we demonstrate highly efficient expression of the GFP reporter protein and upregulation of NeuroD1 in hPPCs upon transfection with GFP and Ngn3 modRNA, respectively. Indeed, our strategy could be of great potential in driving expression of transcription factors during development and to direct differentiation of human progenitor cells for transplantation.

2 Materials

2.1 Tail PCR

1. KAPA high fidelity PCR kit (KAPA Biosystems).
2. Custom forward and poly-(T) tail-containing reverse primers, 10 μM.
3. Linearized ORF-containing plasmid, 40–80 ng.

4. DNase-, RNase-free PCR water.
5. PCR purification kit.

2.2 In Vitro Transcription

1. Tail PCR product: diluted to 100 ng/μL.
2. MegaScript T7 transcription kit including T7 RNA polymerase, (10×) T7 transcription buffer, ATP (75 mM), and GTP (75 mM) (Ambion).
3. m7GpppG RNA cap: reconstituted to 60 mM stock (New England Biolabs).
4. Me-CTP (100 mM).
5. Pseudo-UTP (100 mM).
6. RNase-free PCR water.
7. RNase-free pipette tips.
8. RNase-free Eppendorf tubes.
9. RNase Zap (Ambion).

2.3 modRNA Processing and Purification

1. Turbo DNase (2 U/μL) (Ambion).
2. MEGAclear transcription clean-up kit (Ambion).
3. 100 % ethanol.
4. Antarctic phosphatase kit, including phosphatase (5000 U/mL) and phosphatase buffer (10×) (New England Biolabs).

2.4 modRNA Transfection in hPPCs

1. Lipofectamine RNAimax (Life Technology).
2. Opti-MEM (Life Technology).
3. RPMI 1640.
4. FBS.

2.5 Analysis of Protein Expression and Cell Survival

1. Paraformaldehyde.
2. Saponin.
3. Normal goat serum.
4. Rabbit anticleaved caspase 3 antibody (1:200; Cell Signaling Technology).
5. Alexa Fluor 546-conjugated goat antirabbit IgG (H + L) antibody (1:600; Life Technology).
6. Hoechst 33342 (1 μg/mL).

2.6 Analysis of Downstream Gene Expression

1. RNeasy mini kit (Qiagen).
2. iScript cDNA synthesis kit (BioRad).
3. iTaq universal SYBR Green supermix (BioRad).
4. Customer primers.

3 Methods

3.1 Tail PCR

1. Prepare the PCR reaction by mixing 100 μL of (2×) KAPA high fidelity PCR mix, 6 μL of custom forward primer (10 μM), 6 μL of custom reverse poly-(T) tail-containing primer, 40–80 ng of linearized ORF-containing plasmid, and RNase-free PCR water up to a total volume of 200 μL.
2. Aliquot the PCR reaction mix into eight PCR tubes with 25 μL in each tube (*see* **Note 1**).
3. Perform PCR using the following program: initial denaturation at 95 °C for 3 min; amplification for 33 cycles, with each cycle including 98 °C for 20 s, 60 °C for 15 s, and 72 °C for 1 min; final extension at 72 °C for 3 min.
4. Check the quality of PCR product by gel electrophoresis.
5. Purify the PCR product using PCR purification kit according to manufacturer's instructions.
6. Measure DNA concentration of the purified PCR product using NanoDrop Spectrophotometer. Adjust the concentration to 100 ng/μL (*see* **Note 2**).

3.2 In Vitro Transcription

1. Clean the working bench and pipettes with RNaseZap solution (*see* **Note 3**).
2. Thaw the required reagents and DNA template on ice.
3. Prepare the NTP mix for five in vitro transcription (IVT) reactions (74 μL) by combining 20 μL of G cap analog (60 mM, *see* **Note 4**), 4 μL of GTP (75 mM), 20 μL of ATP (75 mM), 15 μL of Me-CTP (100 mM), and 15 μL of Pseudo-UTP (100 mM).
4. Mix the content thoroughly by vortex and spin down briefly.
5. Prepare the master mix for five IVT reactions (200 μL) by mixing 80 μL of purified PCR product (100 ng/μL), 74 μL of NTP mix (from **step 3**), 6 μL of RNase-free water, 20 μL of (10×) T7 transcription buffer, and 20 μL of T7 RNA polymerase.
6. Mix the content thoroughly by vortex and spin down briefly.
7. Aliquot 40 μL into each of five PCR tubes.
8. Incubate IVT reaction mix at 37 °C for 3–6 h using PCR thermocycler (*see* **Note 5**).

3.3 ModRNA Purification

3.3.1 DNase Treatment

1. After IVT reaction completes, remove PCR tubes from the thermocycler. Add 2 μL Turbo DNase per IVT reaction (40 μL).
2. Continue incubation at 37 °C for 15 min.
3. Purify the IVT reaction product using MEGAclear kit according to manufacturer's instructions (*see* **Note 6**).

3.3.2 Dephosphorylation

1. Prepare phosphatase reaction by mixing the following components: 100 μL of purified IVT reaction product, 11 μL of (10×) phosphatase buffer, and 2 μL of Antarctic phosphatase.
2. Split the reaction mix into four PCR tubes (i.e., 28 μL each).
3. Gently mix the samples by vortex and spin down briefly.
4. Incubate PCR tubes at 37 °C using PCR thermocycler for 1 h (*see* **Notes** 7 and **8**).

3.3.3 Column Purification

1. Purify the IVT reaction product using MEGAclear kit according to manufacturer's instructions.
2. Measure modRNA concentration using NanoDrop spectrophotometer. The average yield is about 50 μg per 40 μL IVT reaction (*see* **Note 9**).
3. Check RNA integrity by running on a denaturing agarose gel.
4. Adjust modRNA concentration to 100 ng/μL with the elution buffer. Store aliquots at −86 °C.

3.4 Analysis of modRNA Transfection and Translational Efficiency

1. Prepare the transfection mix in two separate tubes. In tube A, mix 12 μL of GFP-ModRNA (0.2–1 μg) and 48 μL of Opti-MEM. In tube B, mix 6 μL of RNAimax and 54 μL of Opti-MEM. Incubate the two tubes at room temperature for 15 min after preparation.
2. Make transfection mix (120 μL) by combining the content of tube A and B, and incubate at room temperature for another 15 min.
3. Replace the hPPC growth medium with 800 μL RPMI 1640 containing 2 % FBS for each well, and add 120 μL transfection mix into each well of a 6-well plate.
4. Incubate hPPCs in transfection mix for 6 h and replace with routine hPPC growth medium afterwards.
5. Translational efficiency of the GFP modRNA can be taken images under a fluorescence microscope 24–48 h posttransfection (*see* **Note 10**). Images can be quantified by software like Image J (NIH) semiquantitatively (Fig. 1).

3.5 Analysis of Posttransfection Cell Survival Rate

Cytotoxicity 24–48 h posttransfection can be evaluated by immunostaining as below (*see* **Note 11**). After immunostaining, the percentage of survival can be calculated semiquantitatively by counting the number of $GFP^+/Casp3^-$ cells among the total GFP^+ population (Fig. 2).

1. Fix GFP modRNA-transfected hPPCs with 2 % paraformaldehyde for 15 min.
2. Wash hPPCs twice with PBS for 15 min.

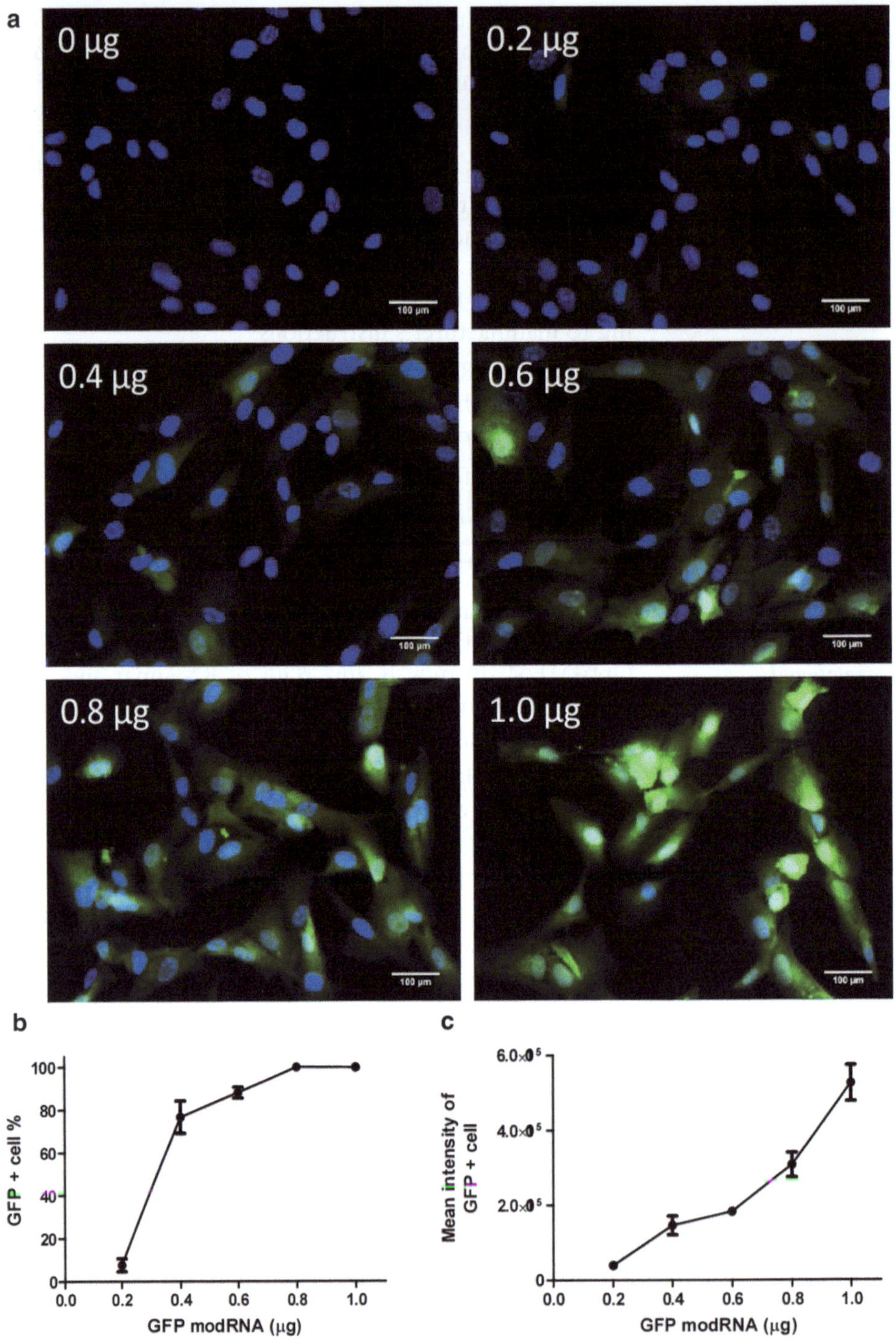

Fig. 1 Efficiency of GFP reporter expression in hPPCs at 24 h posttransfection. (**a**) Representative images of GFP fluorescence in hPPCs at different modRNA doses. Nuclear counter stain by Hoechst 33342 (2 μg/mL). Scale bar = 100 μm. (**b**) GFP^+ cell percentage in total cells examined at different modRNA doses. 3–4 fields as taken in panel (**a**) were quantified. Error bar = ±SE. (**c**) Mean GFP fluorescence intensity per GFP^+ cell at different modRNA doses. Three to four fields as taken in panel (**a**) were quantified. Error bar = ±SE

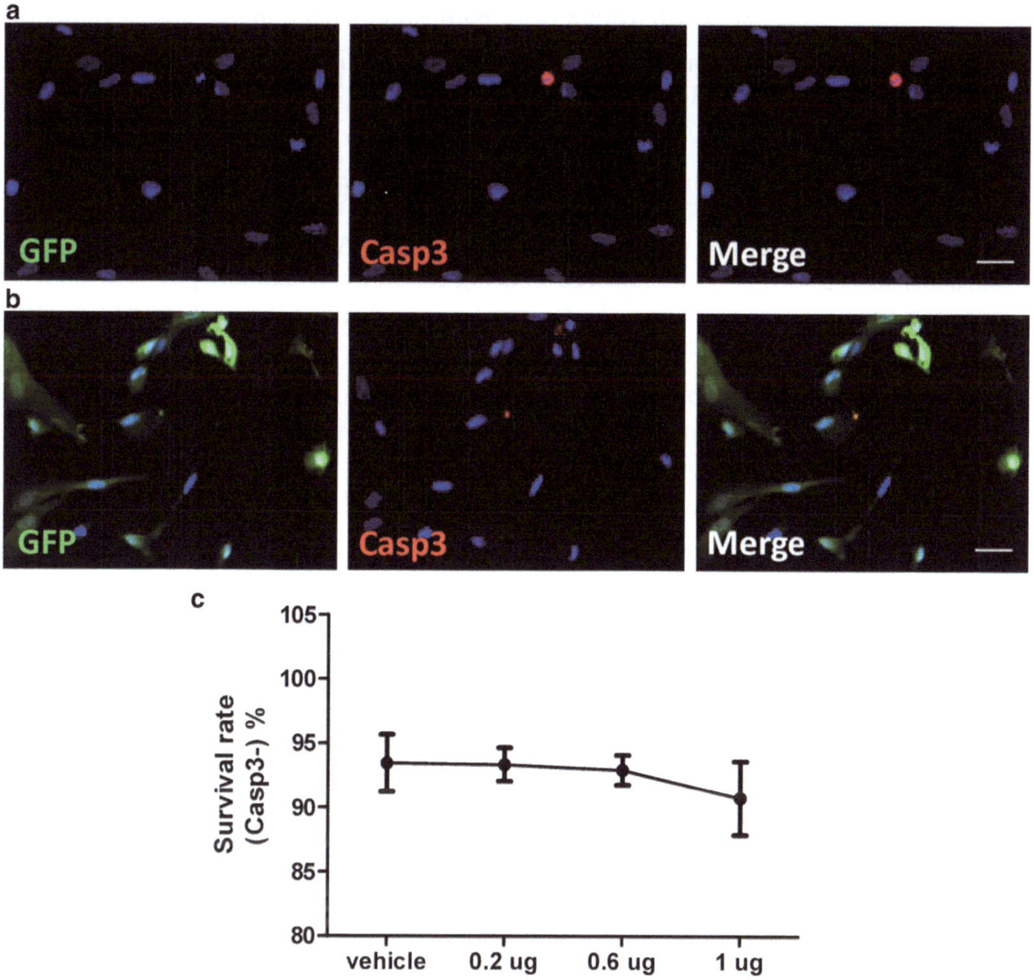

Fig. 2 Cytotoxicity associated with GFP-modRNA transfection. (a, b) Representative images of cleaved Casp3 staining at 24 h posttransfection. Nuclear counter stain by Hoechst 33342 (1 μg/mL). Scale bar = 100 μm. (**a**) Transfection by vehicle control (lipofectamine only). (**b**) Transfection by GFP-modRNA (1 μg per well of 6-well plate). (**c**) Cell survival rate at different modRNA doses. Survival % = 1 − (number of Casp3$^+$ cells/total cells examined) × 100 %. Three to four fields as taken in panel (**a**) were quantified. Error bar = ±SE

3. Incubate hPPCs in 0.1 % saponin and 2 % goat serum for 30 min.
4. Incubate hPPCs in 0.1 % saponin and 2 % goat serum with rabbit anticleaved caspase 3 antibody at 4 °C overnight.
5. Wash hPPCs twice with PBS for 15 min.
6. Incubate hPPCs in 0.1 % saponin and 2 % goat serum with the secondary antibody, Alexa Fluor 546-conjugated goat antirabbit IgG (H + L), for 30 min.

7. Wash hPPCs twice with PBS for 15 min.
8. Incubate hPPCs in 0.1 % saponin and 2 % goat serum with Hoechst 33342 (1 μg/mL) for 15 min.
9. Wash hPPCs twice with PBS for 15 min.
10. Image analysis under a fluorescence microscope.

3.6 Analysis of Innate Immunity and Downstream Gene Expression

1. Extract mRNA from modRNA-transfected hPPCs 24–48 h posttransfection using the RNeasy mini kit.
2. Convert mRNA to cDNA using iScript reverse transcription kit.
3. Dilute the reverse transcription reaction product as template and perform PCR or qPCR using customer primers (SYBR Green supermix).
4. Analyze results as in Figs. 3 and 4.

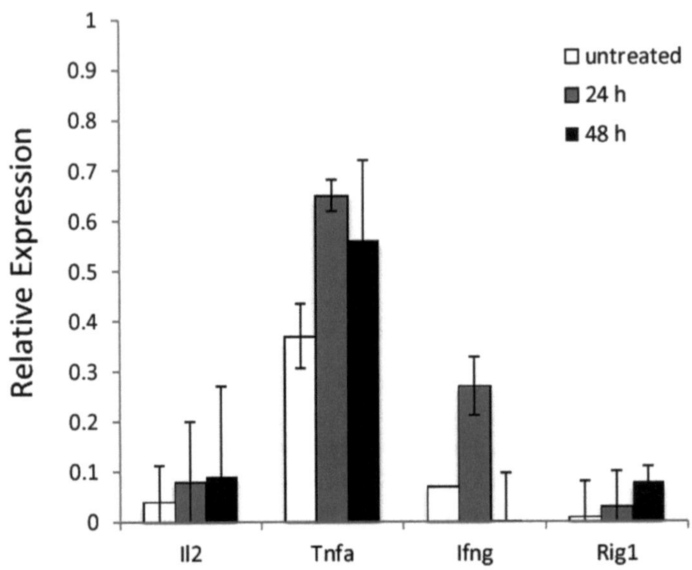

Fig. 3 Innate immune response of hPPCs after transfection with GFP-modRNA. RT-qPCR for genes involved in innate immunity: IL2, TNF-α, IFN-γ, RIG-I. hPPCs were transfected with 1 μg GFP-modRNA per well of 6-well plate or equal amount of lipofectamine as used in modRNA transfection (vehicle control), and cells were harvested for analysis at 24 and 48 h posttransfection. *White bar*: vehicle control at 48 h. *Gray bar*: GFP-modRNA at 24 h. *Black bar*: GFP-modRNA at 48 h. *Error bar* = ±SD

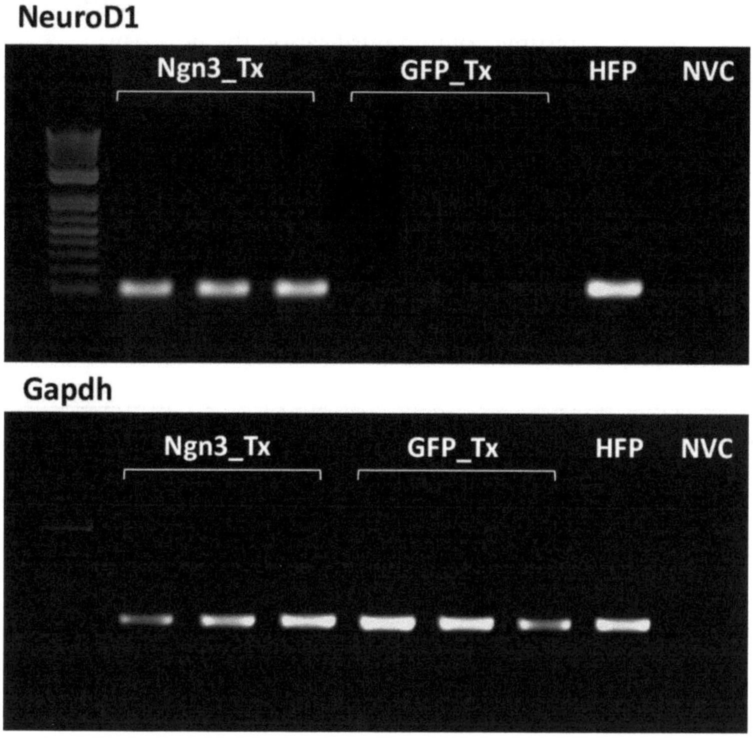

Fig. 4 Upregulation of NeuroD1 at 24 h posttransfection with Ngn3-modRNA. RT-PCR for NeuroD1, a direct downstream target of Ngn3. hPPCs were transfected with either Ngn3-modRNA (Ngn3_TX) or GFP-modRNA (GFP_TX) at 1 μg per well of 6-well plate. Cells were harvest 24 h posttransfection and each lane represents one independent experiment. GAPDH was used as an internal control. *HFP* human fetal pancreas cDNA as positive control, *NVC* no cDNA as negative control

4 Notes

1. A minimum of eight tubes is needed to ensure sufficient DNA templates generated for subsequent 5–10 IVT reactions.
2. The PCR product can be stored in aliquots at −20 °C. The average yield is about 3 μg per tail PCR reaction after purification.
3. Use RNase-free pipette tips and Eppendorf tubes.
4. G cap analog is reconstituted with 20 μL DNase- and RNase-free water.
5. Do not incubate in water bath to minimize contamination with RNase during incubation.
6. Both elution options 1 and 2 in the user manual will give a similar yield.
7. Phosphatase treatment removes 5′ triphosphates and prevents RIG-I mediated innate immunity.

8. After this step, one can store the modRNA at −80 °C and resume column purification the next day.
9. The concentration of modRNA generated per IVT reactions after purification should be sufficient for 50× in vitro transfection.
10. Transfection with modRNA can be repeated [21, 23], according to the concentration and duration required for expression of a particular protein.
11. Unless otherwise specified, all procedures are done at room temperature.

Acknowledgments

We thank Dr. Ke Xu for his advice on molecular cloning. We also thank Dr. Kathy Juan Liang for sharing her experience in hPPC cultures. The authors declare no conflict of interest. This work was supported by Research Grants Council-Early Career Scheme (24110515 to K.O.L.), Health and Medical Research Fund (03140346 to K.O.L.) the Croucher Foundation (K.O.L.), CUHK Strategic Recruitment Grant (K.O.L.), and CUHK Faculty of Medicine Postdoctoral Fellowship (S.L.).

References

1. Warren L, Manos PD, Ahfeldt T et al (2010) Highly efficient reprogramming to pluripotency and directed differentiation of human cells with synthetic modified mRNA. Cell Stem Cell 7 (5):618–630. doi:10.1016/j.stem.2010.08.012
2. Elcheva I, Brok-Volchanskaya V, Kumar A et al (2014) Direct induction of haematoendothelial programs in human pluripotent stem cells by transcriptional regulators. Nat Commun 5:4372. doi:10.1038/ncomms5372
3. Hansson ML, Albert S, Gonzalez Somermeyer L et al (2015) Efficient delivery and functional expression of transfected modified mRNA in human embryonic stem cell-derived retinal pigmented epithelial cells. J Biol Chem 290 (9):5661–5672. doi:10.1074/jbc.M114.618835
4. Lui KO, Zangi L, Silva EA et al (2013) Driving vascular endothelial cell fate of human multipotent Isl1+ heart progenitors with VEGF modified mRNA. Cell Res 23 (10):1172–1186. doi:10.1038/cr.2013.112
5. Creusot RJ, Chang P, Healey DG et al (2010) A short pulse of IL-4 delivered by DCs electroporated with modified mRNA can both prevent and treat autoimmune diabetes in NOD mice. Mol Ther 18(12):2112–2120. doi:10.1038/mt.2010.146
6. Kormann MS, Hasenpusch G, Aneja MK et al (2011) Expression of therapeutic proteins after delivery of chemically modified mRNA in mice. Nat Biotechnol 29(2):154–157. doi:10.1038/nbt.1733
7. Zangi L, Lui KO, von Gise A et al (2013) Modified mRNA directs the fate of heart progenitor cells and induces vascular regeneration after myocardial infarction. Nat Biotechnol 31 (10):898–907. doi:10.1038/nbt.2682
8. Barton FB, Rickels MR, Alejandro R et al (2012) Improvement in outcomes of clinical islet transplantation: 1999-2010. Diabetes Care 35 (7):1436–1445. doi:10.2337/dc12-0063
9. Pagliuca FW, Millman JR, Gurtler M et al (2014) Generation of functional human pancreatic beta cells in vitro. Cell 159(2):428–439. doi:10.1016/j.cell.2014.09.040
10. Leung KK, Liang J, Ma MT et al (2012) Angiotensin II type 2 receptor is critical for the development of human fetal pancreatic progenitor cells into islet-like cell clusters and their potential for transplantation. Stem Cells 30 (3):525–536. doi:10.1002/stem.1008

11. Leung KK, Suen PM, Lau TK et al (2009) PDZ-domain containing-2 (PDZD2) drives the maturity of human fetal pancreatic progenitor-derived islet-like cell clusters with functional responsiveness against membrane depolarization. Stem Cells Dev 18 (7):979–990. doi:10.1089/scd.2008.0325
12. Jennings RE, Berry AA, Kirkwood-Wilson R et al (2013) Development of the human pancreas from foregut to endocrine commitment. Diabetes 62(10):3514–3522. doi:10.2337/db12-1479
13. Pan FC, Brissova M (2014) Pancreas development in humans. Curr Opin Endocrinol Diabetes Obes 21(2):77–82. doi:10.1097/MED.0000000000000047
14. Benitez CM, Goodyer WR, Kim SK (2012) Deconstructing pancreas developmental biology. Cold Spring Harbor Perspect Biol 4 (6). doi: 10.1101/cshperspect.a012401
15. Grapin-Botton A, Majithia AR, Melton DA (2001) Key events of pancreas formation are triggered in gut endoderm by ectopic expression of pancreatic regulatory genes. Genes Dev 15(4):444–454. doi:10.1101/gad.846001
16. Gu G, Brown JR, Melton DA (2003) Direct lineage tracing reveals the ontogeny of pancreatic cell fates during mouse embryogenesis. Mech Dev 120(1):35–43
17. Swales N, Martens GA, Bonne S et al (2012) Plasticity of adult human pancreatic duct cells by neurogenin3-mediated reprogramming. PLoS One 7(5):e37055. doi:10.1371/journal.pone.0037055
18. Van de Casteele M, Leuckx G, Baeyens L et al (2013) Neurogenin 3+ cells contribute to beta-cell neogenesis and proliferation in injured adult mouse pancreas. Cell Death Dis 4:e523. doi:10.1038/cddis.2013.52
19. Yang K, Wang Y, Du Z et al (2014) Short-reactivation of neurogenin-3 and mesenchymal microenvironment is require for beta-cells differentiation during fetal pancreas development and islet regeneration. Rom J Morphol Embryol 55(2):305–311
20. Chien KR, Zangi L, Lui KO (2015) Synthetic chemically modified mRNA (modRNA): toward a new technology platform for cardiovascular biology and medicine. Cold Spring Harbor Perspect Med 5(1):a014035. doi:10.1101/cshperspect.a014035
21. Lui KO, Zangi L, Chien KR (2014) Cardiovascular regenerative therapeutics via synthetic paracrine factor modified mRNA. Stem Cell Res 13(3 Pt B):693–704. doi:10.1016/j.scr.2014.06.007
22. Suen PM, Zou C, Zhang YA et al (2008) PDZ-domain containing-2 (PDZD2) is a novel factor that affects the growth and differentiation of human fetal pancreatic progenitor cells. Int J Biochem Cell Biol 40(4):789–803. doi:10.1016/j.biocel.2007.10.020
23. Lui KO (2014) VEGF-A: the inductive angiogenic factor for development, regeneration and function of pancreatic beta cells. Curr Stem Cell Res Ther 9(5):396–400

INDEX

Robert E. Rhoads (ed.), *Synthetic mRNA: Production, Introduction Into Cells, and Physiological Consequences*, Methods in Molecular Biology, vol. 1428, DOI 10.1007/978-1-4939-3625-0,

L

M

N

O

P

R

S

T

U

V

W

MIX
Papier aus verantwortungsvollen Quellen
Paper from responsible sources
FSC® C105338

If you have any concerns about our products,
you can contact us on
ProductSafety@springernature.com

In case Publisher is established outside the EU,
the EU authorized representative is:
Springer Nature Customer Service Center GmbH
Europaplatz 3, 69115 Heidelberg, Germany

Printed by Libri Plureos GmbH
in Hamburg, Germany